ESSENTIAL ADULT CARDIOLOGY

A Textbook for Aspiring Cardiologists

Dr David H. Dighton

MediCAUSE

2025

About the Author

& Copyright

Dr David H. Dighton

Formerly,

British Heart Foundation Research Fellow,
St. George's Hospital, London.
Lecturer in Cardiology,
Charing Cross Hospital, London.
Chef de Clinique, Cardiologist, Vrije Universiteit, Amsterdam.

Director: Cardiac Centre, Loughton, Essex, UK.
Chigwell Cardiac Centre, Essex. UK.
www.daviddighton.com
email:
david@daviddighton.com

Dr. David H. Dighton qualified at the London Hospital Medical College in 1966 with MB and BS (London) degrees. In 1970, after some time spent in NHS general practice, he became a British Heart Foundation Fellow in Cardiology at St. George's Hospital, Hyde Park Corner, London, working with cardiologists Dr. Aubrey Leatham and Dr. Alan Harris. In 1973, he became a MRCP(UK), and later a Lecturer (London University) in Medicine and Cardiology at Charing Cross Hospital, London.

In 1980, he became Chef de Clinique (Assistant Professor) at the Vrije University Hospital, Amsterdam. After returning to the UK in 1982, he worked both in his own private medical and cardiac practice in Loughton, Essex (The Loughton Clinic, established in 1973), and at the Wellington Hospital, London. In 2000, he started a private cardiac diagnostic centre, specialising in heart disease prevention and the early detection of heart and artery disease (The Cardiac Centre, Loughton). This closed after the PSA directed the GMC to withdrawn his licence to practice (see the details in 'The NHS. Our Sick Sacred Cow.' 2023). As an independent private cardiologist and general physician, he disagreed with them about UK medical practice bureaucracy, and those most qualified to devise, regulate and supervise it; opinions he based on having been a medical student, general physician and cardiologist for sixty years.

In 2003 and 2006, he wrote two books on food and the heart, and between 2023 and 2014, six books on medical and cardiac subjects. He is currently engaged in forming Chigwell Cardiac Centre, Essex, UK., and continuing to publish books on both medical topics and his research relating to secondary cardiac prevention.

One result of his interest in the frontier between art and science, is one small book of poems, haiku and senryu, on inter-personal relationships. His magnum opus, *The Art and Science of Medical Practice*, details what he learned from practising the art and science of medicine for sixty years.

This is his ninth book, but he has other interests. He draws (all the illustrations in this book) and paints in oils on canvas. One of his paintings features on the cover of his book, *The Art and Science of Medical Practice*. He has played the guitar and piano for many decades, for his own amusement. He has composed several simple melodies, one of which introduces his YouTube series on understanding heart problems (Dr. Dighton interviews), another of which he played live (on Facebook) for his now much missed patient and friend, June Allpress.

For further information go to:
www.daviddighton.com

First published in 2025 by MediCause.
The copyright© belongs to Dr David H. Dighton ™

All rights are reserved.
No part of this publication may be reproduced or transmitted in any form or by any means.

British Library Cataloguing in Publication Data
A CIP catalogue record for this title is available from the British Library.

ISBN: 978-1-0683597-1-2

Dedication

Those who gain knowledge rely on the work of many others.

I dedicate this book to all those who graciously taught me cardiology:

Dr. Alan M. Harris.
Dr. Aubrey Leatham.
Dr. Michael Davies.
Dr. Keith Jefferson.
Mr. Geoff Davies.
Dr. Alan Gelson.
Dr. Paul Kligfield.
Dr. Peter Nixon.
Dr. Keith Woollard.
Dr. Pim de Feyter.
Prof. J.P. Roos.
Dr. David Lipkin.
Mr. Stephen Edmondson FRCS
and many nurses and cardiac technicians at
St George's Hospital Hyde Park Corner, London SW1,
The New Charing Cross Hospital, London W6,
the Vrije Universiteit, Amsterdam,
and The Wellington Hospital, London NW8.

Other Books by the Author

Eat to Your Heart's Consent. The diet and lifestyle for a healthy heart (2003). HeartShield. ISBN: 0-9551072-0-2

HeartSense. How to look after your heart (2006). HeartShield. ISBN: 0-9551072-1-0

The NHS: Our Sick Sacred Cow; Causes and Cures. (2023) Paperback. ISBN: 978-1-3999-6027-4 (e-book available from stan.store/drdhd001001)

How to Become Heart-Smart. A User's Guide to Heart Health and Heart Disease Prevention (2023) 1st Ed. 2nd Ed 2024 ISBN: 978-1-3999-7461-5 (e-book available from stan.store/drdhd001001).

Who Loses Wins. Winning Weight Loss Battles: A 'Fat' Mentality v a 'Fit Mentality'. (2024). ISBN: 978-1-7385207-1-8. (e-book ISBN: 978-1-7385207-2-5 from stan.store/drdhd001001).

Doctors and Nurses. How to Survive Medical Practice. (2024). Paperback ISBN:978-1-7385207-5-6; ebook available from stan.store/drdhd001001.

The Art and Science of Medical Practice (2024). Hardback ISBN: 978-1-7385207-7-0; Paperback ISBN: 978-1-7385207-3-2; e-book from stan.store/drdhd001001.

Poems for Recycling Lives (2024). Paperback ISBN: 978-1-7385207-8-7; ebook from stan.store/drdhd001001.

Author's Journal Publications

Dighton, D.H. (1974) Sinus Bradycardia : Autonomic Influences and Clinical Assessment.
British Heart Journal. 36: 791-797.

Dighton. D.H. (1975) Sino-Atrial Block : Autonomic Influences and Clinical Assessment.
British Heart Journal : 37: 321-325.

Dighton. D.H. (1975) Complete Heart Block : Studies of Atrial and Ventricular Pacemaker Site and Function
British Heart Journal. 37(11):1156-1160.

Dighton, D.H. (1976) Autonomic Features of Supraventricular Tachycardia.
British Heart Journal. 38 (3) 319.

Nixon, P.G.F., Dighton, D.H. (1976)
Meditation and Methyldopa.
British Heart Journal 2 (6034):525.

Dighton, D.H. (1980) Sleep Therapy (On Sleep, Homeostasis, and Cardiovascular Problems).
General Practitioner.

Dighton., D.H. (1980) Sinus Bradycardia.
The Practitioner. 224 : 261-266.

Dighton, D.H., Golding, R., de Feyter, P., (1982)
Post-Pneumonectomy Pericardial Effusion.
Chest. 82 : 389-390.

Katona, P.G., McLean, M., Dighton, D.H., Guz, A. (1982).
Sympathetic and Parasympathetic Cardiac Control in Athletes and Non-Athletes.
J. Applied Physiology 52(6): 1652-1657.

de Feyter, P., Van Eenige, M.J., Dighton., D.H., et al (1982).
Prognostic Value of Exercise Testing, Coronary Arteriography and Left Ventriculography 6-8 Weeks after Cardiac Infarction. Circulation 66 (3): 527-536.

de Feyter, P.J., Van Eenige, M.J., Dighton, D.H., Roos, J.P. (1983)
Exercise Testing Early after Myocardial Infarction
Chest. 83: 853-859.

de Feyter, P.J., Van Eenige, M.J., et al, including Dighton, D.H., (1982)
Experience with Intracoronary Strepto-
kinase in 36 Patients with Acute Evolving
Myocardial Infarction.
Eur. Heart J 3:441, 1982.

Dighton, D.H. (1986) Vasovagal Syncope. Lancet 1: 982.

List of Illustrations

CREATED BY DR. DAVID DIGHTON

Fig 1: Anterior Chest Wall. c = costal cartilages and costo-chondral joints. A = Aortic Area. P = Pulmonary Area. M = Mitral Area. **Page 343.**

Fig 2: Schema of a normal (A) and boot-shaped heart (B) on CXR. **Page 344.**

Fig 3: Auscultation diagrams of sounds made by MS, MI, AI, AS. **Page 345.**

Fig. 4: Anterior Left Ventricular (LV) overlapped by the Right Ventricle (RV). As on a PA CXR. **Page 346.**

Fig. 5: Combustion engine cylinder, intake and exhaust valves. Page 347.

Fig. 6: Position of mitral (inlet) valve and aortic valve (outlet) and LV. Ao is aortic valve; MV is mitral valve; LV is left ventricle. **Page 348.**

Fig. 7: Heart valves, all in the same plane. Seen from above. AoV is the aortic valve; PV is the pulmonary valve; MV is the mitral valve; TV is the tricuspid valve.
Page 349.

Fig 8: The coronary Arteries and their major branches. Ao = Aorta; D1 and D2 = Diagonal branches; LAD = Left Anterior Descending; LCA=Left Coronary Artery; Cx = Circumflex; M1 & M2=Marginal Branches; PA=Pulmonary Artery; PDA=Posterior Descending Artery; RMA=Right Marginal Artery; RV=Right Ventricle; SVC=Superior Vena Cava; SA=Sino-Atrial Artery. **Page 350.**

Fig 9: Inter-Atrial Conduction Tracts. **Page 351**

Fig. 10: LV Pressure -Volume Cycle. **Page 352.**

Fig 11: The Frank Starling Relationship. A: Changes of Cardiac Output (CO) with heart rate. B: The effects of higher heart rates on CO. **Page 353.**

Fig 12: Pull Back Pressures. **Page 354.**

Fig 13: Multi-electrode Catheter and His Bundle Recording. **Page 355.**

Fig 14: Einthoven's Triangle. **Page 356.**

Fig 15: An Intra- (I) and Extra-cellular (E) Electrograms. **Page 357.**

Fig 16: P-Wave Morphologies. A is a normal P-wave; B is a bifid P-wave; C is P-pulmonale; D is P-mitrale. **Page 358.**

Fig 17: Sinus Rhythm (A); Atrial Fibrillation (B); Atrial Flutter (C); VF (D); VT (E). **Page 359.**

Fig 18: Various Degrees of AV Block. A is normal conduction; B is 1st degree block; C is Wenckebach block; D is 2:1 AV block; E is complete AV block. **Page 360.**

Fig 19: A is a normal QRS; A is normal; B a delta-wave; C a widened S wave in RBBB. **Page 361.**

Fig 20: STEMI and NSTEMI ECGs. **Page 362.**

Fig 21: Myocyte T-tubules. **Page 363.**

Fig 22: Stylised Phonocardiogram in mitral stenosis; patient in SR with a pliable valve. **Page 364.**

Fig 23: Carotid atheroma (ultrasound image). **Page 365.**

Contents

Introduction	1
PART 1: CLINICAL BASICS	5
1. History Taking	6
2. Clinical Examination	31
3. Cardiac Investigation	44
4. Cardiac Diagnosis	60
PART 2: HOW THE HEART FUNCTIONS	71
5. Essential Cardiac Anatomy and Physiology	72
6. Cardiac Imaging and Physiological Measurements	84
7. Understanding ECGs	103
PART 3: CARDIAC CONDITIONS	116
8. Rhythm Problems	117
9. Valvular Heart Disease	136
10. Congenital Heart Disease	155
11. Atherosclerosis	165

12.	Heart Failure and Cardiomyopathy	184
13.	Myocarditis and Systemic Disease	198
14.	Heart Disease in Pregnancy	212
15.	Cardiac Tumours	220
PART 4: CARDIAC TREATMENT		223
16.	Pharmacology and Therapeutics	224
17.	Surgical Interventions	269
PART 5: PREVENTATIVE CARDIOLOGY		280
18.	The Early Detection of Heart Disease	281
19.	The Inheritance of Heart Disease	293
20.	The Prevention of Heart Disease	305
Figures & Illustrations		342
Abbreviations Used		366
Bibliography		369
INDEX		395

Introduction

Many medical students and junior doctors find diagnosing and managing cardiac conditions more difficult than most other specialties. The same often applies to specialists practicing in other fields of medicine and surgery. The need to explain the key information about cardiac problems, and to provide the evidence needed for sound cardiac management, is the aim of this book. Although aimed at medical students and junior doctors, and those undertaking training in cardiology, some nurses and paramedics should find explanations within that could help them better understand cardiac problems.

Knowing and understanding all this book has to offer will not a cardiologist make; especially not one capable of complex cardiological assessment, clinical judgement, and invasive intervention. It will, however, provide the essential elements for the foundation of safe and effective cardiological practice.

Managing a disturbed cardiovascular system from every physical, emotional, and functional point of view requires some talent. Through knowledge contained within and much hands-on experience, a cardiologist with enough clinical wisdom to benefit cardiac patients could emerge.

Cardiology has developed rapidly over the last one hundred years. The history of cardiac discovery, the development of new techniques, and the advances derived from many clinical trials are all necessary studies. Within, I have summarised the history of cardiology, modern advances in cardiac intervention, and the key references to important therapeutic trials. All are necessary for an appreciation of current cardiac management and essential for those wanting to create a positive impression among their peers and colleagues.

Cardiologists like to quote relevant trials as acronyms, so some of the game-changing and controversial ones appear throughout the text. The extended bibliography provided will enable dedicated students to take their study of cardiology further; to an expert level, should they wish. Those who want details of cardiac surgery will not find enough detail here. They should consult dedicated surgical textbooks on the subject.

Although impressive to be able to quote pertinent scientific journal references, one must be able to discuss the perspective they offer, together with the devils in the details. Many meta-analyses, for instance, lump selection criteria together as best they can, with devils in the details being overlooked for the sake of providing a neat conclusion. From the forest of information available on any topic, experts will decide which research they want to favour, and which they will allow to influence their opinion. These sometimes reveal personal biases, uncertainties, and surprising leaps of faith. It can cause informed experts to appear unsure of what is best to do. Perceiving this can be disconcerting for the inexperienced, biased to think that their teachers and peers must know best (the halo bias). To be competent, one must decide for oneself. To this end, I have provided many key research references which the inquisitive will want to look at in detail, thereafter undertaking their own research.

A few patients referred to cardiologists have no significant cardiac problems, mainly because patients and doctors alike get anxious about heart problems. Many fear missing coronary artery disease and conditions with fateful consequences. Some doctors will refer patients to cardiologists simply to avoid undue patient risk or any risk to themselves.

An area of difficulty for many students and doctors is cardiac examination, especially auscultation of the heart. Another area of difficulty is 'reading' ECGs. Few now appreciate that it was over nine decades ago that cardiologists first started to develop the skills needed to detect and assess congenital and valvular abnormalities using only their eyes, ears, and hands. Using feedback from cardiac catheterisation and surgery, these skills could be developed further. Except for those working on desert islands and in impoverished third world countries, the highly specialised skills of cardiologist doyens like Paul Wood and Aubrey Leatham might now seem antiquated, but they are not. Echocardiography, MRI, CT, and PET scanning are now seen as more useful and reliable. Although it may be fashionable to do so, only a fool throws away reliable clinical tools, however old.

Since the basics are crucially important to diagnosis and management, I have included chapters on cardiac history-taking and physical examination, supported by a chapter on essential cardiac anatomy and physiology.

ECG interpretation is difficult because it requires pattern recognition of abnormalities, subject to many variations. By describing how ECGs are generated, I hope to impart a better understanding of both normality and abnormality.

The two commonest causes of cardiac morbidity are atherosclerosis and arteriosclerosis. It is therefore essential to understand how atherosclerosis affects the coronary and other arteries, and how arteriosclerosis relates to left ventricular hypertrophy and hypertension. The information included aims to explain these processes at clinical, histological and molecular levels.

Patients are most troubled by shortness of breath, angina, heart attacks, and rhythm problems. To consider them fully, one must consider valvular and adult congenital heart disease, the reasons for heart failure, and its optimal management. Less common conditions like myocarditis, cardiomyopathy, cardiac tumours, and heart disease in pregnancy are reviewed briefly.

I have introduced the way forward by discussing cardiac screening, early heart disease detection, and the interventions now suggested for primary and secondary heart disease prevention. After a while, one can feel inadequate as a physician dealing only with established heart disease and symptom relief. That is why, early in my career, I accepted a mission to help prevent or ameliorate atherosclerosis and arteriosclerosis. Important to this are cardiac screening and the inheritance of heart disease. Later chapters briefly cover both.

I have attempted to give those tips and tricks, gained from over fifty years of cardiac practice, that might be most helpful to the inexperienced. These should be of some help to those seeking cardiological competence. For those with some cardiac skills already, what I have described should quicken their progress.

The book includes twenty-three illustrations, drawn by me to avoid copyright and other issues. I apologise for those that are cruder than readers might expect. In days gone by, I would have drawn such diagrams for medical students on a blackboard. I have put them together at the end because they will be found more easily than by searching through the text.

PART 1: CLINICAL BASICS

Chapter One

History Taking

Holding supreme for the solving and management of difficult clinical cases is one epizeuxis: History. History. History.

The Pre-History Evaluation

With new patients, doctors find themselves in a 'blind date' situation. Both the patient and doctor have their aims. The doctor aims to make diagnoses that will lead to successful management if they are correct; patients mostly hope to get medical help. Patients want answers to their problems that will improve their condition and prognosis.

Before any conversation begins, a mutual exchange of non-verbal information will always take place. Before taking a history, a doctor can observe the signs of health and ill-health, or the robustness and frailty of their patient.

A few cardiac conditions will change the patient's appearance, but many do not. Even those who look robust and healthy, can harbour serious cardiac problems. Eventually, most of those with serious illnesses will look ill and have facial changes typical of weight loss (disciplined dieting does the same), if left untreated. Some will appear cyanosed, have breathin difficulties, be in pain, have palpitation or have experienced syncope. Most manifestations of illness take time to develop; some being obvious only in the later stages of the natural history. All those who look ill need rapid appraisal; some will need emergency intervention. The assessment of clinical priority is a crucial art.

Non-verbal behaviour, some of it fleeting micro-behaviour, can be important. Transient, or more permanent facial expressions can reflect pain or nervousness, while observed breathlessness is always clinically important. In noticing the patient's state of dress, age group, gender, gender orientation, and race, clues about how best to handle them can become obvious. One might try to guess the patient's cultural, financial, social and educational status, and that of those who may accompany them. This meta-information is often useful when deciding patient management.

Tense facial features and eyes that seldom blink, allow one to presume worry, pre-occupation, or pain. Any such 'tension', needs explaining. If the excuses given do not make intuitive sense, and the patient tries to change the subject, one must resist being diverted from pursuing answers. In the absence of any 'ring of truth', patients could be hiding important issues. One will risk missing a diagnosis or failing to advise the most appropriate clinical management. If possible, allow patients' answers to emerge without undue pressure. Naturally delivered answers, usually reveal more truth than forced ones. Similar to childbirth, facts often emerge in a natural fashion, although they often need a lot of time to corroborate. Sometimes, clarity will emerge only after several consultations.

From the start of every consultation, make a mental check-list of further questions to ask about unresolved issues and discrepancies. Disinterested doctors will not do this, and may try to avoid solving discrepancies for fear of creating extra work. Detachment from patients is never appropriate, but it happens to doctors under stress, and those with personal issues they take to be more important than those of their patient. They should not continue in their role as a doctor until they have resolved their problems.

Doctors who look at their watch every minute, are less likely to develop patient trust! Patients need time and often our patience. No doctor should tolerate working for any organisation that allows insufficient consultation time and discourages patience. I had to create a practice of my own, to escape the ill-informed dictats of bureaucrats whose job it is to maintain efficient corporate functioning, rather than the sanctity of doctor-patient relationships.

The electro-mechanical nature of cardiac problems allows an algorithmic approach. Unlike general medical practice, one can practise cardiology with little regard to patient understanding. Because physical rules govern the heart and circulation, some cardiologists have ignored humane considerations like patient circumstances, emotional issues, psychology, and socio-economic background. Those who see themselves as mechanical interventionists can come to regard such information as irrelevant. Cardiology thus attracts those happy to focus on physical processes and mechanical functioning. Problems may arise for them, when patients become resistant to treatment, and a patient's background and meta-information become critical to their improved management.

Cardiology has remained removed from William Harvey's original quest, when in 1628, he wrote *'Exercitatio Anatomica de Motu Cordis et Sanguinis in Animalibus'*. He tried to understand why the pulse changed with anxiety and emotion. Unfortunately, he found no answers. He knew nothing of the autonomic connections between the brain and heart; a subject that would wait another two centuries for discovery.

When William Heberden first described angina in 1772, he referred to it as occurring with exercise and emotion (Heberden, W. 1772: *Some account of a disorder of the breast*.

Medical Transactions. The Royal College of Physicians of London. 2: 59-67). Like Harvey, he also cared to take the human condition into account.

History Taking

The primary aim of history taking is to discover diagnostic clues. It should also aim to discover pertinent meta-information. Some clues lead directly to correct diagnoses, and to facts that will later improve patient management. Traditionally, we start with present history, proceed to past medical history, and then to family, social and therapeutic history. There is no need to change this sequence, although one may need to change the order, or bypass some aspects when expedient.

Although there are only a few significant symptoms in cardiology, most need specific definition if they are to foster diagnostic accuracy. The most important symptoms are chest pain, shortness of breath, syncope, palpitation and ankle swelling (a symptom and sign). I will consider them in the detail required for diagnostic relevance.

Chest Pain

True cardiac pain is important to identify correctly (angina or that from a dissecting aortic aneurysm). It can be difficult for patients to describe, their description often merging with that of musculoskeletal pain or indigestion. Heberden's description of pain arising from the heart (angina) needs no alteration.

As Heberden described it, angina is a tight feeling across the chest, coming with exercise or emotion, and relieved by rest. It is more frequent after meals, in cold weather; with arduous exercise (walking uphill or carrying a load), especially when the patient is being stressed by their circumstances. Perhaps because of increased cardiac work, and a greater need for myocardial oxygen, angina then comes all the quicker.

Ischaemic cardiac 'pain', like that felt across the anterior chest, can occur only in the upper arms, face, and upper back and neck. If caused by cardiac ischaemia, all are worse on exercise. Angina can be severe enough to stop patients in their tracks, sometimes inducing facial pallor as a vasovagal reflex response progresses (bradycardia and a low BP).

A key point. Patients rarely describe angina as 'pain'. Because many describe what they feel as tightness, some will deny having chest 'pain' when asked directly. Typically, true angina will disappear a few minutes after stopping exercise.

If you ever encounter a patient with constant, tight chest pain, accompanied by signs of distress (sweating, pallor, low BP, difficulty in breathing), immediately assume that a heart attack or aortic dissection is in progress. If you observe these signs, cut your history taking short, take the BP, get an urgent ECG, take blood for troponin and enzymes levels, and start emergency investigations (CT, for coronary calcium and evidence of dissection) and organise any intervention necessary.

Other causes of less acute chest pain can have other characteristics—sharp, dull or nagging like toothache. These pains can be constant or intermittent (under certain circumstances). For those with persistent pain, ask if it is worse with torso movement or hurts when pressed (as in commonly occurring, musculoskeletal chest wall pain). Breathing may exacerbate it (sometimes pleuritic, resulting from pericarditis), or it may occur only on exercise (pain from particular muscle groups).

One key consideration for those diagnosed with pericarditis or pleurisy, is whether they are ill and febrile. A well person with pain on inspiration is most unlikely to have pleurisy.

One symptom that can confound a correct diagnosis of angina is indigestion. Hiatus hernia and oesophageal spasm can cause tight, central chest pain, not unlike angina. Indigestion, unlike angina, is often relieved by standing erect, exercise and medication (antacids and H_2 receptor inhibitors). Ask for a history of acid regurgitation (usually, but always, in obese subjects), and a history of antacid medication, known to have helped in the past. Never forget the possibility of the patient having both angina and indigestion

A common, misleading form of chest pain, arises from inflammation of the anterior chest wall cartilages. Sometimes referred to as Tietze's syndrome (described in 1921 by German surgeon, Alexander Tietze), chest pain can arise from costal cartilage (costochondritis, or costochondral joint inflammation). The costochondral joints extend bilaterally, along a line from the mid clavicle, downwards and laterally. Because it is usually chronic and persistent, diagnosis is often straightforward. Pressure over the painful areas (lateral to the sternum or over the costochondral joints) confirms it. There is no known cause. I often found the pain to be relieved by weeks of specific exercises (press-ups, and exercises that involve the pectoral muscles).

During the American Civil War, many soldiers developed chest pain of uncertain origin. Stress was perhaps the cause. Based on an analysis of around three hundred soldiers in a dedicated wartime hospital, Dr. Jacob Mendez Da Costa, a Philadelphia physician named it 'soldiers' heart' or 'irritable heart'. It is mostly of musculoskeletal origin, associated in my experience, with the long-term exposure to stress: situations that have consumed the patient's energy, requiring an unusually high level of coping. Such situations are often unpleasant, inescapable, and potentially long-lasting. Losing control can be central to the cause, especially for patients who are more anxious or sensitive than usual to external pressures. Sometimes it will feature as part of PTSD. Sometimes such patients, get referred to cardiologists. The chest discomfort they have, is usually unrelated to exercise or breathing. Some will exhibit hyperventilation and have anxiety-related atrial and ventricular ectopic activity or sinus tachycardia.

Dyspnoea

Dyspnoea is the sensation of shortness of breath. A key question is whether it occurs at rest, or only with exercise. It has many non-cardiac causes that include unfitness, obesity, hyperventilation, anxiety, anaemia, asthma, COPD and the effect of some drugs.

Breathlessness that occurs specifically on lying flat (orthopnoea), is pathognomonic of left heart failure, until proven otherwise.

It is useful to grade the severity of breathlessness; the more severe it is, the more urgent it will be to make a correct diagnosis.

The New York Heart Association (NYHA) classification of heart failure, provides a practical template for grading shortness of breath. We can apply it to mitral stenosis, even though mitral stenosis does not cause breathlessness from left ventricular heart failure. Between doctors, the classification directly communicates severity and the need for some urgency.

This is the NYHA grading system:

Class	**Patient Symptoms**
I	No limitation of physical activity. Ordinary physical activity does not cause dyspnoea.
II	Slight limitation of physical activity with shortness of breath.
III	Marked limitation of physical activity because of dyspnoea.

IV   Unable to carry on any physical activity without dyspnoea.

(Adapted from Dolgin M, Association NYH, Fox AC, Gorlin R, Levin RI, New York Heart Association. Criteria Committee. *Nomenclature and criteria for diagnosis of diseases of the heart and great vessels.* 9th ed. Boston, MA: Lippincott Williams and Wilkins; March 1, 1994).

Original source: Criteria Committee, New York Heart Association, Inc. *Diseases of the Heart and Blood Vessels. Nomenclature and Criteria for diagnosis.* Sixth edition Boston, Little, Brown and Co. 1964, p 114.

Even mild angina (caused by cardiac ischaemia) can cause breathlessness. Ischemia, as a mismatch between the demand and supply of myocardial oxygen, will cause some ventricular muscle to become temporarily weaker; in effect inducing transient heart failure. This is more often associated with anterior descending coronary artery stenoses, than with right coronary and circumflex stenoses.

Other important cardiac symptoms are syncope, palpitation and ankle swelling.

Syncope

Diagnosing various types of syncope is fraught with difficulty because patients find it difficult to describe the events; after all, they may have experienced altered consciousness. Although an oversimplification, consider the three main types as if they were clearly distinguishable (in practice they are rarely so).

With **vasomotor syncope (fainting)**, the patient usually gets a forewarning. The progression of fainting can take several minutes, although patients mostly feel unwell, only once their blood pressure drops. Some will also feel nauseated. Those accompanying the patient, may notice facial pallor. Although the patient may feel the need to lie down, many will feel too embarrassed to do so. When examined, their BP will be low and their pulse, sometimes difficult to feel (described as 'thready'). The pulse will usually be slow.

Some faint at the sight of blood, others with injections or when in pain. Many faint more easily once a febrile illness has developed. Others experience their first faint in the early stages of pregnancy. These conditions lower both the blood pressure, and the threshold to fainting.

Fainting patients can injure themselves while falling. Because most patients faint slowly over one minute or two, they will usually have enough time to sit down or lay flat. Just occasionally, deep unconsciousness occurs within seconds; most, however, become semi-conscious at worst. If called to assist, immediately lay the subject flat (wherever they are). Then elevate their legs, getting them to pump their calf muscles to enhance venous return. Apart from pallor, low BP and bradycardia, those with vasomotor syncope often take a long time to recover and feel better.

In all cases of syncope, interview any witnesses. Circumstances, and some knowledge of previously occurrences can aid diagnosis. Known proneness to travel sickness, and some sensitivity to too much alcohol, are common factors. Many who faint, have had a long-term proneness to it (active vagal reflexes). For reasons I do not fully understand, I have observed vasomotor syncope more frequently among those experiencing stress.

In **Adam-Stokes syncope**, the onset of unconsciousness is either instant or fast. In older adults, complete heart block (CHB) is the commonest cause. Sinus arrest and runs of ventricular tachycardia or fibrillation, are causes in younger patients. The pallor patients typically exhibit, quickly disappears once sinus rhythm returns (replaced by facial flushing). If no help is at hand, and asystole has occurred, ventricular fibrillation could intervene. Patients with CHB need a pacemaker and/or an implanted defibrillator.

Epilepsy

Syncope associated with epilepsy is not associated with facial pallor. A short warning or premonition is common. In *petit mal*, the patient may become 'distant', but there is no tremor or convulsion. The absence of any features of vasomotor or Adam-Stokes syncope, help the diagnosis. The diagnosis can be difficult, even with EEGs taken awake and during sleep recorded after the event. Vagal epilepsy was a concept once postulated (causing sinus bradycardia and hypotension). I have seen no convincing evidence for it, although I accept the theoretical possibility. Since vagal nerve stimulation is now accepted as a treatment for difficult-to-control epilepsy, depression and some stroke rehabilitation, there is no consistency to the idea.

Palpitation

Palpitations come in many forms. They all entail an awareness of the heart beating. Extrasystoles (a common cause of palpitation) and tachyarrhythmias, are all but impossible to diagnose from questioning alone, since most patients are indefinite when describing them.

Palpitations can be sustained or intermittent. Extrasystole is the commonest cause of intermittent palpitation (both atrial and ventricular). Many complain of 'missed beats', although some are more aware of the exaggerated beat that follows an extrasystole after a compensatory pause (these beats have a larger stroke volume). The extrasystole itself, causes little forward flow and often goes unnoticed.

Extrasystoles occur with anxiety, especially after first lying down (due to increased venous return and right atrial expansion). For the physically unfit, they are common after exercise; something rare in athletes.

Important tip. Extrasystole occurring during exercise, should suggest structural heart disease (coronary artery disease and cardiomyopathy), as a cause.

Atrial fibrillation is the commonest cause of transient and persistent palpitation in those over 60-years of age. In younger patients, drug taking (alcohol and cocaine) are commoner causes. Others may have inherited electrical pathway defects, like WPW syndrome, based on the re-entry phenomenon. An ECG at rest can reveal them, although 24-hour ECG recordings, more often reveal the cause. Sometimes, electrophysiological studies will help diagnose them, and define whether ablation is appropriate.

All cardiologists get referrals for palpitation. This is because they can induce trepidation and a fear of death, when they are mostly of little clinical consequence.

Ankle Oedema

This can concern patients who find it difficult to put on their shoes. Prolonged sitting is mostly the cause: gravitational oedema is common in older adults sitting for long periods. It will occur in right heart failure, renal dysfunction and some metabolic disorders, but is also associated with lower leg venous insufficiency, varicose veins and phlebitis (or thrombophlebitis). One-sided oedema should always prompt the diagnostic possibility of a deep leg-vein thrombosis (DVT).

There is an art to the practice of medicine that develops from experience. The way each doctor uses the science of medicine as a tool to benefit patients, expresses it. Doctors proficient in this art, can change their history taking style in response to each patient's background and personal characteristics. Now consider some different styles.

The Executive Style

As the fictional US army officer, Dr. Major Charles Emerson Winchester III, said to the character Radar in the TV series 'MASH', when starting a discussion: "Be quick, and be gone!"

Most doctors prefer this style. It requires specific questions, is business-like, formal, and to the point. If there are significant psycho-social aspects to the patient's condition, they are unlikely to get exposed; indeed, they may be avoided on purpose. Most cardiologists I have known, had no wish to expose personal issues. Doctors who limit themselves to this style, waste no time on small talk, and can miss important aetiological factors outside of their usual frame of reference. There are patients who prefer doctors to restrict themselves in this way, hoping to limit the depth of enquiry and any undue revelation. By respecting these wishes, key issues can remain hidden. Incorrect diagnoses and ineffective management can follow.

A distant, anonymous and detached approach, is a fully workable position for most cardiological cases, even though it may bypass the fullest of evaluations, and the best of clinical judgements. It may explain why some clinicians cannot manage complicated cases successfully, and why some patients prefer more engaging doctors. To be disinterested in a patient's personal life, is to relegate them to being pathophysiological tissue only (a number, not a person). Detached doctors using this approach, happily populate cardiac centres of technical excellence, where the desired focus is pathophysiology, anatomy, and mechanical correction, rather than humanity. A combination is possible of course!

The Focussed Style

This is a sub-type of the executive style, useful in cardiology and other specialties with circumscribed symptoms. Aubrey Leatham adopted it with questions that were often limited to: "Are you breathless walking up stairs. How many flights of stairs or steps can you manage before feeling short of breath? Can you lie flat? If you walk quickly, does it cause chest discomfort? Are your palpitations regular, or irregular? Have you had rheumatic fever? Did your blackout come without warning?"

The information gained, helped him grade the severity of coronary heart disease and mitral stenosis, and to guess the type of syncope. Clinical expertise depends a great deal on the ability to weigh the significance symptoms correctly in relation to any diagnosis. Specialists in all fields of medicine have their favourite questions. Experience has taught them which are sensitive and specific enough to provide the positive and negative diagnostic accuracy they seek. This is an art.

The Interplay Style

As Judge Judy once said to a meandering defendant: "I don't need your story from 'Once upon a time'!"

Withholding information, giving too much information, and giving selected snippets of information, can all occur as the doctor-patient interaction proceeds. This interaction, one could regard as a transactional 'game' (as Eric Berne defined it in, *The Games People Play*). When we suspect that someone is holding top secret information, or information too sensitive or too important to be revealed, it can be beguiling. A common inter-personal game is *'poor little me' (I am just someone nobody can help)*. There are many attention-seeking and other gambits, but beware most of being duped, and a failure to make a diagnosis caused by withheld or falsified information. If you ever suspect a patient-doctor game in progress, revert to the succinct executive style.

The Discursive Style

I used to talk too much, but then I always had the time with patients I needed. It helped patients relax and allowed some to 'open up' and provide me with significant insights into their problems. It annoyed others. As an interview technique, this requires one to be relaxed about technical clinical issues, and to have time enough for individual personal considerations.

Through chatting, and having an open discussion, many patients will sense that one has their best interests at heart. With enough time to assess you, they might find reasons to trust you. From this privileged position, helping them is easier. As trust develops, you might venture to put yourself in their place, and even preface your advice with, "If I were you . . ." This forthright approach is unacceptable to many professionals, be they a doctor, lawyer or accountant. A willingness to share their identity, is a step too far for them.

This style aims to define a patient's inscape — the distinctive design of their mind that makes up individual identity (Gerard Manley Hopkins. Poet and Jesuit priest:1844-89). The medieval philosopher John Duns Scotus (1266-1308), referred to this unique human property as haecceity.

The discursive style resembles a fireside chat. It is most suitable for introvert patients; many extroverts prefer the executive style. The discursive style will allow some patients to forget their shyness and to express themselves freely. They are then more likely to reveal what is really bothering them. Doctors and patients lacking inter-personal skills will find this style awkward.

Most patients' complaints arise from the insufficient explanation of medical matters given by medical staff (usually doctors).

For those patients with significant psycho-social problems, nurses and a social work team may be of more help than a doctor. In my early days, ward sisters and lady almoners accepted this role as an integral part of their work.

One can of course, blend history taking styles, and employ them in hard or soft ways. Colloquially, one can choose to 'put the boot in' or walk on eggshells. To use both consecutively, mirrors the 'good cop, bad cop' technique, used by police when dealing with incommunicative captives.

Resistance to questioning is common among those patients who have been sent to doctors; perhaps by their boss or partner. Because someone else has required their attendance, they may choose non-compliance. Some will expect results that are not in their best interest.

During the process of history taking, awkward and embarrassing moments, emotional reactions, and *faux pas*, can all occur. There is an art to handling each of them sensitively.

In complicated or difficult to resolve cases, it is often best to get contextual, coexistent meta-data, aiming to complete the clinical picture from a fly-on-the-wall perspective. As history taking proceeds, be sensitive to the possibility of missing information, and the reasons for its omission. Assessing the factual and emotional relevance of what patients say, or try to say, is essential when asking supplementary questions to define their circumstances.

Forensic History Taking

Barristers employ this style. It is an infrequently needed medical style of questioning, but one well worth knowing about when dealing with those who withhold information, those giving mis-information, and the uncommunicative (for reasons best known to themselves).

Language Use

'16% of adults in the UK are 'functionally illiterate'. National Literacy Trust (2020)

Educated people have a larger vocabulary than the poorly educated. Choice of words, and the correctness of grammar, accurately reveal educational status in the UK. They do not, however, accurately predict socio-economic status or intelligence. Many of the most intelligent and successful business people I met, functioned well in business without too many words. Some were illiterate and needed a companion to complete forms for them.

Without a large vocabulary, clarity of thought and communication suffer. Some patients cannot express their ideas and or their intended meaning well enough for others to fully understand. Those who cannot say what they mean, are liable to get frustrated and angry when challenged.

I have met many whose vocabulary barely exceeds 300 words; they often use colloquial expressions and slogans that are emotive, but lack descriptive clarity: "It's a no-brainer!" (it's obvious); "Not on your Nellie!" (never to be considered); "He went mental!" (he lost control) "What are you like?" (I can't believe you did that); "one-hundred percent!" (I totally agree). Although most doctors will probably guess what patients mean, not all patients will understand what doctors mean; not unless they choose their words carefully in order to be understood. To communicate well, one must use the language and idiom appropriate to the patient. Words from the relevant local dialect can help, but this requires local knowledge and experience.

Masters of history taking will choose the style that best suits each patient: one style to get the facts, another to put relevant leaves on the data tree. Like a talented actor, doctors sometimes need to change the language they use, and their style of delivery (tone, volume, character, and expression), manner (respectful, or not; sympathetic, or unsympathetic), and demeanour (supportive, or dismissive), to best serve communication. Ideally, one would speak Cockney to Cockney's; Liverpudlian to Liverpudlians; Geordie to Geordie's, and Scots to Glaswegians. Attempt these dialects only if fluent. Sharing cultural identity helps, but being thought patronising, obsequious or ridiculing, could permanently damage a doctor-patient relationship.

There are other aspects of linguistic diplomacy. Try not to practice your Greek on Turks; your Turkish on Greeks, or your Russian on Ukrainians, Poles and Czechs. Memories of national conflict have lasted generations. It is less of a problem than once it was, but such *faux pas* can still embarrass. Ask for the patient's mother tongue and what other languages they speak, before launching your linguistic talents. Unlike the average Brit, our European compatriots (now ex-compatriots), often speak two or three languages. Because many want to learn English, they will mostly forgive British linguistic ineptitude.

If you are not sure of what to say in a difficult situation, say nothing! Your body language will, however, not go unnoticed.

Past Medical History

There are many sources of past medical history. The most usual are the patient, their family, friends, workmates, and the clinical records. Much of the patient's past history will be irrelevant, but one or two pertinent facts can sometimes prove crucial.

The diagnostic relevance of some information can fade with time. From a cardiological point of view, there is a now less need in Europe to enquire about diphtheria and rheumatic fever. This is not the case in some 3rd world countries where some diseases, once common in the UK before the 1950s, are still present.

Rheumatic fever was the major cause of heart valve problems seen by European and US cardiologists, prior to the 1970s. Fewer now have post-rheumatic heart valve problems (thickened or calcified valves that are stenosed or incompetent). One concern I have, is that rheumatic fever, mastoiditis, tonsillar abscess, etc. could all recur, now there are strict limitations placed on prescribing antibiotics like benzyl-penicillin and erythromycin for minor conditions. The aim, however, is worthy: to prevent the world-wide growth of antibiotic resistance. Here, worthiness is no measure of clinical appropriateness.

A small percentage of patients with sino-atrial block, or complete heart block, once had diphtheria. Diphtheria toxin is an anti-cholinesterase (a ubiquitous enzyme in the cardiac conducting system). With the enzyme inhibited, acetylcholine will persist, possibly resulting in atrial bradycardia and AV conduction delay.

The history should include details of past travel. I saw my first and last case of diphtheria, ten years ago. The patient had just returned from a holiday in Venezuela. The grey film covering the back of his throat was pathognomonic.

In my whole career, I saw only two cases of acute rheumatic fever. I had to discharge one young patient from a private hospital, in order to treat her appropriately. She was 8-years-old, and rules created to prevent Reye's Syndrome, restricted aspirin use for those over 12-years-old. Having unsuccessfully prescribed an NSAID, I discharged her to be treated at home with aspirin alone (with the fully informed consent of her parents). It helped immediately. Aspirin remains the mainstay of treatment for uncomplicated rheumatic fever. Thirty years later (still my patient) she had not developed clinically detectable renal, cerebral, or cardiac damage.

In the latter case, my judgement (having tried the patient on other anti-inflammatory drugs) was to ignore the vanishingly small risk of Reye's Syndrome, in favour of gaining an immediate clinical improvement with aspirin (as happened). Fixed, arbitrary rules can often conflict with experienced clinical judgement. When this happens, it is understandable that most doctors will feel obliged to follow an inappropriate therapeutic rule to the letter (to avoid regulatory action).

And now for a brief political aside, some may wish to skip.

Doctors are right to fear regulators Many doctors are driven to conform, even if their actions are clinically inappropriate. Experienced shepherds who never lose sheep, are the only ones trainee shepherds should follow. Unfortunately, there are too few experienced shepherds available. I wonder if actual shepherds must tolerate farm regulators: those who know nothing much about farming or sheep, yet have the authority to tell experienced shepherds how best to manage their flocks.

One should always question generalised rules. While most often they can provide intelligent advice, they should not cause us to abandon our clinical experience and judgement. The ability to use one's experience and judgement is a major aspect of clinical mastery. Beware of using your judgement to adapt a guideline; it could expose you to regulatory action.
When I was at medical school, only those who could think for themselves were eligible to become doctors. To-day, I wonder if the medical students now chosen are those most likely to conform to sets of rules and regulations, rather than practice their judgement. Medical bureaucracy has a mission to outlaw personal clinical experience and judgement. Only in this way, is standardisation and medical practice regulation possible. Because their work is large scale, corporations need standardisation to control the management process. Although a mission the corporate NHS has set itself, it is impossible to standardise doctors, nurses, and the illnesses patients suffer.

In this age of supposed improved communication, access to patients' records should be easy (for those with the patient's permission). It is not so, despite the many promises made over decades by IT professionals. The IT firms employed by the NHS, have earned fortunes from their promises, despite their many failures. My older foreign patients had the perfect answer: they carried their written notes with them at all times!

Should we trust the State with any patient information, especially if it could make the difference between a patient's life or death in an emergency? Even now, electronic record systems have a long way to go before they can match the security, completeness, and accessibility of hand-written notes. Some of my patient's files were 50-years old when I digitised them. Since I gave each patient a summary of their condition (after an in-depth consultation and investigation), they were always free to present my findings and opinions, to any other doctor of their choice.

In order to brief myself, before seeing any patient, I could always rely on my previous summaries. After spending time with patients, and uncovering relevant clinical information, I committed all the details to a coloured summary sheet (pink paper was easily spotted within a thick A4 file). Most of my colleagues achieved the same end, by writing letters with a header summarising the clinical facts of each consultation. Summary sheets are most useful when they list all relevant diagnoses to date, treatment, previous interventions and the thinking behind each opinion and action. With a detailed history summary, one can ask patients more pertinent questions (patients will use this to grade doctors); they also help to avoid the repetition of investigations and repeated wrong diagnoses.

Being well briefed is something every intelligent business executive expects from his accountant, lawyer or surveyor. Why should doctors be exempt? Being well briefed, denotes personal engagement and a mark of professionalism. To give the best advice possible, one must have access to each patient's present, past, family, social, and therapeutic history.

Fail to prepare and you must be prepared to fail.

Many patients will judge clinical efficiency from the number of specific and personalised questions a doctor asks. For instance: "How has your shortness of breath changed since your CABG operation three years ago?" and, "You don't respond well to NSAIDs, do you?" Patients take such questions as signs of personal engagement. Knowledge of a patient's intimate details, and their personal story, will characterise you as an attached, caring doctor; not one who wishes to be detached and to remain anonymous.

Mastery of the clinical facts, fosters patient trust. Losing a patient's notes may be understandable, but you and the system you work for, risk being judged incompetent. All doctors should question the wisdom of working for any organisation incapable of providing patient notes when required.

Family Medical History

Family history enquiries might need to extend to great grandparents and cousins. Since only a few patients have any in-depth medical knowledge, the accuracy of family history information is likely to suffer from hearsay, misinterpretation, omission, distortion, and modification of the Chinese whisper sort. There are conditions, such as syphilis and suicide which some families will choose to forget.

I have often asked patients to enquire among their relatives for family history verification. As with all other answers to medical questions, one must view each as positive, negative, false positive or false negative. In this way, Bayesian considerations can assess their likely diagnostic accuracy. Ask for copies of hospital notes and death certificates, in order to reveal the evidence, although, these are also subject to error. If you watch detective dramas, you will know about devils in the detail and how persistence can make or break an enquiry. The same principles apply to scientific and medical enquiry.

A positive family history can signify an increased antecedent probability of the same medical condition recurring. Inherited conditions are as diverse as coronary heart disease (CAD), hypertension, allergy, psoriasis, osteoarthritis, prostate cancer and migraine.

From a clinical point of view, coronary artery disease and hypertension are inherited predictably. The likely genetic contribution, has yet to be identified and verified scientifically, but from experience it is in the order of 95%. In individuals, one can use family history as both a positive and negative predictor. If a patient has one parent with these conditions, their risk of developing the same will be at least 25%. By the same token, a completely negative family history makes any similar occurrence unlikely. Such predictions are especially useful when assessing the significance of chest pain, and the true nature of hypertension (true hypertension as opposed to 'white coat', labile hypertension).

A simple Mendelian 'rule of thumb' is useful in predicting autosomal, dominantly inherited conditions (hypertension and CAD). If one parent has the condition, expect one in four of their offspring to develop it. If both parents have it, expect at least 75% of their offspring to develop it. These are short odds, readily confirmed from experience (despite relatively small numbers). From clinical experience alone, the chance of occurrence (4 : 1), is useful diagnostically.

The inheritance of both diabetes and cancer is much more complicated, and far less predictable. Even with 50+ years of clinical experience, I have only rarely seen cancer inherited (breast and prostate cancer). The few examples I saw, could easily have been coincidental.

It is useful to appreciate that inheritance can result from a matrix of genes. In straightforward Mendelian inheritance, a 2 x 2 matrix can exist (two genes from the father; two from the mother). In conditions with a complicated inheritance, many more genes may be involved. There could be a 16 x 16 or 32 x 32 matrix of relevant genes, each contributing to the phenotype. If you imagine a matrix as a crossword puzzle with many boxes, the number of 'boxes' ticked (the genotype), and the strength of their influence (their penetrance), will help to determine the clinical features (phenotype).

I had a friend with severe type 2 diabetes, CAD, and peripheral vascular disease. He had no retinopathy or peripheral neuropathy as one might expect. Obviously, patients have their own genetically pre-determined set of clinical features. The complete genotype for Duchenne muscular dystrophy is now known. If all the boxes in the relevant matrix are 'ticked', the patient will have the full-blown condition; with only one box ticked, the patient might just have weak legs, and wonder why they never won races at their school's sports day.

Fifty years ago, few people knew much of their family history. With improved availability of family information aided by longevity, this no longer applies to many young patients. Many have living grandparents and great grandparents. A family history has thus become more reliable. Whenever able to check a family history for morbidity and mortality, I found accuracy to be poor. Relatives more reliably report events (CABG and artery stenting) than diagnoses.

Both my father and his mother, told me that my grandfather had died of alcoholism. My father, 18-years-old in 1922 when his father died, remained a lifelong non-drinker. My alma mater, the London Hospital, had my grandfather's admission notes on microfilm, so I checked my grandfather's clinical data. He actually died of pneumonia, although on admission, his breath did smell strongly of alcohol. Looking through his clinical notes, written in an impressive copper-plate style and reproduced as if photo-copied (rather than musty old originals), his treatment surprised me. They gave him Tincture of Lily of the Valley and other inefficacious medicines. A strange urge overcame me. I felt the need to telephone the hospital and suggest they prescribed him penicillin. My call would have been 78-years too late, and 19-years before penicillin was first given!

Social History

The patient's social history should include alcohol use, cigarette smoking, drug use, occupation, social status, relationships, lifestyle, living and working conditions. Significant life changes (divorce, bereavement, bankruptcy) can be important; they can influence cardiovascular morbidity and mortality, and help to determine the prognosis. In 1967, Holmes and Rahe were first to publish their work on life changes and the attendant risks. (Holmes, T.H, Rahe, R.H. *The social readjustment rating scale.* Journal of Psychosomatic Research. 1967; 11:213–218). Life changes can cause depression, and that is associated with CVS risk (Hare, D.L, Toukhsati, S.R, et al. (2014). *Depression and cardiovascular disease: a clinical review.* Eur. Heart J. 2014;35:1365–1372).

Liberal, politically correct thinking in 2024, would have us resist recognition of distinct socio-economic classes, race, and their cardiovascular consequences. On the street, in every society, both 'insiders' and outsiders' are recognised, regardless of establishment political correctness. The 'outsiders' often need to withstand social discrimination.

Doctors born outside of the UK might find British attitudes to class, somewhat baffling; especially if, in their country, there is a President and not a royal family. As one example of cultural difference, one might ask who has the highest social status in the UK? Is it a village priest, teacher or billionaire? In many countries, it will be the billionaire; in the UK, nurses, teachers and priests may not be rich, but they have the higher social status. Citizens of countries where wealth defines class, will often find UK social politics quaint and anachronistic. It may not be obvious, but in UK society, class and education still define the social strata much more than wealth. Wherever one works, every nurse and doctor will have to contend with many differing cultural presumptions, beliefs, values, educational backgrounds and behaviours held by colleagues and patients.

When medical management is being considered, social differences are of practical significance. They can challenge our ability to value and understand a patient's symptoms, our ability to make an argument understood, and our sensitivity to the resources available to them. A patient's ability to understand and accept evidence-based reasoning, is sometimes of life and death significance. Together with lifestyle differences, these factors partly explain the health divide (the three to five-fold difference in morbidity and mortality between the social strata) that exists in all nations.

For further consideration of these factors, read my book on *The Art and Science of Medical Practice* (2024).

Therapeutic History

There is more to therapeutic history taking than listing a patient's drugs. Therapeutics is a vital branch of clinical knowledge, too large to summarise here. Instead, I offer some tangential views of drug use, to show the breadth of experience required to achieve therapeutic success while minimising risk.

Forms of therapy other than pharmacological can be useful. Talking therapies (cognitive therapists, psychotherapy and psychiatry), manipulative interventions (physiotherapy, osteopathy), and lifestyle changes when offered as 'treatment', can all have benefits. Those with mechanical cardiac problems primarily need plumbers (cardiac surgeons) and electricians (electrophysiologists), not psychologists!

Many patients see themselves as health experts, so the therapeutic history must include the non-pharmaceutical preparations they believe in and consume regularly. One must consider the lifestyle and dietary habits they have adopted for health reasons. Vitamin and health food supplements now represent a 315 billion-US dollar industry; one that is growing rapidly. Many of the mixtures some take every day, are potions concocted from fruit, vegetables and supplements (minerals, vitamins, amino acids, ginseng, garlic, etc.). Perhaps they need to be swallowed with two teaspoons of mythology, a tablespoon of pseudo-science, and a whole packet of scientific evidence. Many now share a nature-based mind-set, readily accepting that sulphurous spring water from spas will fortify health; that ginseng, garlic, soya, vitamins, lots of fresh air and running naked through snow-clad forests, will prevent disease. Powdered parts of animals, like tiger, shark, elephant, and rhino, still hold their place as medicines in some parts of the world.

When health rather than disease is the issue, belief in these therapeutic alternatives is of less concern; what patients believe in, can affect how healthy they feel. When disease prevention and cure are the issues, there is rarely enough evidence to support the many beneficial claims made for alternative medicines, except for a placebo effect. Unfortunately, those who believe in snake oil, often choose to ignore scientific evidence. However cynical one becomes, however, one must never underestimate the therapeutic power of a placebo (20% of those who take them, report a benefit). To experience the benefit of a new mitral valve or a pacemaker requires no belief; the physical benefits that follow become all too obvious.

Running counter to the ready acceptance of alternative therapies, are the double-blind controlled trials and meticulous safety-checking required of the pharmaceutical industry when they introduce new drugs. Nevertheless, mistrust in pharmaceutical drugs persists. The uptake of COVID vaccinations in some population sub-groups was evidence of this. Long before that, the Victorians would omit their medications on Sundays; it allowed the presumed toxic effects of their drugs to 'clear the system'. This was a good thing for those taking digitalis.

Patients harbour many attitudes and beliefs about pharmaceuticals. Coming to know what they are when taking a therapeutic history, can help when prescribing, altering or withdrawing medication. A few drugs have caused havoc in the past (like thalidomide). These changed patient attitudes to medication.

We all deserve validated assurance about drug efficacy and compatibility, but it has led to something unfortunate—the medico-legal necessity of listing every drug side-effect in information leaflets. Although correct and essential, these have not fostered patient confidence. We will have to await genetic profiling before we can better predict individual responses to each drug.

Drugs of addiction are now a major problem. Every doctor needs to know about the latest street drugs, together with those that have been used for centuries. One should never dismiss the effects of alcohol, ephedrine analogues and cocaine, in cases presenting with palpitation and hyperactivity.

Not To Be Missed

In urgent cases, only a few essential facts will be pertinent; a full history may be unobtainable, irrelevant, or unnecessary. How much information do we need to deal with a severed hand, a bleeding nose, a CVA or an unconscious, syncopal patient? One glance will often decide the action needed.

Whenever a diagnosis is elusive, retake the history before ordering batches of further investigations. If the diagnosis remains elusive, take the patient's history in the presence of their friends or relatives; those they trust and know well. I have found this an invaluable strategy. When flummoxed, unsure, or bereft of diagnostic ideas, resist getting second opinions and ordering more investigations; re-take their history again. General conversation and questioning, often flushes out important information. Based on the 'horses for courses principle', the time, place and style of further questioning, should be appropriate. In difficult to solve cases, always ask the patient (or his relatives) what *they* think is the problem.

There are several classes of unhelpful histories: fraudulent and false ones (encountered in some insurance, and legal cases); those that are deliberately misleading (to avoid embarrassment, etc.), and those that are absent (dementia, or convenient omissions for the sake of pride or politics). You may need a relative or friend to supply the information.

One can often sense when information is missing from the picture being painted by the patient. Why might someone want to withhold information? Do they have an undeclared agenda? Are there facts they do not wish to declare? It may be important at this stage to convey the importance of getting information to patient prognosis.

I have sometimes used a simple metaphor while trying to put prevailing psycho-social stresses into perspective. An elastic band left on a desk, never snaps of its own accord. I would follow this by asking, "What has stretched your elastic (or made it snap)?" Some patients will be forthcoming; others will look blank, remain in denial, or choose to evade further questioning. Some will suggest that their symptoms are a mystery, and that they make no sense, perhaps implying, 'I have successfully defeated many doctors with these suggestions. Now it's your turn to try!'

With detailed personal knowledge of a patient, usually comes the ability to predict their acceptance of advice, be it to take a statin drug or submit to a coronary stent or coronary by-pass. Your acquaintance with patients should not, however, interfere with your judgement and advice, or become detached from conveying your commitment to their best interests.

For diagnostically impenetrable medical cases (diagnosis not obvious, even after several encounters), you may need to employ forensic history taking. This is especially important if the patient is 'acting dumb', being secretive (hiding embarrassing facts), or playing hard to get (for reasons that remain elusive). Occasionally, it might become expedient to cross-questioning their friends, relatives, and work associates.

Chapter Two

Clinical Examination

A nursing home nursing manager once called me to examine an elderly lady; she was unwell and fast deteriorating. The patient's GP had visited, found her with oedematous legs, and prescribed diuretics for what she presumed to be heart failure.

The observation of dependent oedema was correct, but did she try to corroborate heart failure? Without a raised jugular venous pressure (JVP), a diagnosis of right heart failure could be incorrect. In borderline heart failure, the JVP might rise only after exercise Also, an effective diuretic can lower the JVP in heart failure, obscuring the clinical diagnosis. Misinterpreting the cause of ankle oedema (the result of sitting all day), especially for elderly patients, often results in inappropriate prescriptions for digoxin and diuretics.

The patient became weaker, albeit with some improvement in her oedema. The combination of dehydration and hyponatraemia, caused her to become weak and thirsty all the time. She had a dry mouth, inelastic skin, and a JVP around zero. Digoxin had caused her to become nauseated (associated with typical T-waves on ECG, and a high blood digoxin blood level). Having stopped all her treatment, I revisited her 10-days later. She was no longer nauseated and was feeling better. She also had a normal JVP (before and after a short walk).

My conclusion was that her original oedema was gravitational, caused by sitting for long periods, and not the result of heart failure.

Diagnosis on First Sight

The examination process starts before the first 'Hello'. Facial features and expression can be valuable if the patient's state of mind is on display. Dress standard, demeanour and cleanliness (hands, shoes, and clothes), can be important. Other easily observable signs are gait, tremor, jaundice and cyanosis of the lips and nose. A distraught demeanour, and difficulty with breathing, are part of a long list of instantly observable, invaluable physical signs.

Your initial observations should direct your history taking, especially if you have already made a preliminary diagnosis. Parkinsonism, acromegaly, myxoedema, arthritis, heart failure, and possibly depression can all be observation-based diagnoses, awaiting verification. One should explore every presumptive diagnosis during history taking and examination. An accomplished observer is always on the lookout; primed not to miss any tale-tale sign of disease or discomfort. Disengaged observers can start an examination, hoping not to find a problem.

I have fond memories of a brilliant teacher. He helped me prepare for the MRCP examination (Membership of the Royal College of Physicians). Dr. Alan Gelson, was tutor to both Dr. Paul Kligfield and myself while we all worked at St. George's Hospital, Hyde Park Corner, London, in the 1970s. When we ventured out into Knightsbridge or Piccadilly, Alan would ask, 'What's wrong with that chap over there?' He expected his pupils to make instant diagnoses from a distance. I still try. Because they so influence one's understanding and appreciation, one will rarely forget a brilliant teacher.

'Chance', said Louis Pasteur, 'favours the prepared mind'.

This is never more so than when needing to recognise significant physical signs.

Not to be Missed

It helps to devise an examination routine. This is one tick-box method worth adopting. Once learned, no further thought is required to avoid omissions. Both positive and negative observations are important; both provide valuable clinical evidence. One problem to remember is that some physical signs take time to develop; many are absent in the early phases of disease development.

Examination requires physical contact. After obtaining permission, hold and examine the patient's nearest hand. Observe its colour and temperature. Is the skin dry or supple? Are the nails bitten off or deformed? Is there clubbing, cyanosis or pallor? Ask the patient whether they are right or left-handed. Ask questions that will establish rapport. Note the hand joints: Heberden's nodes on distal finger joints (osteoarthritis); enlarged MP joints in rheumatoid arthritis. Some processes causing inflammatory heart disease, also affect the skin (SLE) and joints (RA).

Now feel the radial pulse. Is it fast or slow? Is it regular or irregular? Record the rate. It is often good enough to approximate this (fast, average, slow). Measure it accurately if appropriate, it could be relevant later when assessing treatment. When fast, think of anxiety, blood loss, febrile illness, thyrotoxicosis, drug intoxication (cocaine etc.). When slow, think of hypothyroidism, and various types of electrical heart block. More common than either are the effects of medication (β-blockers and digoxin).

With experience one should be able to detect the character of the pulse waveform. It can be slow rising with aortic valve stenosis and sharp with aortic incompetence. If in doubt, go to the neck to feel the carotid pulse. A slowly rising carotid pulse waveform from significant aortic stenosis, and a short, sharp pulse with aortic incompetence, only occur when the valve is significantly abnormal. By 'significant', I mean bad enough to be considered for surgical intervention.

When the pulse is irregular, there are a few immediate considerations. In younger patients, repeated extrasystole is often the cause. If this is the case, the beat will always return to synchronise with the previous regular rhythm (clumsily referred to as 'regularly irregular'). In older adult patients, think of atrial fibrillation (AF). In AF, the rhythm never resynchronises with the regular beat (as if metronomic). Because it is never 'in sync', we refer to it as 'irregularly irregular'. Because AF can sometimes seem regular, feel the pulse for at least one minute before deciding. Any true irregularity should then be apparent.

Age-related, subendocardial atrial fibrosis (the commonest cause), is a common cause of atrial fibrillation. Thyrotoxicosis, ischaemia and atrial dilatation are other causes, not to be missed. For corroboration, look for sharp tendon reflexes in thyrotoxicosis (quicker than normal contraction and relaxation phases), and when hypothyroidism is being over-treated. In hypothyroidism, under-treatment with thyroxine, causes the reverse. The relaxation phase of reflexes will often be noticeably slow. The tissue effects of two thyroid hormones, T4 and T3 are the cause, but blood levels do not always reflect their tissue effects. Remember that T3, usually has more pronounced effects (exophthalmos and pulse rate) than T4.

One cannot afford to miss colour abnormalities: skin and conjunctival colour, lip and buccal cavity colour, fingertip and nose colour. Note any rashes, skin haemorrhages, and spider naevi; they are always of significance. Diminished peripheral blood flow can cause cold blue hands and a cold nose.

In Shakespeare's Henry V, the Hostess recalls John Falstaff, asking her to lay more clothes on his feet. Falstaff was dying in his bed. She observes, 'So a' bade me lay more clothes on his feet: I put my hand into the bed and felt them, and they were as cold as any stone'. Henry V. Act II, Scene III.

Either a poor cardiac output or a slowed peripheral perfusion (Raynaud's or peripheral artery disease) are the cause. If the lips and hands are blue and warm, the patient may be centrally cyanosed (from COPD or blood flowing right to left, perhaps through an atrial septal defect). Note the patient's breath. It can have a distinctive odour (foetor) in dehydration, ketosis, renal and hepatic failure.

Obesity, frailty, and wasting, should all prompt enquiries about diet. Weight loss and gain over time, are useful for assessing the stage of illness. In the neck, never forget to feel for lymph nodes, if only for completeness.

Always try to integrate your examination findings with the historic data. Give thought to any discrepancies.

It is important to learn from pathological feedback. From every confirmed, final pathological diagnosis, one can learn in retrospect which points from the history and clinical examination, held most and least diagnostic value. All clinical presumptions need evidence-based confirmation. History taking and clinical examination are searches for evidence, with our biases, assumptions, presumptions, and personal interests, continually tempting us to value evidence wrongly. Sometimes, the accuracy of our diagnoses will depend on how well we deal with these biases. Rush your history taking and examination then jump to conclusions, and you will make diagnostic errors.

Examining the Heart

When examining the heart, the patient positioning is important. Several positions are useful:

- Reclining at 45º on a couch or chair.

- Lying flat.

- Sitting, torso inclined forward.

- Lying on the left side, left arm above the head.

When first examining the heart, the patient can be sitting. In this position, one should first observe lip colour for pallor or cyanosis. Observe the chest movements with breathing. Is there hyperventilation (deeper and / or faster)? Is breathing laboured (intercostal muscles and the supraclavicular spaces drawn inward)? Do both sides of the chest move equally? In obese patients, note the form of the neck. A short neck, with an expanded circumference, could imply obstructed breathing during sleep (sleep apnoea).

Next observe the JVP and carotid pulse. With the patient reclining (45° is traditional), observe the top edge of the jugular venous pulse (internal or external jugular). Indirect lighting in a darkened room can help; bright lights sometimes make it more difficult. Incline the patient's sitting angle in order to see the top of the JVP waveform. It is the height of the JVP above the atrium (sternum) that counts, not the sitting angle. This height equals the blood pressure in the right atrium.

After that, the patient should lie at 45° on their left side, with their left arm raised, and their left hand behind their head. In this position, one can detect LVH and mitral murmurs are most easily heard. After that, sit the patient upright and lean them forward. With the patient holding their breath, after exhalation, one can usually hear the murmur of AI, either in the aortic area or at lower left sternal edge. The flat diaphragm of the stethoscope transmits the high-frequency sounds best (the mid-diastolic murmur of MS is low-frequency, and best heard with the bell of the stethoscope).

With the patient lying flat, one might feel a right ventricular heave at the lower right sternal edge or beneath the xiphisternum, as evidence of RVH.

Measuring the JVP

The JVP is a measure of right atrial pressure, as a column of blood. It will be raised in right heart failure and diminished in dehydration; the combination can normalise it. Observing the waveform can help to diagnose atrial fibrillation and tricuspid incompetence.

Two situations make its measure difficult: one when the JVP is high—above the angle of the jaw (in heart failure at rest, or mediastinal obstruction); the other is when it is low or negative, in dehydration (excessive diuresis perhaps).

Only with the patient lying flat, can a low JVP be measured. When normal or high, measure it with the patient sitting upright in a chair (especially if high), or more usually when inclined at 45° on a couch. The object of the exercise is to visualise the top of the venous wave-front, and to measure the right atrial pressure in centimetres of blood above the right atrium (which lies behind the upper sternum), whatever the position of the patient.

There are some important technical points to observe. With the patient sitting at 45°, and chin elevated, turn the patient's head away and look for the fastest inward neck movement (the 'y' descent of the venous pulse); this will be visible, somewhere between the clavicle and the earlobe. Change the patient's elevation and head position when necessary, in order to visualise the inward pulsation. The key question is: does what you see have a fast **outward** waveform component (carotid artery pulse), or is it predominantly **inward** (the jugular pulse)? The internal jugular, is visible even in obese people with short necks (candidates for sleep apnoea). You do not have to rely on viewing external jugular pulsations; the internal jugular is more reliable.

KEY TIP: the JVP is better seen using indirect light. It is best seen (if found difficult) in a darkened room with a single light source playing across the skin, rather than in a room with bright diffused light.

By slowly elevating an outstretched arm, one can sometimes observe at which point the dorsal hand veins collapse. This is far from being a reliable technique for measuring high central venous pressure, but try it sometime.

Look just below the angle of the jaw to view the highest of jugular venous pressures. You can sometimes lower this pressure to make it visible, by asking the patient to inspire deeply while sitting erect. This will help to explain ankle oedema, paroxysmal nocturnal dyspnoea and orthopnoea (congestive heart failure, caused by left heart failure leading to right heart failure) – the syndrome we call congestive heart failure..

The jugular pulse may, or may not, move up and down with breathing. If no movement, there could be mediastinal obstruction (retrosternal thyroid, etc.).

The JVP wave can reveal tricuspid valve incompetence. A prominent upward, jugular 'v' wave with each right heart contraction, will synchronise with the right ventricular apex beat. Use the apex beat to time it. If you can distinguish it, a prominent 'a' wave suggests pulmonary hypertension, obstructive pulmonary emboli, mitral stenosis or emphysema. One can diagnose atrial fibrillation by observing 'fibrillation waves' (fast vibrations), superimposed upon the jugular venous pulse.

Detecting Ventricular Hypertrophy

With one hand, find the most lateral position of the left apex beat (the left ventricle). If it reaches the anterior axillary line, the heart is most likely enlarged (usually LVH). How heavily does it beat? Is there a sustained outward impulse? In left ventricular hypertrophy (LVH), it will not tap, but heave outwards in a sustained manner. When differentiating true, from labile hypertension, detection of LVH provides clinical evidence for an increased left ventricular after-load. This occurs in true systemic hypertension, aortic stenosis, and HCM., but not always in labile BP cases.

Interestingly, the diagnosis of LVH may not appear as increased voltage changes on an ECG (increased QRS voltages = $SV_1 + RV_5 >$ 45mms or

4.5mVs). The ECG is an electrical device that unreliably reflects physical/mechanical changes. QRS voltages on an ECG (associated with ventricular thickness), can remain normal until advanced LVH develops. ST changes (T-wave inversion) in the chest leads, usually occur first. One can usually corroborate a clinical diagnosis of LVH using a PA chest X-ray. The shape of the heart can come to resemble a boot, with a prominent left heart border (Fig. 2). Echocardiographic measures can provide further physical evidence. On any desert island, with no technical equipment available, a trained hand alone may suffice!

On either side of the lower end of the sternum, or in the upper epigastrium, one can sometimes feel the outward heave of right ventricular hypertrophy (RVH). Use your thenar eminence and only light pressure over these points to detect it.

Auscultation

One can only learn to auscultate the heart with reliable diagnostic accuracy, only after clinical-pathological feedback. An apprentice working in a cardiac unit should eventually be able to match heart sounds to valve pressure gradients, and be able to gauge various degrees of valve stenosis and regurgitation (as seen on echocardiography, or revealed by cardiac catheterisation). Except for making cardiac diagnoses on desert islands, many now regard cardiac auscultation as obsolete. Echocardiography, introduced in the 1970s is responsible. When developed to a masterly degree, however, auscultation can be diagnostically accurate. Unfortunately, few doctors now wish to achieve such proficiency.

Advanced Cardiac Physical Signs

Aubrey Leatham's landmark book on cardiac auscultation, describes all one needs to know about heart sounds and murmurs. His apprentices, me included, learned to diagnose the stenosis and incompetence of each valve, even when occurring together. We also learned to detect developmental abnormalities like ASDs, VSDs, coarctation, Fallot's Tetralogy, and the degree of pulmonary hypertension using auscultation alone.

One can use a simple schema to denote cardiac sounds and murmurs: (Fig 3).

Reading and learning Leatham's book on auscultation, and becoming proficient at it, are different things. The only way to gain diagnostic accuracy is at the bedside, with the various heart sounds being pointed out by an expert; in my case it was usually cardiologist, Dr. Alan Harris. We once used to confirm our auscultatory findings using phonocardiography (and later, with echocardiography). With these recordings in hand, we could revisit the patient and listen again. In the early 1970s, we had many patients who had rheumatic heart disease, so it was easier to become proficient at valvular diagnosis using auscultation.

To auscultate the heart with diagnostic accuracy, one must adjust one's perception: from hearing usual everyday sounds, to listening for almost inaudible low-frequency sounds ≈20 cps, (like 3rd and 4th heart sounds); then to higher frequency clicks (prolapsing valves), opening snaps (mitral stenosis) and loud murmurs (VSD and mitral incompetence).

Using a stethoscope, one should be able to discern the two high-frequency components of the second heart sound: the first caused by aortic valve closure (A_2), the second (following immediately) caused by the sound of pulmonary valve closure (P_2). A_2 is exaggerated in systemic hypertension, and P_2 in pulmonary hypertension (the loudness of P_2 relates directly to the pulmonary artery pressure). In normal 2nd heart sound splitting, respiratory inspiration delays P_2, increasing the time difference between A_2 and P_2. With larger ASDs, the splitting can be fixed -unchanged by respiration. In LBBB, A_2 can be delayed enough to coincide with P_2; sometimes being delayed beyond it. In the latter case, the second sound is reversed; the sounds merge on inspiration rather than separate. To the practiced ear, listening to the components of the first heart sound (mitral then tricuspid closure), can have diagnostic value.

All students should be able to discern the difference between a pan-systolic murmur (mitral, and tricuspid incompetence), and an ejection systolic murmur (AS in adult cardiology), but this is not as easy as some textbooks suggest, or some teachers imply. Diastolic murmurs are more challenging because of their lower frequency (being produced at lower pressures). An early diastolic murmur (EDM), usually from AI, but also from PI, is best heard in the aortic area (to the right of the upper sternum), or at the lower left end of the sternum. It is easier to hear with the patient leaning forward, and breath held in expiration. Mitral stenosis can cause an opening snap (if the valve is pliable), followed by a late diastolic murmur (LDM) in the mitral area (lower left chest; mid-clavicular to anterior axillary line, with the patient turned on their left side). The 'snap' will not be easily audible if the mitral valve leaflets are thick, stiff, or calcified. The length of the LDM can indicate the degree of moderate stenosis. In both severe and mild stenosis, the murmur length can be short.

An important clinical question concerns the severity of each valve defect. Another key question when dealing with valve defects, is whether they have caused chamber dilatation or hypertrophy. Although the first question might get answered with auscultation, and the latter with ventricular palpation, both will now need echocardiography for verification.

Along with the auscultatory findings, one can always return to the waveform of the arterial pulse, best felt in the neck (carotid artery) or groin (femoral). From these (and the radial or brachial pulse), it is sometimes possible to deduce the severity of AS (slow to rise), and AI (sharp to fall), although one must never expect to diagnose minor degrees of aortic valve disease in this way.

The Lungs

Lung examination is essential for detecting both cardiac and pulmonary disease. Before rushing into lung auscultation, observe the shape of the chest (is there hyperinflation?) and the movements associated with breathing. Remove sufficient clothing to expose the chest (if allowed). Are the respiratory movements equal on both sides? Is breathing laboured or restricted; when laboured, are the intercostal muscles and supraclavicular fossae being drawn in. Look for finger nail bed clubbing, cyanosis, pallor and nicotine staining. Before auscultation, get the patient to breathe deeply through an opened mouth: listen for wheezes and rhonchi, magnified by the airways. Repeat this after the patient has coughed. Listen to their cough: is it dry or wet and rattling with sputum? Does coughing induce wheezing or does it remove wheezing? Any wheezing resulting from pulmonary oedema in left heart failure, will not usually be removed by coughing.

Next percuss the chest to assess resonance. Dullness, a lower frequency sound like percussing a brick wall, suggests a pleural effusion or a thickened pleura. A more than normally resonant sound (higher frequency) will suggest air-filled space. This might result from emphysema or a pneumothorax (hyper-resonant), especially in situations where the patient finds breathing difficult.

With the patient sitting upright, use your stethoscope to auscultate both the front and back of the chest. Remember that the lungs are above the nipple-level in men. Are there crepitations on auscultation at the bases (in the back, at the lowest level of the lungs), and do they remain after coughing? They will remain in pneumonia, lung consolidation, and pulmonary oedema (left heart failure, and mitral stenosis). With the patient still leaning forward, after posterior lung examination, carefully tap the length of their spine for tenderness (osteoporosis, metastases) with the ulnar edge of your clenched fist. First carefully, then more forcefully, if there is no tenderness. Palpate the renal areas (below the last ribs) for tenderness (always important in women who more commonly suffer from pyelonephritis).

One could regard such examination observations as irrelevant, even nostalgic. They are not. Also, they are not just an attempt to keep old clinical skills alive, useful only on desert islands or in places where no investigation equipment is available. They have proved indispensable to me. I could rely on these simple techniques wherever I was. As important as anything else, this level of contact with a patient, creates an impression of involvement, caring and thoroughness; evidence that the examiner not only knows what she is doing, but is also engaged with the patient. In a world where anonymity is fast becoming the accepted norm, such contact may help to define a doctor; one trusted to expose a patient's body and trusted to examine them. Sending a patient for a chest X-ray, without examining them, will add little to a patient-doctor relationship and further foster an anonymous approach.

Leg Examination

Leg examination can have cardiac diagnostic implications. One can detect atherosclerotic problems (reduced foot pulses) in those suspected of having claudication (ischaemic calf pain on walking), and reduced sensation in those with early peripheral neuropathy (in diabetics, and pre-diabetics). I never routinely performed straight leg raising (to confirm lumbar nerve root compression), looked for varicose veins, or tested co-ordination, strength, and reflexes in cardiac patients, unless the patient's symptoms suggested a need. It is often essential, however, to detect calf tenderness, and to look for a positive Homan's sign (calf pain with ankle flexion), especially when unilateral ankle swelling is present or there is any suggestion of pulmonary embolism.

The Integration of Symptoms with Signs

As much as possible, symptoms should prompt the search for physical signs. A complete examination, looking for every physical sign (most of which will be irrelevant) is useful only for the inexperienced. Although complete examination never goes amiss, too much information can distract focus.

One can always return to examine the patient again; a tactic of unquestionable value, both for challenging cases and those with an evolving condition. The time dimension of repeated examinations is sometimes crucial, especially if one suspects cardiac infarction, appendicitis, stroke, meningitis, myocarditis, or septicaemia. One can examine patients too early, before any significant physical sign develops. Dependent on the circumstances and the disease process involved, one must decide when a second examination might be appropriate, be it in an hour or two, or a day or so later. The point is not to miss the disease process evolving. Often, the second examination reveals most. The point is to make an accurate diagnosis as soon as possible and to treat the patient early; one might thus prevent disease progression and some complications.

Every doctor will have their own perspective when examining patients. This perspective can be too focused (missing the big picture), or too wide (missing the detail). Knowing which focus is appropriate for each patient is an art; a matter of experience and personal judgement. Some say it can't be taught or easily learned.

Having fully examined a patient, you might find yourself confused, or unsure of the diagnosis. There are reasons for this. First, a diagnosis based on the evidence available, may not be possible. Too many conflicting findings can be problematic for the inexperienced. It is always of value to make one diagnosis, although many older adult patients will have multiple conditions, some of which could inter-react. Now patients are living longer, multiple diagnoses are becoming commoner. One should aim to gain a clearer perspective of each diagnosis and its relationship to any other conditions. Diagnostic clarity is essential when deciding treatment priorities.

It is crucial to recognise red herrings early and to discard them quickly.

Beware of investigation details that inappropriately divert your attention.

Back in the day, when many more complex heart valve cases presented to doctors (in western countries), making a balanced decision about each valve could be challenging. For instance, was the patient's mitral incompetence more important mechanically than their aortic incompetence? One simple principle still applies. First, treat the more life-threatening condition, then those causing symptoms. They can be the same, but are not always. In a case of severe coronary artery disease and severe hip pain, it might be expedient to undertake a CABG, before risking hip replacement. Consider the clinical priorities before deciding patient management.

Chapter Three

Cardiac Investigation

Over time, more non-invasive investigations have become available. What were once unpleasant techniques, involving dye injections and radiation, have given way to no-touch imaging techniques that reveal both the anatomy and physiological functioning. The painless nature and improved results of imaging developments, has steadily changed the focus away from clinical appraisal.

Traditionally, investigations were used to provide evidence for the diagnoses made after history taking and physical examination. The most judicious tests reliably prove or disprove a diagnosis. Using Bayesian terms, they have a high diagnostic accuracy positive or negative.

When using diagnostic investigations, there are important considerations:

- **Diagnostic Uncertainty**. The need for any investigation increases with diagnostic uncertainty and the likelihood of clinical error.

- **Diagnostic Accuracy and Relevance**. The predictive value of every test depends on its specificity and sensitivity, and its then calculated diagnostic accuracy (positive and negative). Every investigation can yield true and false-positives, as well as true and false-negative results. One can use these to calculate diagnostic accuracy (Bayes' Theorem. See later). One can evaluate every symptom and sign similarly.

- **Interpretation.** Test results need to be interpreted for clinical relevance.

- **Appropriateness.** For every medical condition, there are appropriate and inappropriate investigations.

- **Urgency.** Test urgency must consider the clinical state of the patient and the resources available.

- **Corroboration** Corroborate the diagnostic contribution of different tests, bearing in mind their individual reliability.

- **Clinical Management.** Which investigations will most help to decide a patient's management? Choose investigations that can grade the disease process (early to advanced), as well as its rapidity of progression or regression (anatomically, histologically, and patho-physiologically).

- **Test Worthiness.** Worthy investigations are those which involve little or no procedural risk and have the potential to influence a patient's management. (Some will want to add low cost to this definition).

- **Test Unworthiness.** Unworthy investigations are those which are unlikely to contribute anything to the patient's clinical management. Some request them for 'research' purposes or just 'out of interest'. Neither low cost, nor low risk, condone their use.

- **Clinical Progress.** To assess progress, repeated clinical examination and investigations are necessary. The time intervals chosen will depend on the initial assessment, and the likely rapidity of change. Repetition needs to be minimised where risk, discomfort and radiation are involved.

- **Medico-legal Investigations.** A distinct set of considerations apply. Some are used to either exonerate doctors from clinical responsibility or implicate them, even when they lack diagnostic value. As doctors have become trusted less, and subjected to more litigation, some investigations have become *de rigueur*.

Before requesting any investigation, review the patient's notes. Avoid unnecessary repetition. Access all the past results and note any change. Many of my patients kept their own notes; some even plotted their test results on a spread-sheet. Is this business-like, or neurotic? Both, of course, but useful for efficient clinical management.

Why Investigate?

The first justification for any investigation is to support a presumptive diagnosis. The second is to refine the diagnosis by defining the pathological anatomy and physiology: its extent, severity, progression or recession (before and after surgery or radiotherapy, for instance). Omitting pertinent investigations can be fatal.

Essential Bayes' Theorem

To understand the relevance of investigations, an understanding of Bayes' Theorem is essential.

The most worthy investigations are those that are diagnostically sensitive and specific for a particular diagnosis. Few tests qualify sufficiently. One of the best, is the pregnancy test that detects placental human chorionic gonadotrophin (HCG). When performed correctly, its diagnostic accuracy, both negatively and positively, is impressive. The timing of the test, however, is important in the earliest days of pregnancy. The test is both sensitive (blood levels become positive soon after implantation of a fertilised ovum, 6-10 days after ovulation), and specific to pregnancy. There are very few conditions that can confound the result (liver disease, and choriocarcinoma, etc.), but this rarely occurs.

Clinicians cannot escape some numerical analysis, even though medical practice relies almost entirely on anecdote, with only one patient being under consideration at any one time (i.e., no cohort analysis). Bayes' theorem provides a useful framework for physicians whose concern is the diagnostic accuracy of every question they ask, every physical sign they find, and every investigation they order.

As applied to every medical condition, each symptom, sign and investigation result, has its own sensitivity, specificity, and diagnostic accuracy. A sensitive test is one that easily detects a specified condition; if too sensitive, false positives will occur. High specificity means the test will be positive only for specific examples of a particular pathology (with little or no overlap with other conditions).

Investigations cannot be too specific, but they can be too sensitive.

Test sensitivity we define as the percentage of only truly positive tests, among all positive results (both truly positive and false positive tests added together).

Test specificity we define as the number of truly negative tests as a percentage of all negative results (both true and false negatives added together).

Early detection requires a sensitive test; perhaps sensitive enough to detect pre-symptomatic disease. Specificity is a measure of the confidence one can have, ruling in or ruling out, any diagnosis.

Physicians rarely calculate these figures. Instead, many estimate them from experience, supported by those research reports which add perspective and can temper our biases. Every physician will soon get to know which tests are prone to false positives, and which are prone to false negatives (useless tests). It is just as important to know which tests usually yield true positives and which yield true negatives (useful tests). We usually refer to them as clinically unreliable and reliable, respectively.

How we weigh the relevance of each test result is critical to diagnostic certainty. When tests are unreliable, we must seek verification and corroboration with other tests. Each car, boat and airplane has its own handling characteristics; so does every investigation we perform. Getting a 'feel' for tests, and their practical usefulness, is an essential aspect of the art of medicine. We must employ this art, each time we 'interpret' the test results of any individual. All machine-generated results, be they from simple devices like thermometers, ECG machines or AI-driven software, all need patient-specific clinical interpretation and evaluation.

Patients often get bothered when test results are outside of the normal laboratory range. Too frequently having to explain a high blood potassium result, can get tiresome (in an otherwise healthy person), caused mostly by red cell lysis rather than renal failure or ACE inhibition. False hyperkalaemia most often results from blood being left too long before analysis. I installed a centrifuge to avoid it, but continued to see a few cases of false hyperkalaemia.

Error Forever?

Every measurement we take is subject to error; the instruments we use, and the techniques and processes we employ, can impose systematic errors. Each person as an observer, can introduce human error. In a physics laboratory, these errors are often small (±0.001%); in a biology laboratory they can be ±10%, or more. Important types of error arise in medical practice:

- *Observational error*: failure to measure accurately, what one intends to measure.

- *Measurement error:* measurement instrument error.

- *Error of perspective*: Like parallax error, not getting the correct view of what needs to be measured.

- *Sampling error*: Errors imposed by data acquisition. Not a clinical problem.

In the clinical domain, there are also:

- *Errors of assumption*: Assumption is a major source of diagnostic and management error.

- *Errors of omission*. Errors cause by forgetting or omitting information.

With echocardiographic measurement, exactly which chords across the left ventricle should one take to calculate left ventricular (LV) volume and ejection fraction (EF)? Even slight changes in the chords measured, will affect the resulting ejection fraction calculation. A major clinical error could lead to the false diagnosis of LV dysfunction or heart failure. An error related to perspective can occur when we accept the wrong view of an entity (like measuring the LV cavity in the short axis, rather than in a four-chamber view or both).

Test Appropriateness

Every test used to diagnose or define a medical condition, has its diagnostic accuracy for positively diagnosing it or excluding it (negative diagnostic accuracy). Physicians and surgeons need to know, for instance, which type of brain scan (MRI, PET, or CT) will more reliably diagnose a brain tumour or multiple sclerosis? Is a Rubidium PET scan better than a stress echocardiogram, when trying to detect deficient myocardial perfusion? In each medical specialty, there are tests which offer best or worst diagnostic accuracy.

The polymerase chain reaction (PCR) has now enabled bacterial and viral identification as a rapid, highly specific process. A few decades ago, we used to send swabs to the pathology laboratory for culture and sensitivity. We waited, then waited some more for some results! A few patients died waiting! Culture and sensitivity (C&S) testing was all we had, and not always practicable for the very sick. We had to employ intelligent guesswork, and decide our management before we received any results. We became used to presumption; a common one was that the commensal bacterium *streptococcus viridans* was mostly caused subacute bacterial endocarditis. ('Viridans' because it appears green—*viridis* in Latin—on agar cultures). It was expedient to send blood cultures to the laboratory, and start intravenous penicillin immediately. We would change the antibiotic regime once we received the culture results.

In the past, we worked with tests that had poor test sensitivity and specificity. Effective judgement honed by experience, allowed us to help patients.

Test Interpretation

Test interpretation is by far the most salient and contentious aspect of clinical investigation. Interpreting some tests, like the ECG, relies entirely on pattern recognition, learned from experience. Tarot card readers, clairvoyants, palmists, and tea-leaf readers, employ a similar form of art. Both ECG and EEG interpretation are ideal for neural network analysis and diagnosis, but their conclusions still need to be interpreted for correct clinical context. For this reason, interpreting test results will remain an art-form. For decades, ECG machines have incorporated diagnostic algorithms; they can yield errors that take no account of clinical context.

Any investigation dealing with electrical phenomena requires an understanding of vectors.

A note on electrical vectors and the ECG: *A sheet of paper seen end-on appears thin. As it rotates, more paper becomes visible, until face-on, we can see the full area. While the area of the paper has remained the same, the direction of view determines how much we can see (and measure). Because all vectors have both force and direction, various ECG lead positions (equivalent to the direction of view) either minimise or maximise the voltage displayed. When measuring a vector, both its direction and force matter. The increased QRS voltages seen in left ventricular hypertrophy (from hypertension and cardiomyopathy) increase or diminish with the anatomical orientation of the heart within their chest, and chest lead positioning.*

In my younger days, those who could 'read' ECGs with confidence, sometimes achieved iconic status. Apart from true abnormalities, it is often the many variations of normal that confuse the beginner.

Machine outputs can hide a multitude of errors. How does each capacitor, resistor, diode or integrated circuit within the machine, shape the analogue signal displayed as output? Electronic engineers do a commendable job trying to remove wave function distortion, but to what extent are they successful? Because we are so used to accepting machines for what they provide us, few question the distorting effects of electronics. At least such distortion will usually remain the same!

Unnecessary Testing

Without full access to a patient's notes, investigations can easily get repeated unnecessarily. Time and money might get wasted, and some repeated tests will not be in the patient's best interest (as when radiation or operative procedures are involved). This never happened to any of my overseas patients—they carried their notes with them. Even to this day, almost no British patient carries their clinical information with them to share. The current UK Secretary of State for Health (2024), Wes Streeting, has devised a plan to allow distant access to NHS notes. A good suggestion, but will it be reliable?

When I was a junior doctor in A&E, it was not unknown for non-urgent patients to be sent for a chest X-ray or a blood test. This had one main purpose—to shorten the queue of waiting patients!

Cardiac Screening Investigations

I have dealt with these separately in Chapter 18, The Early Detection of Heart Disease. They include blood lipids and inflammatory markers, ECGs, ECG exercise testing, cardiac ultrasound and the direct identification of atheroma in the carotid and coronary arteries using ultrasound and CT scanning.

Worthy and Unworthy Investigations

A clinically worthy test is one that has the antecedent potential to change a patient's clinical management and outcome. An unworthy test is one that lacks this potential and will raise questions about why one requested it. Clinically unworthy tests are done, either 'out of interest' (an attempt to be complete), for research or medico-legal reasons. Patients deserve a full explanation.

Other aspects of investigation worthiness, relate to risk and cost. Ideally, a worthy test has a level of risk that matches the clinical need for the information it can provide. One must also consider the clinical risk of omitting a test.

The worthiness of a test is not always clear cut; some tests being done to exclude a diagnostic suspicion. Ruling-out diagnoses can be a justifiable strategy in challenging cases. Even if unlikely, we have a duty to ensure that we are not dealing with hepatitis B or C; syphilis, HIV and AIDS, or TB., so tests for these conditions can be justified, even when they are unlikely. Diagnosis by exclusion is sometimes the only option. *'Are you sure he hasn't got . . . ?'* is the favourite rhetorical question of armchair physicians. Although one cannot prove a negative, only the brave and foolhardy, dismiss such questioning completely. If any of your seniors is an enthusiastic hobby horse rider (spanophiliac); has a special interest in genetics, the gut biome or some rare condition, gather lots of evidence before challenging them.

To challenge opponents effectively, know their arguments better than they do!
Charlie Munger. Former Vice Chair of Berkshire Hathaway.

In some places, like the NHS (never in my private practice), the financial cost of testing can define their acceptability, availability, and worthiness. This can apply to private patients with limited means and to limitations imposed by private medical insurance cover. Patients may need to get prior confirmation from their insurer, before a test is 'allowed'. Medical insurance companies only occasionally cover screening tests. Medical insurance companies are mostly generous, and rarely stand in the way of clinical management. They do, however, retain the right not to renew a patient's contract at the end of its annual term.

As a junior doctor, I did whatever investigations I thought necessary, including some done for academic interest. Fifty-years ago, the dividing line between tests that were clinically justifiable, and those done for research, was not always clearly explained to patients. I did harmless physiological manoeuvres, like carotid artery massage and Valsalva manoeuvres, while running an ECG rhythm strip. I used these to test cardiac pacemaker autonomic responsiveness. At other times, my research involved arterial puncture and the administration of drugs like atropine, propranolol, prostigmine and isoprenaline.

In the 1960s and 1970s, the public regarded doctors as mostly altruistic, honest, and trustworthy, and patients were content to think that our research would benefit others, even if they did not fully understand its relevance. Many patients gave *carte blanche* to testing, regardless of personal risk or benefit. This was especially so if they thought it might advance a junior doctor's career. Only rarely did the trust between doctor and patient get questioned. Any suggestion that we might still carry out such practices today, is likely to make modern medical bureaucrats and regulators apoplectic.

Investigation: Risk and Benefit

The worried parents of a 29-year-old man, from a wealthy Asian business background, brought their son to see me as an emergency. He was feeling unwell and had palpitations. He had been drinking more than one bottle of vodka every day for several years. When he arrived, he was semi-comatose and jaundiced. He had multiple spider naevi and an enlarged liver. I admitted him as an emergency to the liver unit of a private hospital in central London (private doctors could not admit NHS patients directly to NHS hospitals). They then transferred him to an NHS liver unit.

Despite his acute toxic state, and a corroborated history of excessive alcohol consumption, doctors in the NHS liver unit decided on a liver biopsy. What chance was there that a biopsy would reveal anything other than alcoholic cirrhosis? Was it done to exclude unforeseen causes of hepatic failure, or to follow a protocol as a useful teaching exercise? They chose a trans-venous biopsy method, rather than a traditional percutaneous one, in order to minimise the risk of blood loss. The patient died 24-hours after a difficult procedure. The junior operator had ruptured his vena cava.

The history of excessive alcohol intake (confirmed by intelligent parents), made any cause of liver failure other than alcohol, very unlikely. Was there any need for an urgent biopsy? Surely, the priority was to detoxify the patient and stabilise liver function.

At the subsequent Coroner's Court, the consultant hepatologist in charge of his case, did his best to deflect any criticism of his decision to biopsy the patient. Having listened to his explanation, I regret to say that both the coroner, and I, suspected his was fulfilling another agenda: to allow a junior doctor to practice the transvenous liver biopsy procedure. The Coroner could have concluded that professional ineptitude occurred, although the public can only expect doctors to 'do their best'. As long as doctors work within published guidelines, regulators will not usually take action against them. Since the junior doctor in question was trying to 'do his best' (albeit not well), the coroner's decision had to be one of unfortunate mishap; one that led to the death of my patient!

All would-be interventionists must start somewhere. Even with an experienced tutor in attendance, a little luck and some talent are required. One can never justify performing a dangerous test, however, without there being some chance that it will change the course of subsequent clinical management. If one ignores this rule, and someone dies, the result will be more than regrettable.

All interventionists in training will come across situations where they do not agree with established protocol, or their superiors. If they do not question decisions, directions and orders (or write their disagreement in the patient's notes), their career will be safe. If they do question them in writing, their career could be in jeopardy. They might then face a dilemma. Should they risk the life of a patient or their career?

In the 'good old days' (1970s and before), three appointed senior consultants (three wise men) would have interrogated the consultant who actioned the liver biopsy mentioned before. The only fair way to review any doctor's actions is to employ doctors with equivalent experience, to sit in judgement. All those who sit in judgement on doctors should have this basic qualification. They should be independent of the case and have enough appropriate clinical expertise and experience. This does not always happen. Coroners have legal and medical degrees, but do not always have enough specific medical expertise or experience. Lawyers and members of the public will hear allegations, aided mostly by a GP, when the MPTS at the GMC, judges doctors, but few will have any specialist medical experience or expertise at all. Third party, expert 'referees' may supply reports, but will they know the patient, the doctor on trial, or his circumstances.

They can easily ignore such valuable meta-information and consider the documentary evidence only. Tribunals may refuse to consider circumstantial issues, without which a true evaluation of the whole clinical situation is not possible. No wonder many doctors live in fear of regulators and the bureaucrats employed to investigate them.

The greater the risk of an invasive investigation, the better must it be justified. Even with complete justification, a few unfortunate patients will die being investigated. Given the need for failsafe rules, GMC (MPTS) tribunals in the UK, composed as they are from those with little clinical understanding or experience, can easily conclude that a doctor acted with a cavalier attitude, was reckless, or lacked insight. Many doctors are now aware of the inexpert way lawyers and lay people can judge them. They are more than aware of the risk bureaucrats represent to their careers. Highly defensive doctors can avoid harming themselves by following every bureaucrat rule and guideline, but what if the rules are not appropriate clinically? What good will they bring to the patient? In emergencies, we need men and women of expertise, action and experienced judgement, aware of the rules and guidelines, but not made catatonic by the thought of taking a risk.

I once performed a routine cardiac catheterisation on a 27-year-old woman in Amsterdam. She had multiple heart valve problems and was awaiting major cardiac surgery. In such cases, our departmental routine was to perform a pre-operative coronary angiogram. Unsuspected coronary disease can complicate major valve surgery, but this was very unlikely at her age. The routine coronary angiogram proceeded well until one image showed dye persisting. The dye did not flow away as usual, and remained static. Something untoward had happened. Soon after, she developed tight anterior chest pain and her ECG showed an ischaemic pattern. With my colleagues, we concluded I had dissected the left coronary artery wall. She was stable, but there was no time to waste. We rushed her to theatre for an emergency coronary bypass. Unfortunately, she died several days later.

After something like this, I needed to review my attitude towards fixed clinical protocols. At her young age, this patient was very unlikely indeed, to have had significant coronary disease, yet I followed the prescribed departmental protocol. While most clinically agreed protocols are sensible, there will always be exceptions. Inexperienced doctors, working with such protocols, will only rarely have the courage to countermand them (the chance of death from a coronary arteriogram was 20,000 to one, at the time). I wish now I had objected to the rules and argued for an exception in her case.

Cardiac catheter technology has improved considerably since the 1980s. Cardiac catheters are no longer as stiff (except in a rotational direction, which helps to guide them into position). The death of this young woman, resulted from using a catheter that was too stiff; stiff enough to puncture the intima of her proximal left coronary artery. After 40-years of performing cardiac catheterisation, this was the only time I experienced coronary artery dissection. It did not stop me catheterising others; clearly a characteristic of a cavalier doctor with no insight into risk! Some detachment, and an ability to take such risks, are necessary prerequisites for anyone considering invasive cardiology as a career. Unfortunately, those called on to sit in judgement on us, do not always have enough relevant experience to appreciate clinical risk.

The Pitfalls of Investigation

Here are some pitfalls of investigation I have encountered:

- Inappropriate investigation.

- Investigation errors and 'red herrings'.

- The wrong patient.

- Investigations lacking sensitivity, specificity and diagnostic accuracy.

- Misinterpreted variations of normal ('peculiarities', not 'abnormalities').

- Overlooked abnormal results.

- Failure to recognise the correct clinical significance of results.

- Overestimating and underestimating the clinical significance of results.

- Results wrongly filed, ignored or overlooked.

- Allowing computer algorithms / outputs, to offer preferential diagnoses.

We sometimes use investigations, based on a false assumption of their significance (a skull X-ray for migraine headache; shoulder X-ray for pain occurring while walking). Some doctors order tests with no more justification than 'just in case we miss something'. It is surprising, however, just how often such poorly justified investigations will uncover important (or erroneous) clues. Because atypical presentations are a reason for missed diagnoses, they will often need investigation.

I previously mentioned false hyperkalaemia, with blood allowed to lyse in collecting tubes after venesection. Other disparities between tissue and blood levels occur with measures of thyroid hormone and digoxin. If you want to assess the tissue levels of digoxin and potassium, look at an ECG (look for the cupped ST segments of digoxin toxicity). If you want an independent measure of myocardial potassium, look at T-wave height and trace any T-wave changes with time. If the T waves on ECG are normal, when blood potassium is reported as high, first believe the ECG, and repeat the blood level of potassium before taking any action.

A raised ESR, WBCs, CPK, or liver enzyme result, can cause both patients and doctors to worry. The results can be of significance, of course, but trying to explain false or insignificant abnormalities to a patient, can be challenging. They may assume that medical science is always exact, and in no need of the art of interpretation.

For decades before I qualified, a raised ESR (now partly replaced by CRP as an inflammatory marker) implied hidden inflammation (like TB, or chronic pyelonephritis). Chasing the reasons for a raised ESR, can often prove fruitless in the short-term. A constantly raised ESR (>30mms/h) in the long term, however, is another matter. I saw this a few times, decades before a slowly progressive, chronic disease manifested. One of my patients, with a constantly raised ESR, later developed rheumatoid arthritis; another had low-grade, chronic pyelonephritis, that lasted decades.

I had a patient who for years repeatedly said: "I never feel well". After many years of reviewing her, and finding nothing abnormal, except for a slightly raised ESR, I discovered microscopic haematuria. This led to the diagnosis of a chronic renal abscess. The ESR remains a sensitive test of inflammation, but lacks specificity.

Sometimes the wrong patient gets tested, or we attach the wrong name to a request form or blood collection tube. Those committing this error need to be open about it.

Some tests lack diagnostic accuracy. The PSA test belongs to this category when used to diagnose cancer; it lacks specificity and is over-sensitive. Although every test that needs further corroboration is inadequate, there may be no alternative. Every patient with a raised PSA needs an MRI scan to evaluate its significance.

An up-and-coming demand for total body screening will yield contentious results. Doubtful results can frighten patients. This was one good reason for doctors becoming unsympathetic towards medical screening. The results of screening, however, mostly reassure patients (although sometimes falsely). One callous counter argument I never used was, *it is better to be worried than dead!*

Routine MRI scans can detect anatomical peculiarities, as well as important pathology. There is an art, however, to discerning the difference. Beware of the sometimes ambivalent, defensive, but understandable conclusions, occasionally made by our specialist imaging colleagues needing to protect their medico-legal position. Insufficient clinical information about a patient, can put radiologists at a distinct disadvantage. Some protect themselves with ambiguity, by being non-committal, and sometimes by being negative. In retreat, they will sometimes unhelpfully state: 'one cannot exclude X, Y, or Z, as diagnoses.'

In first world countries, chest X-rays (CXRs) are now used less than when I was a junior doctor. It has always been the subject of many normal variants; some 'normal' features being wrongly misinterpreted as pulmonary oedema, pneumonia, TB, or early lung cancer. CT scanning improved the situation, but tissue diagnosis remains the ultimate gold standard.

Failing to recognise a meaningful abnormality and its clinical relevance, can arise from both a lack of knowledge and inexperience. It is important to attend as many clinical-pathological meetings as possible; it is there that the discussion of clinical data and its interpretation takes place, and where one can review clinical judgements once we know the actual pathology. Because pathologists can help to supply most of the answers, it should be obligatory to attend.

A relative of mine had a cervical smear which revealed evidence of cervical cancer. Unfortunately the gynaecologist involved, did not review the result because his secretary filed it before he saw it. The diagnosis came to light one year later, during her next scheduled visit. Her cancer had spread widely by then, and the filing error proved fatal.

Chapter Four

Cardiac Diagnosis

Diagnostic Steps

To emphasise the relevance of clinical data, it may help students to walk through points raised by a cardiac history featuring chest pain, shortness of breath (SOB), ankle swelling and examination findings.

With PRESUMPTIVE DIAGNOSES as **'P'**, and
RESOLUTION as **'R'**,
we can review each symptom and sign individually.

Chest pain

'P': Is it angina or costochondritis (Tietze's Syndrome) or is it indigestion or muscle injury (common)?

'R': Ask: Is it worse at rest or only on exercise? Is there tenderness of any costal cartilage or costochondral joint? If not, consider an exercise ECG, a stress echocardiogram or MIBI scan.

Consider a carotid artery ultrasound for the detection of atheroma. If atheroma is present, a CT cardiac calcium score or angiogram becomes essential. We usually reserve coronary angiography for those with definite (classic, Heberden's) angina. Their suitability for PCI, stenting or CABG needs to be assessed.

Shortness of Breath

'P': Is it caused by obesity or unfitness? Is it a pulmonary or cardiac problem? It can be the first symptom of IHD or heart failure. Could asthma, pulmonary emboli or severe anaemia be causes?

'R': Is it at rest, or only on exercise? Observe the pulse for AF or some other form of tachycardia. Is the JVP raised? Observe exercise performance. Is the JVP then raised? Are there signs of smoking or asthma? Is the chest hyper-resonant on percussion (emphysema)?

Swollen ankles

'P': Is it dependent oedema? (commonest). Varicose veins can cause it in hot weather. Could it be heart or renal failure?

'R': Is the lifestyle sedentary? Are there signs of varicose veins or any suggestion of DVT? Are creatinine and electrolyte levels normal? Is the JVP raised?

You must ask further questions and seek the relevant physical signs prompted by the history. With experience, the history should link to the physical signs you seek.

The duration of symptoms is important. Long-term SOB is more likely caused by obesity, COPD or heart failure. Acute SOB is more likely caused by the sudden onset of tachycardia, pneumothorax, asthma, or pulmonary emboli (consider the context: post-RTA, or cancer). The length of history is always an important weighting factor.

Now consider the physical signs with INFERENCE 'I' and RESOLUTION 'R'.

SOB at rest

'I': Is it asthma, pneumonia, pneumothorax or severe left heart failure? Is it caused by pulmonary emboli, anxiety or hyperventilation (common).

'R': Staring, unblinking eyes, could suggest increased catecholamine drive (anxiety/ hyperventilation) or thyrotoxicosis. Look for tachycardia

(AF, etc.), and a raised JVP. In asthma, COPD and left heart failure, accessory chest muscles can be used. On auscultation, listen for wheezing and crepitations. Percuss the chest for evidence of pneumothorax, pleural effusion and emphysema. A CXR, ECG or echocardiogram can help to confirm the diagnosis.

Raised JVP

'I': This implies raised PA, RA, and RV pressures.
'R': Is there Right Ventricular Hypertrophy (RVH)? Detect it with your thenar eminence, with your hand over the lower left sternal edge. Is there over-hydration (sometimes seen in CCU), right heart failure, multiple pulmonary emboli or mitral stenosis? Observing the JVP can help detect TI, and restrictive pericarditis – the JVP rising with inspiration.

Cyanosis

'I': Poor oxygenation: 'blue bloater' (COPD), or a right to left cardiac shunt.
'R': Common with lung disease in adults; life-threatening asthma in children, and sepsis associated with failing cardiac output. Is there cardio-pulmonary failure? Are there cardiac septal defects? Consider, gas diffusion, other lung function tests, and echocardiography. Consider blood culture.

An exhaustive list is not possible, but with experience, we should all develop our own lists of presumptive diagnoses and ways to resolve them as history taking and examination proceed.

It can be critically important to assess the severity of any problem discovered, and the urgency needed for their management. If someone with asthma is struggling to breathe, don't waste time requesting a CXR. A CXR in asthma will only rarely contribute to the immediate management, unless you have reason to suspect a pneumothorax.

Your most valuable instruments are your hands, eyes, and ears.

The Diagnostic Value of Surveillance

Does the evidence support an improving or deteriorating clinical picture? Never forget to ask the patient for any changes they have noticed. Make your own judgements. Don't just look at charts, laboratory results, or listen to the comments of nurses' and colleagues; not unless you are completely confident of their clinical acumen.

One golden rule for effective business management is to observe yourself, what is going on at all levels, and do it frequently. Business owners risk failure when they decide not to visit the shop floor. The Field Marshal who never talks to his front-line troops, will lose battles. During the second world war 'Monty' (Field Marshal, Bernard Montgomery), never lost touch with his front-line troops.

All golf professionals will advise something similar: never take your eye off of the ball! Sick people need the same focus. The discovery of a sick, toxic, or seriously ill patient, must trigger urgent action, whatever the diagnosis. All doctors and nurses must learn to recognise such clinical states as a basic requirement. Because they will know them well, the friends and families of patients will usually be the first to recognise deterioration. Never ignore or underrate their observations; check the patient for confirmation. A quick call to the ward asking, *'How is Mr. Q?'* will not always result in a reliable clinical evaluation. If you choose not to review your patients regularly, you will make a mistake eventually. Your patient, your responsibility. Junior doctors should involve their senior colleagues once they have identified a problem and taken steps to resolve it. Involve them, the instant you feel unsure about a diagnosis or the required management.

Dr. Hadiza Bawa-Garba found herself in this situation when, in February 2011, no consultant was available to review the case she was dealing with: a sick, six-year-old boy with pneumonia. As a junior doctor needing back-up, she could find none. Because the patient died, the GMC wanted to strike her from the medical register. The Medical Practitioners Tribunal Service (MPTS) disagreed, preferring to suspend her for one year. Nottingham Crown Court later found her guilty of gross negligence manslaughter. She lost her job with a young child to support. UK law sanctions the use of vindictive retribution as a punishment, even for those skilled enough to save lives.

So many other examples of doctors suffering injustice from bureaucrats now exist, the content would fill a few books.

For some, calling for help is the same as declaring incompetence. Some cultures find any loss of face completely unacceptable. This itself, is an important cause of clinical incompetence. It may prove difficult, but doctors must always consider personal intervention when a patient, other than your own, is suffering from a lack of medical expertise (nurses are often privy to this, and some will report it to you). Remember that no patient belongs to any doctor, regardless of how any organisation allocates them. Patients have the final say about who they want to entrust with their lives. Those whose culture includes avoiding the loss of face as a priority, will strongly resist accepting help. The correct response to this is to insist on intervention.

Whenever you delegate to others, check their clinical competence. Never assume the competence of colleagues. Several times in my career, I have encountered doctors unable to cope with certain procedures: their inability to read an X-ray or ECG; to perform cardiac catheterisation or undertake a pacemaker implantation. It can even apply to simple procedures like venesection, arterial puncture and venous cannulation for iv infusion. Some incompetence is apparent to all. Nurses and doctors should openly discuss these matters, and be free to report incompetence when medical intransigence or recidivism risks harming patients.

The need for immediate action will elude some doctors and nurses; they may be unsure of their ability, or try to avoid tempting providence. Some who are able, have an in-born error: a surfeit of inertia. Others will follow dubious protocols (bureaucrats can insist that their priorities should supersede clinical priorities). Bureaucrats may insist that form filling and office administration, should take priority over a patient's needs. To get their way, they may cite restrictions: too few appropriately trained staff, too many patients, no equipment, a hospital ward too far away, a work shift about to end; staff hand-over in progress, etc., etc. No excuse is ever good enough. My clinical teachers, many of whom experienced wartime, rarely made or accepted excuses. They knew how to side-step bureaucratic red tape, and how to take action.

Be on the lookout for obsessive bean-counters. You will find them motivated to finish whatever they are doing, regardless of any clinical priority.

Some will resist being disturbed, and continue to complete their paperwork, rather than assist a patient in need.

Difficult and Elusive Diagnoses

All presumptive diagnoses must stand firm against academic challenge. Your evidence must be solid and your justifications firm, because wrong diagnoses lead to incorrect management.

My close friend, the late Dr. Alan Gardiner, was a GP famous for his signature diagnosis in cases he found difficult to resolve: it was 'GOK' or 'God Only Knows!' Having given his patient this diagnosis, he would then refer them to a more knowledgeable doctor!

The usual reasons for diagnostic failure are:

- Failure to get a pertinent history.

- Failure to examine the patient fully and meaningfully (with the history in mind).

- Making incorrect assumptions, based on dis-information, misinformation or no access to reliable information.

- A lack of knowledge.

- A lack of data.

- Misjudgement. This can arise from an inability to choose the most relevant clinical features from an array of symptoms, signs and investigation findings. An ability to pick the best fruit from an orchard of trees is similar. This is the eclectic functioning, I previously referred to as nebula thinking (*The Art and Science of Medical Practice*. 2024). The essential art is an ability to apply the correct weight of clinical significance, to each piece of available clinical evidence. Both talent and experience are crucial to this indispensable ability: a basic component of clinical mastery.

Consider a few cases to illustrate misinterpretation:

During the Second World War, RAF men needed incentivising to accept being posted to the island of Uist. They were told that there was 'a pretty girl behind every tree'. What they were not told was, there were very few trees on Uist!

My friend Don Russell had been a nurse for many decades. He fell and subsequently developed cellulitis around one ankle. He continued in pain for many weeks after the fall. Since cellulitis rarely causes pain, the pain had to have another cause. Antibiotics did not help, but colchicine, a treatment for gout quickly relieved him.

Mrs. M. presented with intermittent, right subcostal colicky pain. She had multiple gallstones, but her pain never resulted from eating fatty food. After gallbladder removal, her pain continued unabated. The cause of her pain was colonic spasm arising from the hepatic flexure of her colon, not gallstones. Treating her IBS was the treatment she needed, not cholecystectomy.

Just as important as insufficient clinical data, is a lack of knowledge about a patient's circumstances. When stumped for a diagnosis, re-interview and re-examine the patient. Consider also, interviewing their relatives and friends, and pursuing them for relevant information. All this allows time to think further, and time to review the clinical evidence afresh. Pursuing theoretical differential diagnoses, without being anchored to confirmed, corroborated symptoms and physical signs, will lead to incorrect diagnoses.

There is another important clinical sense—an awareness of something missing (or being withheld). One missing 'clue', can sometimes solve a difficult case. All doctors will encounter a eureka moment, as one pertinent fact comes to light. The phenomenon is common amongst experienced researchers, adventurers, and police detectives (real or fictional). As in the cases solved by the fictional detectives Columbo and Sherlock Holmes, vital clues were often in plain sight, or buried in the detail.

Mrs. Cook was a patient in her eighties; the matriarch of a well-known, East London family; a family known for their Pie and Mash shops. She experienced constant itching, but no skin rash. She was being treated for hypertension, so I wondered whether her medication might be the cause. There was no history of allergy and anti-histamines gave her no relief. I interviewed her husband (who was being kept awake every night by her scratching), hoping that he might provide a relevant clue. He knew nothing. Desperate to find an answer, I reviewed her extensive records.

When I reviewed her haematology, the answer jumped out. Her latest haemoglobin was 19g/l. Since this was unusual for an 80-year-old woman, I looked back at her previous results to find a gradually rising haemoglobin: from 12 to 19g/l over a five-year period.

I was then sure of the diagnosis: Polycythaemia Rubra Vera (PRV). I telephoned one of my haematologist colleagues and was promptly told that, based on odds, my diagnosis was most likely incorrect. Statistics have their place in defining large group characteristics, with random selection in play. There was nothing random about this patient's rising haemoglobin levels, progressive hypertension and constant itching. Although PRV is rare, radioactive studies confirmed my diagnosis. My hubris was short-lived: PRV then had a poor prognosis.

Professor of Medicine at 'the London', Clifford Wilson (from 1946 to 1971), once told my student group, 'Never forget this case (of PRV), you will never see another one!' My guess is that few of my contemporaries at medical school, ever saw a case. I diagnosed two further cases within three years. Even with odds of forty-five million to one, someone will win the UK National Lottery. A few have won it twice!

Presenting Your Diagnosis

Most doctors have to expose their diagnoses to the comments of others. Presenting cases intelligibly, is therefore an important skill. There are some golden rules:

- Be succinct, and focussed.

- Avoid irrelevant information.

- Never force feed others with information.

- Offer sequential bits of pertinent information, never disjointed reams. Those who have ever fed pigeons, ducks and swans, will know that one small morsel is enough to attract wide interest.

- Allow your audience to form their own conclusions from the data you present. Editors of news media know that the conclusions people reach, depend on how they select, edit and present information (the framing bias). They can easily warp public opinion with their choices, omissions and exaggerations, and so can you.

- The inexperienced often need to be led to the correct conclusion.

Imagine presenting the case of a 60-year-old woman, who had experienced the sudden onset of shortness of breath. She had rheumatic fever at age 15-years. Your audience should now think: she might have rheumatic heart disease, with the sudden onset of atrial fibrillation (AF) or perhaps pulmonary emboli (PE). She had no recent operations or accidents (perhaps not a deep vein thrombosis and PE), but noticed palpitation (perhaps sudden onset AF).

On examination, there was no pronounced opening snap (either severe or minimal mitral stenosis), a long, mid-diastolic murmur audible in the mitral area, but no pre-systolic accentuation (AF removes this). They should have now concluded that she has rheumatic mitral stenosis, AF and pulmonary oedema. If your audience has failed to make these deductions from the data given, they cannot be cardiologists! You might have to conclude by detailing the significance of every symptom and sign.

Diagnosis in the Future

Medical science is moving on rapidly to define the genomic associations for each disease. Disease stratification using karyotyping, will add a level of diagnostic engagement, we will have to get used to. Transcription factors (the transcriptome), gene sequences, and mutations, relate to the phenotypic expression (clinical features) of inherited diseases. We might find that the subtypes so identified, will respond differently to treatments (molecular guided therapy), or be treatable with novel medicines. This technology is already in use for some leukaemias and lymphomas. Further technological developments, will personalise medicine, but will not replace traditional personalisation that uses age-old, doctor-patient relationships, to better understand patients and their individual circumstances.

A word of caution. From the corporate bureaucratic perspective (little respect for doctor-patient relationships), genetic evidence-based personalised medicine, could help relieve the need for doctors consulting with patients. One bureaucratic benefit would come from blocking any need for the art of medicine, and the time it takes. Without it, however, we will never maximise the benefit each patient perceives.

If AI and computer aided diagnoses, act as useful investigatory processes, they will inevitably become heroes of NHS executives. The reasons are obvious. Once bureaucrats see the potential for corporate financial savings (employing fewer doctors), computerisation will be promoted. AI will improve with heuristic, neural network programming, but there are real limits to useful data collected (perhaps limited meta-data) and processing. Computers are most useful for probability calculations (providing odds for any diagnosis), but they cannot think independently, weigh common sense or practical experience, or form balanced judgements. In a world that progressively assigns the brightest of halos to AI, and denigrates human wisdom, discouragement for using our brains for judgment, will surely grow.

Computers use big data, and odds-based analyses, to make diagnoses. They have yielded some interesting results, although the current vogue for collecting big data has a problem: it lacks clinical wisdom and specificity. It is, however, easy to use, cheaper than employing humans and fast. The accepted dogma is: if you search enough haystacks, you will surely find a golden needle. This runs parallel to the inductive idea that, since there are so many stars and planetary systems, alien life must exist somewhere in the Universe. Both are odds-based, inductive thinking; both bolster fallacies borne of human aspiration. There is nothing wrong with human aspiration, as long its conclusions remain ephemeral.

Frank Blake of Cornell University, attempted to calculate the likely number of alien civilisations in the Universe. He devised the equation $N = R^* f_p n_e f_i f_l f_c L$., where N is the likely number of technical civilisations that exist in *our galaxy at the moment. Some suggest we would need to look at 10,000 planetary systems to find one with life. This is guesswork with a gloss of mathematics applied to make it acceptable. Those interested should Google the other terms of his equation.*

What will members of the public gain from the vain idea of the 'digital self'? Other than the meditational value of a daily routine, the obsessional satisfaction of collecting data by weighing stools, recording pulse and breathing rates, and counting the number of steps we take every day, none of it is likely to yield any new clinical insight.

Collecting personal data is now big business, with those selling the idea, making substantial profits. The process has gone beyond any need for scientific enquiry. As a narcissistic lifestyle choice, it is almost a necessity for those who can afford it and who enjoy a distracting pastime. It will adequately feed the neuroticism of every body-centric, self-possessed, obsessive neurotic and probably result in their addiction to it. It will also allow the tech companies who design the apps, to make money and directly influence their customer's behaviour.

PART 2: HOW THE HEART FUNCTIONS

Chapter Five

Essential Cardiac Anatomy and Physiology

As I write (Oct 2024), an improved understanding of 3D cardiac anatomy has emerged. It has come from Hierarchical Phase Contrast Tomography (HiP-CT scanning). It offers resolution down to the cellular level (2-3 microns)(Link:https://doi.org/10.1148/radiol.232731). It relies on wave shift (refraction), not on X-ray absorption. Even the conducting pathways are visible. Such minute detail will enhance our understanding of cardiac anatomy and pathological anatomy. Unfortunately, this is not an *in vivo* application:a fourth-generation synchrotron is necessary to produce images.

First consider the cardiac anatomy (the ventricles, valves and coronary circulation). I later cover some key clinical aspects of cardiac physiology and two headings: electrical and mechanical.

Viewed from the front, the heart sits much higher in the chest than many expect, resting as it does on the diaphragm, well above the lower end of the sternum.

The heart sits on the diaphragm, in what resembles a plastic bag, formed from a thin layer of pericardium. The heart, great blood vessels and the oesophagus, occupy the central mediastinal space between the lungs.

The Ventricles

When viewed from the front (in an anterior-posterior direction on CXR), the right ventricle partly overlies the left ventricle (Fig. 4). On a CXR, the right ventricle forms the lower left border of the cardiac profile; the lateral left ventricle forms the lower right edge. It is sometimes possible to palpate both ventricles independently. The left forms the usual apex beat (anterior and posterior to the anterior axillary line), felt with the patient lying on their left side. One can also sometimes partly feel the right ventricle below the xiphisternum. In left ventricular hypertrophy (LVH), the lower right edge of the cardiac contour seen on CXR, may expand laterally to make the heart contour look like the shape of a boot (Fig 2). In right ventricular hypertrophy (RVH), the right ventricle can be seen to protrude to the left on a CXR.

In the anterior view, the right atrium lies above the right ventricle (as seen on CXR)(Fig 4) and the left atrium lies above the left ventricle. Both can protrude a little on chest X-rays, when dilated. Some of these features are subtle, but someone versed in the art of cardiac radiology, can make surprising diagnostic deductions. One such person, whose assessments of CXRs I was privileged to experience, was Keith Jefferson, author of *Clinical Cardiac Radiology.* Butterworth (1973).

The Valves

In a combustion engine, the cylinders each have an inlet and an exhaust valve, lying next to one another (Fig 5). The mitral (inlet) valve, and the aortic valve (exhaust or outlet) are positioned similarly in relation to the LV (Fig 6). The mitral valve (the inlet valve from left atrium to left ventricle) lies next to the tricuspid valve (inlet valve for right ventricle from the right atrium). The aortic and pulmonary valves (the outlet valve to the pulmonary artery) are also adjacent. All four lie near to one another, in the same plane above the ventricles (Fig. 7).

Coronary Arteries and Veins

Since no organ in the body can survive without an adequate supply of oxygen, every artery (and the diseases that afflict them) are of key importance. Atherosclerosis is the principal arterial disease, reducing blood flow and oxygen delivery; the process plays a major role in diseases of the heart, brain and limbs.

Why are 'coronary arteries' so called? The coronary arteries ('artery' is a word meaning windpipe or tube) were first drawn by Leonardo da Vinci (1452–1519). He saw them encircling the base of the heart (around the aorta and pulmonary arteries), like a crown with radiant attachments (κορώνη (koróni) in Greek, and '*corona*' or crown, in Latin). Leonardo referred to them as 'vessels' (arteries and veins were not yet distinct entities). He also described and drew the heart valves. With his interest in hydrodynamics, his conjecture was that they supplied the heart muscle with blood.

Only rarely is the part played by coronary veins mentioned. They accompany the arteries and collect venous blood into the coronary sinus and right atrium. When catheterising the right heart, catheters will sometimes pass unintentionally into the coronary sinus. From there, they cannot progress. Not much research has questioned coronary vein function or dysfunction. As such, we know little about their contribution to heart disease.

Both coronary arteries originate from the aorta, close to the aortic valve (see Fig. 8). There are rare, but important variations of origin, referred to as anomalous arteries. The coronary arteries arise normally within one centimetre from the aortic valve, each from a slight depression called a coronary sinus. Each coronary sinus lies behind an aortic cusp as it opens during left ventricular systole. Diseases that cause inflammation of the aorta, distal to the valve (syphilis, lupus erythematosis and atherosclerosis), can affect their origin. After a short single artery (the left main stem), the left coronary artery divides into two— the anterior descending artery (associated with a poor prognosis when narrowed or blocked), and the posterior or circumflex artery. The right coronary does not divide similarly, but has many side branches.

The right coronary artery supplies oxygenated blood to the right atrium and ventricle, although some of its branches can cross-feed the left coronary vascular bed, forming an important (potentially life-saving) collateral circulation. A sudden blockage of the anterior descending coronary artery that would threaten life, would be benign in the presence of sufficient collateral arteries from the right or circumflex coronary arteries. Learn the names of all the most important branches of the coronary circulation (Fig. 8) and more about coronary collateral circulation. An important area of research is now angiogenesis—the growth of new coronary blood vessels.

It is usual for the right coronary and the left posterior descending to be of a reciprocal size (cross-sectional diameter. A small right coronary will usually be associated with a dominant posterior descending artery, and *vice versa*.

Further learning points:

- Learn which systemic diseases affect coronary arteries.

- Study the anatomy of the main branches of each coronary artery, and parts of the heart they supply.

Jim was a sixty-year-old man with a young girlfriend. They seemed very much in love. She wanted him to have a medical check-up, prior to their marriage. His chest X-ray showed something I had only seen before in textbooks: an extensively calcified ascending aorta. This once meant one thing only: tertiary syphilis. For this reason I added tests for syphilis to his blood screening package. I did not discuss my fears with him initially; I wanted the result before mentioning anything. Unfortunately, the tests for syphilis were strongly positive, and I had to confront him with bad news. (I acknowledge that today, patient permission would have to be sought before performing tests for syphilis etc.).

I told him we had known for decades, that patients with aortic syphilis could die suddenly after being given penicillin. The Herxheimer reaction can cause rapid swelling of infected areas, most importantly, around the coronary ostia. If I treated him with penicillin he might die from cardiac infarction (as the coronary arteries swelled and blocked). Leaving him untreated, meant that his brittle aorta could burst, causing his sudden death. There seemed no way out. What happened next was a surprising tragedy.

Next day, the Police came to question me about the patient. He was dead after shooting himself in the head. His girlfriend was nowhere to be found. It is possible that he had told his girlfriend about his syphilis and how his life was in jeopardy. She may have decided not to marry him and to leave immediately.

During our consultation he told me he had served in the Royal Navy, and had contracted syphilis while on shore-leave in Hong Kong. A naval doctor had given him a long course of painful intramuscular penicillin injections. He thought this had cured him, but had never been checked further.

The major branches of the coronary arteries run superficially, the left anterior descending running in the inter-ventricular groove, between the right and left ventricles. The anterior descending artery is 'vital' because its many branches, supply the interventricular septum with blood. Heart attacks in its territory often have a limited, 50% five-year prognosis.

The right coronary artery follows the right border of the heart until it reaches the base of the heart, close to the diaphragm. Its occlusion will usually result in an inferior infarction (right coronary thrombosis does not much increase morbidity; dissimilar to those without infarction). The right coronary supplies blood to both the sinus and A-V nodes. The posterior descending branch of the left artery, courses posteriorly between the left atrium and left ventricle then turns to course along the lateral aspect of the left ventricle. The prognosis of posterior infarction lies somewhere between that of the other two artery branches, although again, the prognosis depends a lot on the presence of collateral arteries.

There are many individual variations of coronary anatomy. Some affect morbidity and mortality.

Both branches of the left coronary artery supply oxygenated blood to the left ventricular muscle; the right coronary artery supplies the right ventricular muscle. It is quite common on angiography, to see some cross-flow between the two, the extent of which will affect survival following coronary artery thrombosis. Cardiac infarction (death of heart tissue following arterial blockage) will not so readily occur when a sufficient collateral circulation exists. The prognosis can also be determined by variations in inherited arterial anatomy.

Essential Cardiac Physiology

I will consider the physiology of the heart under two headings: electrical and mechanical.

Basic Electrophysiology

Beneath the endocardium, within all four chambers of the heart, lie specialised tracts formed from Purkinje cells, that preferentially conduct electrical wave fronts between the sino-atrial node and AV node, and then within the ventricles via well-defined right and left bundle branches. Rapidly conducted action potentials within the atria and ventricles, allow a coordinated and effective contraction sequence.

Within the atria there are many potential pacemaking cells. The sinus node in the upper right atrium, mostly takes the lead. Like the lead instrument in an orchestra, it sets the pace. Pacemaking cells within the sinus node and in other places, possess electrically unstable cell membranes which depolarise at a regular rate. When detached from all autonomic influences, as in transplanted hearts, they beat at their 'intrinsic rate'.

I was once involved in researching the intrinsic rate of athletes. The question we wanted to answer was: does either a reduction of sympathetic stimulation or increased parasympathetic stimulation, account for the slower heart rate of athletes? We showed that the intrinsic rate was slower, and that the autonomic influences were normal. (Katona, P.G., McLean, M., Dighton, D.H., Guz, A. (1982).

The parasympathetic (vagal) nerves innervate the sinus node, the atria and AV node, but not much beyond. Acetylcholine released from parasympathetic nerve endings, slows the sinus rate (diminished sympathetic drive will have the same effect). In fainting or vasomotor syncope (increased parasympathetic activity), the pulse rate and blood pressure diminish together, reducing cerebral perfusion. The effect is to reduce or cease consciousness, because cerebral circulation blood flow is pressure dependent.

Depolarisation spreads out from the sinus node as a wave front; first through the right atrium, then through the left. P-waves on ECGs result from atrial wave front propagation. Two separate parts of the P-wave are sometimes visible. The initial phase of the P-wave results from right atrial depolarisation; the latter part from left atrial depolarisation. With left atrial hypertrophy (as in mitral stenosis), the second part of the P-wave may have a higher voltage; a bifid P-mitrale, can appear on ECG. In theory, separation of the two P-wave peaks might be evidence of inter-atrial conduction block. I have never seen a case.

Between the sinus and AV nodes, there are indistinct Purkinje fibre tracts that allow preferential conduction (Fig.9). These were once controversial and much debated, but are now accepted (Cavero, I., Holzgrefe, H. et al. 2023).

The AV node slows the progression of the atrial depolarisation wave front. The delay allows passive atrial filling of the ventricles (sometimes associated with a 3rd heart sound), and at the end of diastole as the atria contract (sometimes in association with a 4th heart sound). What lies beyond the AV node is a small structure, the Bundle of His, lying close to the tricuspid and mitral valves. This divides into the right and left ventricular bundles. In the ventricular endocardium, the wave front first travels down the larger left bundle, and 40 milliseconds later, down the smaller right bundle.

The QRS complex of the ECG represents the spreading wave of ventricular depolarisation; first within the endocardium, then outward through the myocardium to the epicardium. The wave front can spread throughout the myocardium in different ways, giving rise to various voltage wave patterns seen in each ECG lead. In whichever direction the resultant, maximum electrical force points, one will see a maximal positive voltage. The pattern seen on ECG depends on the orientation of the heart, and any myocardial damage, dilatation or hypertrophy. The voltages appearing on an ECG are determined by vector considerations involving both the voltage and direction of the wave front.

If all atrial pacemakers fail (because of widespread idiopathic atrial fibrosis), or if there is complete heart block due AV nodal damage or fibrosis, part of the AV node will take over the predominant pacing function, at a rate much slower than the sinus node. Like the sinus node, sympathetic and parasympathetic fibres also innervate it. Those with congenital heart block, often have a heart rate comparable to normal.

When AV block occurs suddenly, syncope may occur (Adam-Stokes syncope). Pacemaker implantation is then necessary to prevent further blackouts and sudden death from the escape rhythm intervention of VF. In VF, the heart shimmers rather than contracts, so there is no effective ventricular pumping.

Geoffrey Davis (1924–2008) was working at St. George's Hospital, Hyde Park Corner, when I arrived as a British Heart Foundation research fellow in 1971. Geoff was a humorous, pipe smoking, electronics expert (called 'boffins' in those days). He was an unusually modest, easily befriended, ex-RAF electronics engineer. Under Dr Aubrey Leatham's direction in the 1950s, he engineered one of the first effective, implantable pacemakers, incorporating stimulus inhibition.

Aubrey Leatham gave me the job of finding out why atrial pacemakers to go slow (sinus bradycardia) and cause syncope. Aubrey Leatham once saw me driving a yellow Lotus Elan 2+2. Thereafter, he referred to me as 'an expert on fast cars and slow rhythms!'

Coincident with the PR interval on ECG, the mitral and tricuspid valves open as the ventricular diastolic pressure drops below the prevailing atrial pressure. The 3rd and 4th heart sounds that occur with passive and active ventricular filling, relate to changes in ventricular muscle elasticity / resistance. Athleticism (causing a pronounced physiological 3rd heart sound), LVH (sometimes associated with a pronounced 4th heart sound), and heart failure (pathological 3rd heart sound), are not always easy to detect on auscultation. After being primed with blood from the atria, the ventricles are ready to contract after being stimulated by the advancing electrical wave front (coincident with the QRS, ECG complex).

Mechanical Physiology

The left ventricle contracts first, causing the mitral valve to close a little before the tricuspid; together they create the first heart sound, described as M_1T_1 (where mitral closure = M_1, and tricuspid closure = T_1). With left bundle branch block (LBBB), the first sound is usually reversed as T_1M_1.

Consider what happens within each ventricle throughout a cardiac cycle of relaxation and contraction (Fig 10):

- Just after systole, the aortic and pulmonary valves close, and **isovolumic ventricular relaxation** occurs.

- The mitral and tricuspid valves then open passively as the ventricular end-diastolic pressure drops below the atrial pressure.

- **Passive ventricular filling** from the atria then occurs, followed by active filling with atrial contraction.

- **Isovolumic contraction** occurs next. The mitral and tricuspid close first, then the aortic and pulmonary valves open.

- Following that, **ventricular ejection** of blood occurs through the aortic and pulmonary valves (the ejection fraction is the percentage of the end-diastolic volume).

Echocardiography can show how the relaxation and contraction phases proceed, but there are questions to be asked when patients have ventricular dysfunction:

- Are all parts of the myocardium contracting and relaxing well?

- Is there any sign of reduced movement in any area (following cardiac infarction, etc.)?

- What is the ejection fraction? If less than 40%, it implies reduced ventricular contractile function.

At the end of systole, the aortic valve closes before the pulmonary. The second sound (using the subscript 2) we refer to it as A_2P_2 (A_2 = aortic valve closure sound and P_2 = pulmonary valve closure sound). In right bundle branch block (RBBB), P_2 gets delayed, with the second sound widely split ($A_2...P_2$.). Breathing in, usually delays P_2, making the second sound split even wider. In LBBB, inspiration can make the split disappear or reverse (reversed splitting of the 2nd heart sound: (P_2 A_2)). With practice, one should be able to detect both types of bundle branch block, using auscultation alone.

An essential function to understand is the relationship between LV filling and stroke volume or cardiac output. The early work linked the initial length of myocardial fibres to the force generated by contraction in frog hearts. Bowditch, Cotes, Cyon, and Frank, while working at the Carl Ludwig Physiological Institute, Leipzig in the late 19th century, detailed it initially (Bowditch, HP. 1871)(Coats, J. 1869) (Frank, O. 1898) (Starling, EH and Visscher, MB. 1926).

Based on this work, what is now called the Frank-Starling relationship, is stroke volume (or cardiac output [CO]) increasing with diastolic volume (and heart rate and diminishing peripheral resistance on exercise) in almost a linear fashion, up to a heart rate of 140—150bpm (See Fig: 11 A). Thereafter, the CO diminishes with increasing heart rates. At any point, the onset of fast AF could cause the CO to reduce catastrophically (See Fig: 11 B).

As the heart fails, it achieves its maximum CO, only at a much slower heart rate. This rate can be critical to patient management in heart failure. In older hearts contractility can be reduced by fewer beta-adrenergic receptors, fewer myofibrils, less effective Troponin C activation by calcium ions, and fewer T-tubules coupling action potentials to myofibril contraction

There are physical / mathematical formulae and principles, that underpin some of our understanding of heart and artery mechanics.

The Laplace Formula

Mathematicians use partial Laplacian differential equations to describe changing space and time functions, but the simpler Laplace tension formula, predicts pressures inside and outside of bubbles. It can help understand the functioning of cavities like the urinary bladder and heart ventricles. For understanding blood flow, Poiseuille's formula can predict flow through tubes, like arteries. Are there applications for cardiologists?

As a bubble grows, the internal pressure exerted lessens. In a similar way, larger ventricles are less able to generate the force required for pumping blood. To visualise it further, try squeezing a broom handle. Now try squeezing a large tree trunk. Which generates the most force? The force generated relates to the inverse square of the object radius.

Poiseuille's formula helps to predict flow through narrowing arteries, and why ischaemia can be late in arriving. If you pump fluid through a rubber tube with a clamp attached, fluid flow will not steadily drop in a linear fashion; it will drop rapidly, once the occlusion of the tube approaches 80%. A narrowing of at least 80% in one coronary artery must exist for angina to occur. Several stenoses can co-exist, but it only takes one narrowed one (without collaterals), to cause angina.

Catastrophe and Bi-stable Situations

The French mathematician René Thom, studied discontinuity and catastrophe theory in the 1960s. Some useful analogies are: stretch elastic enough and it will snap; juggle one ball too many and risk dropping them all; push a sailing boat too far and it will keel over.

When stress has 'stretched' a patient mentally, one more stress could cause a mental breakdown (psychosis or an inability to cope). As the atria slowly dilate in heart failure, the sudden onset of AF can occur, sometimes causing the cardiac output to fall dramatically.

Chapter Six

Cardiac Imaging and Physiological Measurements

Cardiac imaging can be **non-invasive** without radiation (ultrasound and MRI), involve radiation (Chest X-rays, CT and nuclear imaging [sometimes combined]), or be **invasive** (cardiac catheterisation and electrophysiology).

Non-Invasive Imaging

Echocardiography

The first use of echocardiography, as we know it today, is credited to Edler and Hertz in 1954. Harvey Feigenbaum, who wrote an early book on the subject, became disenchanted when he could not get his early work published in major journals, reflecting a dismissive disinterest in cardiac ultrasound as a clinical tool (Feigenbaum, H. *Echocardiography*. Philadelphia, Pa: Lea & Febiger; 1972).

I was a British Heart Foundation cardiac research fellow when echocardiography was first introduced to the UK (early 1970s). Feigenbaum's book was then our reference work. Before then, apart from our hands, eyes and ears, our only investigations were chest X-rays, ECGs, phonocardiography and invasive cardiac catheterisation.

Echocardiography first presented us with a small oscilloscope image of various signal peaks of ultrasound reflection that indicated tissue interfaces. The display moved with valve and cardiac muscle movement (A-scans). Graham Leech, who later became a doyen of clinical ultrasound, was an electronics engineer working with us at St George's Hospital, London. It was he who created equipment that first displayed ultrasound reflection peaks as they moved in time. Fifty-years later, we have digital echocardiographic images in 3D, automatic measurements and diagnostic AI.

Using echocardiography, one can measure the dimensions of each chamber and estimate various pressures. Increased atrial dimensions relate to the degree of mitral or tricuspid leakage, and help predict the likelihood of AF and its return after DC reversion. One can measure the thickness (with likely calcification and pliability) of the mitral and aortic valves, to help decide on the most suitable type of intervention or surgery. Using Doppler imaging, one can assess blood flow, including regurgitant flow and flow through septal defects.

One can get better images (especially of the atria) from a transducer placed on the end of an oesophageal probe, placed just behind the LA (transoesophageal echocardiography or TOE). One can more readily see detailed valve structure, and even the valve vegetations of subacute endocarditis.

In the early days, I used to record echocardiograms and heart sounds (phonocardiography) simultaneously. This allowed us to confirm which sounds coincided with which heart valve and muscle movements. The low frequency third and fourth heart sounds, each coincide with an outward movement of the LV wall during the passive and active ventricular filling phases, respectively.

How Efficient is the Heart as a Pump?

A valuable and practical clinical test can be to walk with your patient. Observe if they are short of breath. Breathlessness has many causes: unfitness (after a prolonged illness and a lack of exercise), obesity, lung and heart disease. Observe the JVP before and after exercise that is a little taxing for the patient. If it rises with exercise, a cardiac problem is present. Back-pressure will cause the JVP to rise (right heart failure, left heart failure or mitral stenosis) if insufficient blood is pumped into the lungs by the right ventricle. This simple test will put the patient's heart functioning into perspective. Observing the JVP after exercise can thus detect sub-clinical, early heart failure. A normal JVP can be misleading. It is not always indicative of normal heart functioning in dehydration and borderline heart failure.

One can measure ejection fraction as a measure of heart pumping function using echocardiography. Computed from 2-D views, it is subject to considerable error. Just as important is the impression gained from simply observing cardiac contractions (eyeball assessment of heart muscle contractility). This is sensitive enough to reveal the reduction of ventricular contractility resulting from beta-blockade. Sometimes every part of the myocardium will contract poorly; sometimes it is specific segments of the myocardium that are defective (myocardial dysfunction), or do not contract at all (akinesia). The poorly contracting areas (dyskinetic or akinetic) usually seen, are the interventricular septum and anterior wall (anterior descending artery disease and anterior infarction), or the inferior wall with right coronary and posterior descending artery disease. Infarcted areas will remain akinetic or dysfunctional, despite exercise or pharmacologically induced tachycardia (stress echocardiography). A hibernating myocardium refers to LV dysfunction at rest caused by reduced coronary blood flow. It can be partially or completely reversed by myocardial revascularization and sometimes by reducing myocardial oxygen demand.

Those patients unable to walk on a treadmill for an exercise echocardiogram, can be given an intravenous infusion of dobutamine to stimulate their heart rate. In coronary ischaemia an abnormality may appear only after exercise or pharmacological stress.

Stress echocardiography is a valuable addition to echocardiography alone at rest when diagnosing IHD. Performed straight after exercise, it is possible to measure the increase in ejection fraction as a measure of functional LV capacity. A maximal ejection fraction of less than 40% with stress, signifies significant LV dysfunction.

The presence of bundle branch block will obscure the ischaemic changes normally seen on ECGs during an exercise test. Stress echocardiography then becomes the preferred technique for functional cardiac assessment.

Cardiac MRI has now advanced to provide faster sampling, and more accurate measures of LV function (including 3-D images). It can also reveal some detail about heart muscle structure (pathological in cardiomyopathy).

Hypertension and Echocardiography

Physicians argue about which measures and markers represent true hypertension, and its association with different levels of morbidity and mortality. Although these are mostly statistical considerations, the primary focus for physicians and cardiologists dealing with individuals, must be the pathological risk assessment of hypertension. One valuable measure of this is the observation and assessment of any LVH present using echocardiography.

LV wall and ventricular septal thickness are the important echocardiographic measures of LVH. If systemic hypertension is pathological, rather than 'labile' and caused by anxiety and stress, it will eventually be associated with left ventricular hypertrophy (LVH) and associated arteriolar medial hypertrophy (arteriosclerosis). Because this is a late stage effect, preventative treatment might have started earlier. While following patients in my practice, I often saw LVH arise within a year or two, in some with labile hypertension. The presence of LVH confirms the need for treatment. Decisions about treating those without LVH will remain contentious without the development of more specific prognostic markers (genetic perhaps).

Clinically detectable hypertrophy can occur late in the natural history of hypertension. In the early phases, one cannot be sure (without improved genetic profiling) whether any high BPs measured, represent low-risk labile hypertension or potentially pathological primary hypertension.

Two studies are of relevance:

In Monza, Italy, only when doctor's office BPs were persistently elevated, did white coat or labile hypertension (found on ambulatory recordings) assume prognostic relevance (abnormal long-term mortality risk). (Mancia, G., Facchetti, R., et al., (2015) *Adverse prognostic value of persistent office blood pressure elevation in white coat hypertension.* Hypertension. 66(2):437-44).

The Framlingham study showed a correlation between the average, long-term, level of systolic blood pressure (and pulse pressure), and the onset of coronary artery disease (Mahmood, S.S. et al. *The Framingham Heart Study and the Epidemiology of Cardiovascular Diseases: A Historical Perspective.* Lancet. 2014; 383 (9921):999-1008).

Known well before I was qualified, was that LVH seen on a chest X-ray in a hypertensive patient, predicted a high stroke risk. Treatment then, only lower blood pressure; it was only after angiotensin converting enzyme (ACE) inhibitors and angiotensin-II-receptor blockers (ARBs) were in use, that LVH and the medial hypertrophy of arterioles, became preventable (and stroke risk lowered, independent of BP measures). LVH can sometimes be reversed using both calcium channel blockers, ACE and ARBs, together with an improved prognosis. (Kawasoe, S., Ohishi, M. *Regression of left ventricular hypertrophy.* Hypertension Research (2024); 47: 1225-6).

What we once called 'essential' hypertension (now called primary hypertension) results from early arteriolar hypertrophy (with associated endothelial dysfunction and arterial stiffness, due partly to atherosclerosis). I strongly suspect that the expression of hypertensive genes causes abnormal medial hypertrophy, increased resistance to arteriolar blood flow, LVH and hypertension.

Clinical Learning Point. Antihypertensive treatment is obligatory once hypertension has caused LVH. The hypertrophy of heart muscle is most likely accompanied by widespread arteriolar medial hypertrophy (arteriosclerosis). Contrasting with this hypertrophy, is cerebral artery thinning in hypertension (there is no room in a tightly packed skull, for hypertrophied arterioles). This risks cerebral haemorrhage.

I advocate that all those suspected of hypertension (certainly those with a strong family history of hypertension and stroke) should have echocardiography annually; the risk of stroke is better assessed and the long-term effectiveness of treatment measured. When no progression of LVH occurs, should one regard treatment as successful, even in the absence of improved BPs? Although not reflected in NICE Guideline 136, this was for decades part of my routine hypertension management policy. This raises an important question. Which is the more predictive of a poor prognosis from hypertension: average or random BPs, or LVH?

Clinical Point: The heart beats approximately sixty times every minute: 60 x 60 times every hour. Every day we will all have approximately 60 x 60 x 24 (86,400) heart beats. Many patients measure their BP only once or twice daily (statistically unreliable sampling). Regardless of this, ballpark figures are workable, if taken to be: always lower than average (< 110/60), always between 110/70 and 140/90, or always elevated.

Blood pressures change every few seconds (seen after arterial cannulation). Therefore, what point is there to expressing BPs to the nearest millimeter of mercury? A good example of false exactitude, perhaps more common among those possessed by an obsessive trait.

Ambulant monitoring can provide averaged blood pressures (systolic, diastolic; 24-hour, daytime or night-time). When significantly higher they predicted:

- A doubling of fatal and non-fatal CVS events, and

- 2.4 times more fatal and non-fatal, cardiac infarction and stroke events.

This doubling of events was seen when the average 24h systolic pressures were compared as:

- Systolic: 123.5 mm Hg. (control group) versus 148.8 mm Hg. on average, and

- Diastolic: 78.8 mm Hg. (control group) versus 90.2 mm Hg. (on average).

These are not large differences. Interestingly, the routine office blood pressures recorded, although not greatly different, were of prognostic significance: 148/91 in the control group and 165/96 in the higher risk group (Clement, D.L., de Boyzere, M.L. et al. (2003). *Prognostic Value of Ambulatory Blood-Pressure Recordings in Patients with Treated Hypertension*. N Engl. J. Med.;348:2407-2415).

The latter study suggests that it would not be safe to accept office BPs beyond 148/91. The NHS current target is below 140/85.

Blood pressure and LVH it seems, are both valuable prognostic factors.

The Detection of Atheroma

We can use the ultrasound imaging of major arteries, like the carotid and femoral arteries, to detect not only the presence of atherosclerosis but also its type (lipid-rich, calcified or fibrotic), and its degree (obstructive or non-obstructive). This is a valuable investigation, given that atherosclerosis causes so many cardiovascular deaths of middle-aged people in the western world. Atherosclerosis causes ischaemic heart disease IHD and infarction; cerebral ischaemia and infarction and peripheral vascular disease. The detection of carotid atheroma is useful, not only because of an association with cerebral emboli but also because carotid and coronary atheroma occur together (Plichart, M. et al. *Carotid intima-media thickness in plaque-free site, carotid plaques and coronary heart disease risk prediction in older adults. The Three-City Study*. Atherosclerosis (2011), (2); 917-924). My own research results (2000 to 2021), corroborate this.

Doctors use blood lipids to assess CVS risk in populations. They could indirectly suggest that atheroma is present, even though the correlation between them in individuals is poor. Given the number of deaths atherosclerosis causes, guessing its presence is hardly good enough. There is a problem in practice. There are those with widespread atheroma and normal lipids, and those with the reverse: no atheroma and raised lipids. My studies showed that 95% of those with proven coronary atheroma (all with coronary angiograms to prove it) had carotid atheroma. HDL levels were usually low in these cases, but the total cholesterol, LDL and triglyceride levels did not correlate.

A paradox exists. Whereas in large populations, the average total blood cholesterol relates to the incidence of heart attacks and strokes, it is of little use when used to predict the CVS risk of an individual patient. This statistical paradox is important, pointing as it does to the impossibility of speculating about individuals from group statistics. It is also erroneous to extrapolate individual data to populations.

Learning Point: Detecting atherosclerosis is potentially the strongest risk factor for assessing cardiovascular morbidity and mortality. With arterial ultrasound, one can identify atherosclerosis, and better define individual risk.

In defence of statistical analysis, one can state that blood lipids provide the best risk guidance doctors have at present; at least they can help governments plan where to build cardiac admission units. Now that ultrasound scanning facilities are available on smart phones, blood lipid measurements are arguably obsolete, so easy is it to detect those individuals with atheroma who should benefit most from prophylactic treatment.

The Use of X-rays

Chest X-ray

Chest X-rays are still of value for cardiac evaluation. The size, position and shape of the heart can be clinically relevant. With a PA chest X-ray it is possible to detect enlargement of each of the four cardiac chambers; the great vessels can also be seen and prominent aneurysmal swelling detected. The overfilling of the pulmonary arteries and veins, as well as cut-off vessels seen in pulmonary hypertension, may also be visible. Horizontal Kerley B lines, seen on CXR, indicate oedematous interlobular septae (seen in the peripheral lung spaces), are typical of pulmonary oedema.

It is certainly worth consulting an atlas of chest radiography to learn the many features of cardiac diagnoses. Keith Jefferson was a senior colleague of mine at St. George's Hospital, Hyde Park Corner, London, in the 1970s. His book on cardiac radiology written with Simon Rees, is a classic text (*Clinical Cardiac Radiology.* Butterworth, 1973). Keith, a mild-mannered, modest man, would amaze us at cardiac conferences with his detailed CXR interpretation and accurate diagnoses. He was to die tragically on the 23rd October 1977.

As a simple example of the clinical value of a CXR. Consider the case of a 40-year-old patient with shortness of breath and a history of rheumatic fever. His CXR showed a normal size LV, a bulging left atrium (LA) and Kerley B lines. These features suggested significant mitral stenosis.

CT Scanning

The discovery of calcium scoring as a measure of coronary atherosclerosis, provided a significant step in the non-invasive detection of coronary artery disease. Arthur

Agatston and Warren Janowitz, published the first technique for scoring coronary artery calcium (CAC, or Agatston score) in 1990. (Agatston, A.S., Janowitz, W. R., et al. 1990).

For the first time, calcific coronary artery disease became detectable non-invasively. Since then volume and mass scores of calcium containing plaque, have been added.

A cardiac CT scan can show calcification, even in small arteries. A cardiac CT scan can provide a computed score of the coronary atheroma burden that correlates well with the carotid atheroma (as seen on ultrasound). There is a problem, however. The more dangerous, lipid-rich —vulnerable plaques — that are most liable to fracture, may be undetectable given their low calcium content. This absence is of crucial importance, since CT scanning may fail to detect vulnerable plaque. Those plaques vulnerable to rupture (ulceration, fracture and the intimal tearing), can provoke clot formation, coronary blockage and cardiac infarction. Fortunately, only a minority of plaques are vulnerable.

I worked in the same cardiac department as Michael J. Davies at St. George's Hospital, Hyde Park Corner, London, when he was working on both the pathology of the cardiac conducting system for his MD thesis, and coronary artery pathology. He later wrote an *Atlas of Coronary Artery Disease* (1998. Lippincott Raven). He made us aware of just how extensive asymptomatic coronary atheroma can be. He clearly described plaque rupture as an antecedent to arterial thrombus formation and cardiac infarction, and described the pathology that justifies coronary artery stenting.

Nuclear Imaging

Positron Emission Tomography (PET) Scanning

F-Sodium Flouride, PET scanning, has be used to detect vulnerable plaque in research. It can detect plaque that is metabolically active and mechanically stressed. These processes may provoke plaque microcalcification (incorporating flouride). (Kwiecinski, J., et al. *Vulnerable plaque imaging using ^{18}F-sodium fluoride positron emission tomography.* Br. J. Radiology 2020. 93 (1113): 20190797).

Functional Heart Scanning

There are several questions about cardiac function that are of clinical relevance:

- How efficient is the heart pumping?
- Is there sufficient oxygenated blood being delivered to the myocardium?
- How viable are the various segments of heart muscle?
- How well are the valves functioning?
- How well are the lungs functioning?

Myocardial Viability and Blood Flow

Various types of radioactive scanning are used to measure myocardial blood flow and viability.

These are:

- **Single photon emission computed tomography (SPECT)** (using Technicium-99m: half-life 6hrs, or thallium 201: half-life 73-hours). Both assess myocardial blood flow.

- **Positron emission tomography (PET**, using rubidium-82: half-life, 75-seconds) can also assess myocardial perfusion and ejection fraction. For myocardial metabolism and viability, 2-fluoro-2-deoxyglucose-F18 (FDG) is used: half-life 110-minutes.

Learning Points:

- PET scanning uses half the radiation of SPECT.
- All nuclear scanning techniques use a concurrent ECG to time the moments of acquisition.
- All assess the myocardium at specific moments during the cardiac cycle (gated measurements).
- All deliver 3-D data.
- As a result of myocardial viability assessment, one can observe several myocardial states: fully functional, dysfunctional, non-functioning or hibernating.

The **hibernating myocardium** has shut down contractile functioning while maintaining myocardial cell homeostasis. PET-FDG is the gold standard for myocardial viability assessment, but intravenous dobutamine and exercise stress echocardiogram testing are more often used. In trying to decide whether a patient will benefit from coronary intervention, assessing their myocardial dysfunction is critical. This can depend on how much fibrous tissue is present, that is not always apparent from myocardial viability testing. The more fibrous tissue, the worse the prognosis. The REVIVED-BCIS2 trial 'failed to show that multivessel PCI, improved event-free survival and LVEF, among patients with severe ischemic cardiomyopathy' (see Bibliography)..

Two to four days after cardiac infarction, a greater than 10% reduction in myocardial perfusion on scanning, strongly suggests the need for re-perfusion.

Learning Points:

- In the presence of coronary artery disease, a normal perfusion scan usually denotes low risk.

- In the presence of a stenosed artery (> 85% narrowed), a normal perfusion scan suggests sufficient collateral perfusion, lowering the risk of cardiac infarction.

PET scanning can support diagnoses such as myocardial sarcoidosis, amyloidosis and endocarditis, but MRI scanning showing tissue features is the more useful imaging technique.

Invasive Imaging

Cardiac Catheterisation

Cardiac catheterisation is necessary for the measurement of intracardiac pressures, measures of blood oxygen content, and for radiographic pictures (still and video) of the heart chambers and coronary arteries. The procedure can include coronary intravascular ultrasound images (IVUS) of coronary atheroma. Separate catheters are used to acquire intracardiac electrical recordings for measurement (electrophysiology).

The insertion of a small tube (catheter) into a peripheral vein or artery, allows access to a blood vessels that will lead back to the heart. All veins lead back to the right atrium and ventricle; all arteries lead back to the aortic valve and LV. One can access the left atrium from the right atrium by crossing the inter-atrial septum (through a patent foramen or by needle puncture).

The pressure measurements of interest are those of the RA, RV, and PA. One can indirectly measure the averaged LA pressure, using pulmonary capillary wedge pressure (PCWP), by wedging a right-sided catheter into a distal pulmonary artery. This closely reflects the average LA pressure.

In Fig. 12, the right-sided pressure waveforms, are shown as a catheter is withdrawn from the pulmonary wedge position, back to the RA via the pulmonary artery and RV.

The two positive waves seen clinically in the JVP, and directly from the right atrium, are the 'a' wave (coincident with right atrial systole), and the 'v' wave, coincident with right ventricular systole as the tricuspid valve bulges into the RA. The 'v' wave becomes prominent with tricuspid valve leakage. The negative 'y-descent', following the 'v' wave, occurs as the tricuspid valve opens; it is often the most obvious component of the JVP when observed in the neck.

Typical normal pressures for each chamber are: RA: 5mm Hg; RV: 25/0 mm Hg (systolic peak), PA: 15/10 mm Hg., Pulmonary wedge pressure (5-10mm Hg)(it will always be lower than the diastolic pulmonary artery pressure). Variations of pressure occur with breathing. On inspiration, blood is sucked into the right atrium as intrathoracic pressure reduces; expiration temporarily holds back the flow and raises the RA pressure.

By catheterising a peripheral artery, the aorta and LV can be entered. During left heart catherization, retrograde passage is usually straightforward, but crossing a stenosed aortic valve can be difficult. Thereafter, crossing the mitral valve is not often possible. The direct LA pressure can be obtained by puncturing the interatrial septum (from right to left), but this has its complications. For this reason, pulmonary artery wedge pressure is used instead.

Cardiac catheters are tubes made from a combination of various specialised materials. The design aim is to avoid damaging the intima of blood vessels, while being stiff enough to allow manipulation and placement. In my early days (1960s and '70s), catheters were unduly stiff and liable to cause damage, especially when used by inexperienced operators. They are now softer, and bend more easily in all directions, except in rotation (useful for guiding catheters into position).

Werner Forssmann undertook the first catheterisation, when he catheterised himself in 1929. André Cournand and Dickinson Richards, introduced diagnostic cardiac catheterisation in the early 1940s. Mason Sones introduced selective coronary angiography in the early 1960s. The Mason Sones coronary catheters I once used were quite stiff, and potentially dangerous. Introducing softer, pliable catheters, lessened the risk of coronary catheterisation.

Learning Point: What are the risks of cardiac catheterisation?

In 1979, there was a mortality rate of two per thousand (Davis, K. et al. *Complications of coronary arteriography from the Collaborative Study of Coronary Artery Surgery (CASS).*Circulation. 1979; 59:1105–1112).

The current risk of mortality is < 0.05% (< one in two thousand cases—four times less than in 1979). Major complications are now < 1%. (Manda, Y.R., Baradhi, K.M. et al. 2023).

Cardiac catheterisation allows us to image the mechanical functioning of the heart and lungs. Direct injection of dye into each coronary artery, with pictures taken in several directions, allows silhouette video pictures to be taken of the blood flow. From the various restrictions and indentations seen, one must deduce what might lie within the arterial endothelium. At present, one can gain this directly, only by using intravascular ultrasound IVUS, although some information is available from CT scanning and coronary artery calcium detection. By inserting an IVUS catheter into a coronary artery, ultrasound images can define the actual atheroma present within the endothelium (too small to be visualised using external ultrasound).

Measuring Cardiac Output

Cardiac output (CO) increases with body size and effort; both increase the oxygen per minute required. CO also varies with age. For a resting middle-aged person, the CO is approximately three litres per minute (4.5 litres/min in children, 2 litres/min in older adults).

We use the **Fick Principle** to calculate CO. The uptake of O_2 by the lungs (oxygen consumption by the body) is equal to the blood flow, multiplied by the O_2 saturation difference, before and after the blood has passed through the pulmonary circulation (i.e. the LA or arterial O_2 saturation (AO_2), minus the RA O_2 saturation (VO_2), measured using blood samples).

O_2 consumption of the body = CO x (the AO_2 - VO_2 difference)

How then do we measure the O_2 consumption of the body? One can assume 3mls of O_2/min/Kg body weight, or measure it after the patient uses a Douglas bag, measuring the O_2 exhaled against that breathed in from the surrounding air. This is a cumbersome physiological procedure, and of far less practicable use, now other methods of estimating CO are available.

The thermodilution method uses temperature changes in the pulmonary artery to estimate CO. A thermistor catheter measures the blood temperature change in the pulmonary artery after an intravenous bolus of normal saline. The temperature of the blood and the saline bolus need to be known accurately. The method is only accurate when there are high-flow states. It is unreliable when tricuspid leakage is present.

Shunt Detection and Evaluation

Before Doppler echocardiography, we used oximetry to detect shunts from left to right: we sampled blood from the inferior vena cava (IVC), superior vena cava (SVC), RA (from 3 places), RV and PA. A step up in O_2 saturation of > 5%, indicates blood flowing from left to right through a shunt (ASD or VSD).

We can calculate how much blood is shunting left to right from the ratio of pulmonary flow (Q_p or pulmonary CO) to the systemic flow (Q_s or systemic CO), using the formula for CO applied to each part of the circulation. For the systemic flow, the AO_2 - VO_2 differences are measured. For the pulmonary flow, we can use the difference between PA O2 and arterial O2.

- A flow ratio of < 1.5, indicates an insignificant shunt;

- A flow ratio between 1.5 to 2, indicates a moderate shunt, and

- A flow ratio > 2.0, occurs in haemodynamically significant shunts.

Shunt evaluation is much easier using Doppler echocardiography.

Vascular Resistance

This is an important factor when assessing congenital heart disease and pulmonary disease. It is useful when considering the appropriateness of operative procedures for congenital heart disease and heart transplantation. It is of little value when assessing systemic hypertension.

One can measure vascular resistance (pulmonary and systemic) by dividing the circuit pressure drop with blood flow (pulmonary [Qp] and systemic [Qs]). The pressure difference between the RA and aorta, divided by systemic flow yields the systemic circulation resistance; in the pulmonary circuit the resistance is the difference between the PA pressure and the pulmonary capillary wedge pressure, divided by pulmonary flow.

Alpha blockers reduce vascular resistance. Sepsis also lowers it (vasodilatation), while in cardiogenic shock it increases (vasoconstriction).

Functional Flow Reserve (FFR)

FFR or functional flow reserve is a useful measure of the flow on either side of a coronary stenosis. It helps as a measure of narrowing, compared to estimates taken from angiography. A wire or a micro-catheter, equipped with a miniaturised pressure sensor, is inserted via a coronary catheter across a stenosis. Its use is justified for patients with angina, when they have what appear to be only 50-70% stenoses. When the cause of a significant reduction in flow, PCI may be needed. Angiograms use a 2-D modality and can poorly estimate reduced flow. To the patient's benefit, FFR measures can detect reduced flow from what would otherwise appear to be minor stenoses.

Electrophysiology

Studies using multiple, multi-electrode catheters, placed in the heart can help to reveal the origins and re-entry pathways of tachycardias. In SA and AV block, one can measure the grade of block.

His bundle electrograms, can be recorded by placing an electrode catheter across the tricuspid valve (Fig. 13). One can measure the time taken for sinus node activation and conduction through the AV node, and the site of conduction block occurring before or after the His bundle. Some re-entry tachycardias involve passage through the Bundle of His, in either an antegrade or retrograde fashion.

Other catheters, used simultaneously, can detect the anatomical origin of a tachycardia, and the pathways taken during re-entry. From this information, one can then plan ablation procedures.

External wave front mapping is also available using multiple external chest electrodes.

Chapter Seven

Understanding ECGs

The Development of the ECG

In 1773, John Walsh managed to get a spark from an electric eel. In 1786, Luigi Galvani noticed a frog's leg twitching when stimulated by an electrical generator. He coined the term 'animal electricity' (later called galvanism). In 1838, Carlo Matteucci observed an electric current accompanying each frog heartbeat. What followed was the development of a sensitive galvanometer.

Two phases of heart electrical activity (later to be named QRS and T-waves) were first observed in 1878, by two British physiologists: John Burden and Frederick Page. It was Augustus Waller, at St. Mary's Hospital London in the 1880s, who came to record the first human ECG with his technician Thomas Goswell as the subject. He used a capillary electrometer (see Besterman, E., Creese, R. *Waller – Pioneer of Electrocardiography*. British Heart Journal, 1979; 42: 61). When only 35-years-old, Waller became a Fellow of the Royal Society.

At the first International Congress of Physiology in Basle, Switzerland in 1889, Willem Einthoven saw a demonstration given by Waller, using his pet dog as a subject. Einthoven developed the hardware further, and named the component waves of the ECG: P, Q, R, S, and T. He received the Nobel Prize in Medicine in 1924 for inventing the first practical electrocardiography system. **(Extracted from Dr Marc Barton. Willem Einthoven and the Electrocardiogram. Past Medical History.** https://www.pastmedicalhistory.co.uk/willem-einthoven-and-the-electrocardiogram/**).**

Einthoven used three pots of saline acting as electrodes, big enough for a sitting subject to put one foot into each, and a hand into the other. He then passed the small human electrical current to a string galvanometer (a fine quartz string coated in silver which twisted slightly as a result). The original device used a small mirror attached to the string with a beam of light directed onto it. The reflected beam was directed to strike light-sensitive paper placed on a slowly turning clockwork drum. Recordings of the ECG were first made this way.

The first commercial ECG machines used thermionic valves in their amplifier circuits; transistors came later. In the early days of their use, some electrical current leaked back to the patient (now thought to have been dangerous). ECG machines now use transistors to amplify the input signals first detected by Waller.

Einthoven invented the way we now connect arm and leg electrodes. They are connected either one to another (bipolar leads), or individually (unipolar) and connected to electrical earth (zero potential). He represented the connections as a virtual triangle made from the right arm, left arm and foot (Einthoven's Triangle) (Fig. 14). Leads I, II and III are bipolar leads, connecting the right and left arm, right arm and foot, and left arm and foot respectively. The unipolar leads (called augmented leads) aVR, aVF, aVL, individually connect the arms and foot to earth. These all record the ECG in the frontal plane; the chest leads record it in the horizontal plane. Every electrode records one aspect of the 3-dimensional cardiac electrical waveform, as it progresses through the heart. The sagittal plane is not used in routine electrocardiography, although it is used in vectorcardiography (rarely used in practice). Taken together, the ECG leads represent the electrical equivalent of walking around a sculpture, viewing it from different angles.

Technical Learning Points:

First imagine a three-dimensional atrial wave front, followed by a ventricular wave front, spreading throughout the heart in about 0.6 of a second. The way it spreads remains constant in normal hearts. The presence of a cardiac abnormality will usually disturb its path and create an abnormal ECG.

ECG leads pick up voltage vectors (force modified by direction): the voltage representation of the wave front increases and decreases, with the position of the electrode. When in the same direction as the progress of the wave front, the voltage recorded will be maximal. It is minimised when it records at right angles to the wave front direction.

The theoretical line that connects the top apices of Einthoven's triangle (representing lead I or a horizontal line across the chest) records only the wave front vector 'seen' in this direction. With the wave front progressing at right angles to this line, the voltage recorded will be minimal.

One can liken the ECG voltage measured, to a piece of paper, the full area of which cannot be seen from every angle. Only when face on, is the full surface area seen. When turned away from full view, one will see only a fraction of the whole area. Turn it 90° from face on, and it will appear paper thin. The voltage observed by any ECG electrode similarly depends on both the wave front voltage and the angle of its approach. This principle applies to every vector, and to every wave on an ECG.

Einthoven assumed his model triangle to be equilateral; in reality, it is just a useful approximation. Different torso shapes and densities will vary the conduction of the advancing electrical wave, so the triangle (assuming the same conduction in each direction) should be curvilinear, with sides of different lengths (like a triangle drawn on a sphere). In the same way we accept Harry Beck's London Underground map, we assume Einthoven's triangle to be equilateral. Although physically erroneous, it is of practical use

What the ECG Represents

A wave of myocardial cell depolarisation spreading through the heart, from the sinus node to the apex of the left ventricle creates the ECG. So how does the ECG relate to
the voltage of cardiac cells as they depolarise and change momentarily from negative to positive, then back to negative again?

A microelectrode inserted into a cardiac cell produces an electrogram 'E' (Fig. 15). The ECG effectively represents the rates of change of voltage (first differential) seen on the electrogram waveform (ie dE/dt). In simple electronic circuitry, a voltage pulse passing through the resistor of a CR (capacitor-resistor) circuit with a time constant shorter than the pulse, will do the same.

As a medical student interested in physics and electronics, I found the work of Alan Hodgkin and Andrew Huxley on the depolarisation of squid axons of much interest (*A quantitative description of membrane current and its application to conduction and excitation in nerve.* J. Physiology. 1952 Aug 28; 117(4): 500–544). It became known that sodium influx caused the rapid depolarisation, and that potassium efflux gradually restored the cell to its resting voltage. Now the fuller nature of ion transport, including calcium ions, is now better understood. An understanding of these mechanisms, has allowed a better appreciation of how cardiac drugs might work on Purkinje fibres.

As the depolarising wave front spreads throughout the heart, it will spread a little differently in children, obese adults and those who are underweight. The less bulky the torso, the more energy (current) is conducted and recorded.

Component Waves of the ECG

P-Wave

Atrial depolarisation most commonly originates in sinus node P-cells. From there, an electrical wave front is conducted throughout the atria. Sinus node pacemaking function is special, being created by a slow calcium-dependent, inward leakage current across P-cell membranes. This is unlike nerve and other myocardial cells, where it is sodium influx that causes the depolarisation. The P-wave depolarising wave front spreads first through the right, and then the left atrium. One can sometimes see two P-wave components arising from each of the atria (a bifid P-wave)(Fig 16). The initial part of the P-wave relates to right atrial depolarisation (when the sinus node is the dominant functioning pacemaker); the latter part, arises from the depolarisation of the left atrium.

Right atrial hypertrophy (as in COPD and pulmonary stenosis) can cause the initial P-wave component to increase in voltage (P-pulmonale). Left atrial hypertrophy (mitral stenosis) can cause the later component of the P-wave to increase in voltage (P-mitrale)(Fig. 16).

If the sinus node fails (sinus arrest or sino-atrial block), other areas in the atria can initiate depolarisation (ectopic origin). This changes the P-wave vector, sometimes making the P-wave negative in leads where it would normally be positive. With a lot of atrial fibrosis present, P-waves may have a low voltage.

P-waves may not appear if the sinus node and other potential pacemaking sites lose their depolarising function (in widespread idiopathic sino-atrial fibrosis for instance). We refer to this as **sino-atrial block** (of various sorts). Sometimes P-waves arise from multiple places in the atria, with each P-wave having a different waveform—referred to as a **wandering atrial pacemaker**.

An important, common disturbance of atrial electrical function in older adult patients, is **atrial fibrillation (AF)**(Fig. 17). Instead of regular P-wave activity, an irregular, continuous waveform, may appear on EC

G. It often causes the pulse to be irregular, given haphazard transmission through the AV node. **Atrial flutter** (Fig. 17) is rarer, faster and more regularly transmitted through the AV node; a saw-tooth atrial pattern replaces the P-waves on ECG. Both waveforms replace normal P-wave activity. Various atrial locations can generate the wave fronts (many of them surrounding the pulmonary veins) which can reverberate within the specialised atrial conduction fibres (localised re-entry)(Fig.9).

The commonest cause of AF is age-related idiopathic atrial fibrosis, although, it occurs in any cardiac condition that causes atrial dilatation (heart failure is the commonest). Coronary artery disease with atrial ischaemia is another cause. When the ventricular rate is fast in AF, slowing the rate may be necessary before atrial fibrillation waves become visible on an ECG (or visible in the jugular venous pressure pulse). One can sometimes achieve this transiently by applying carotid sinus pressure while running a simultaneous ECG rhythm strip.

A common clinical question arises when examining patients with AF. Is the rhythm sinus (regular) or AF (irregular, but sometimes seeming to be regular)? The pulse sometimes needs to be felt for 30–60 seconds, before any irregularity is discernible.

The PR interval

The PR interval represents the slowed conduction of the depolarising wave front through the AV node. The normal delay is between 120 and 200msecs (each small square on an ECG = 40msecs. with a paper speed of 25mm/second). At this speed, one larger square on the ECG = 0.2 seconds or 200mscs.). We refer to any greater delay as **AV block** (Fig. 18). When the delay is maximal, P-wave fronts cannot pass through the AV node and no QRS complexes will follow P-waves (**complete heart block**). Here, the atrial and ventricular pacemakers beat independently and QRS complexes arise from the AV node (at a slow intrinsic rate). Complete heart block is either acquired—from idiopathic fibrosis or ischaemia— or derived from a congenital defect.

In partial forms of AV block, the PR interval can gradually lengthen until a ventricular beat is dropped (**Wenckebach phenomenon** (Fig. 18). This causes the pulse to be intermittently irregular; not consistently irregular, as in atrial fibrillation.

Some patients are born with extra conduction pathways that bypass the AV node (Fig.9). The PR interval is then shorter than normal (< 120msecs). The atrial wave front can take a different course through the atria and myocardium, and be associated with a slurred R-wave. Resulting from pre-excitation, it is called a delta wave)(Fig 19).

It is important to know the main types of pre-excitation. An AV bypass pathway called a **bundle of Kent** (multiple bundles can occur), causes Wolff-Parkinson-White syndrome (**WPW**). In **Levine-Lown-Ganong (LLG) syndrome**, the bypass pathway directly enters the bundle of His, so no delta wave occurs. Different AV bypass pathways, have a different proneness to re-entry tachycardia (the depolarisation wave, rapidly reverberating back and forth between the atria and ventricles).

Kent first described an extra pathway in 1913 (Kent, A.F.S: *The structure of the cardiac tissue at the auriculoventricular junction*. J Exp Physiol 47:193, 1913-1914). The two main extra pathways that occur, lie next to the mitral and tricuspid valves, close to their fibrous annuli.

The QRS Complex

The QRS complex occurs as the depolarising wave front spreads throughout the ventricles. Having left the AV node, the wave front first passes to the bundle of His. It then passes down to the left, and then to the right bundle branch. The wave front progresses through the ventricular endocardial Purkinje fibres, and then into the myofibrils where contraction is electrically stimulated. Every electrode will detect some of the resulting electrical vector forces. What we see as the QRS complex, is always the net result of all the voltage vectors occurring in time, as the wave front spreads throughout the chest.

The ventricular wave front mostly follows a fixed pathway, but the pattern will change with individual cardiac anatomy and the presence of pathological changes (as in bundle branch block). The ECG patterns of an individual are like a signature—changing little throughout life. Much the same can be said of patterns seen on chest X-rays.

Someone murdered one of my patients. His assailant set fire to his house, with him in it. The patient's post-mortem CXR, matched the CXRs I had in my store. These helped to identify him. The bony structures seen were identical to those seen on his previous CXRs.

How the wave front spreads through the ventricles, can give rise to several distinct ECG patterns. The more hypertrophied a ventricle, the more current is generated. Right and left ventricular hypertrophy each have distinctive patterns.

Bundle branch block, lengthens the transit time of the depolarising wave front through the ventricles. The result is a longer QRS than normal (normal =< 120msecs) (one of the smallest squares on an ECG = 40msecs, with the ECG recording running at 25mm/second).

The path of the wave front in bundle branch block (BBB) will differ from normal, and the QRS changes accordingly. The following considerations apply:

In RBBB, the wave front stays longer in the LV, then crosses the septum into the right ventricle. This can cause a pronounced R-S-R´ pattern in the V1 chest lead (Fig 19). The wider, prolonged later 'R'' of RBBB, should be the focus when diagnosing RBBB (often seen best in V6). Because the wave front first depolarises the left ventricle (the right fascicle being blocked), a small initial 'R' wave is usually seen. Because the resulting wave front only later swings rightward, right axis deviation will be seen in the frontal plane (leads I, II, III, aVR, aVL, aVF), and the horizontal plane (in chest leads)(Fig. 20).

In LBBB, the wave front first travels to the right ventricle, and thereafter swings leftward. This increases the positivity of the lateral chest leads (V_4-V_6) since they are closest to the leftward progress of the depolarisation wave.

Because of heart shape, size and position, the main electrical axis (resultant vector) may shift in each plane: in the frontal plane as seen in leads I, II, III, aVR, aVL, aVF, and in the horizontal plane (chest leads). Standard ECGs do not utilise the sagittal plane.

Axis deviation, as seen in the frontal leads, varies a lot. Left axis deviation (as the wave front spreads from right to left) means that the resultant vector has swung towards the same positive direction as lead I (maximising it), but away from the direction of lead II, minimising it, or making it negative (see +ve and –ve points of Einthoven's triangle Fig. 14).

In right axis deviation (with or without RBBB), one should see a predominantly negative wave in lead I, as the wave front aligns with it in a negative direction. There may be a positive deflection in aVF, as the wave front moves from left to right, but this is unreliable, given that the position of the heart affects the pattern seen. In those with a normal BMI, the heart will usually sit vertically on the diaphragm. The ventricles can be pushed to the left in the obese, resulting in left axis deviation (the same happens with left ventricular hypertrophy).

Learning Points:

Unless it is seen to change with time, one can make too much of axis deviation. It rarely adds much clinical significance to the diagnosis and evaluation of heart disease.

To keep the detection of axis as simple as possible (minimising error), regard a positive deflection in lead I, with a negative or minimal deflection in III, as a sign of left axis deviation. Regard a negative wave in lead I, with a positive wave in lead III, as right axis deviation (knowing which electrodes are negative and which are positive is important. See Einthoven's triangle. Fig. 14).

Key clinical questions can arise when considering the QRS complex:

Is bundle branch block present? If LBBB, think of organic heart disease. Is there evidence of left or right ventricular hypertrophy (systemic or pulmonary hypertension)?

The ST Segment

In Fig 15, the intracellular action potential will plateau following the initial rapid depolarisation. The plateau phase of the electrogram underlies the ST segment (i.e. no rate of change).

ST segment changes can be of critical clinical significance. When myocardial cells are being damaged (as in cardiac infarction), ion leakage can disturb the electrogram plateau, in one part of the heart more than another. The wave front spread will change, resulting in either ST depression or elevation, depending on the ECG lead viewed. ST changes are used to categorise acute cardiac infarction, one of which is **STEMI** (an acronym for **ST E**levation **M**yocardial **I**nfarction). Because the ST segment does not always change with acute infarction, we also recognise **NSTEMI**, or **N**on **ST** elevation **M**yocardial **I**nfarction.

The T-wave

As myocardial cells repolarise (by potassium being pushed out of cells), the intracellular electrogram returns to the baseline voltage (Fig. 15). A negative rate of change will create an inverted T-wave (after the plateau in the extracellular electrogram). Because the cellular process of repolarisation is potassium dependent, hyperkalaemia can be associated with larger than normal T-wave voltages, while hypokalaemia is associated with smaller than normal voltages.

U-wave

Intracellular electrograms can show small rebound waves, directly following cellular repolarisation. These can give rise to U-waves, following T-waves on an ECG. Both relate to extracellular potassium levels.

Learning Points

Measures of blood potassium and sodium are unlikely to represent either extracellular potassium or total body potassium levels. I suspect that tissue potassium at least, is better represented by the T-waves of an ECG. Unfortunately, this is unlikely ever to be standardised.

The depolarisation of pacemaker cells depends partly on changes in calcium-related, cell membrane conductance. This depends on genetic factors encoded by the HCN2 and HCN4 genes.

Electro-Mechanical Coupling

So far, I have mentioned only the delivery of the electrical wave front to the heart muscle. Lastly, consider how these couple to action the main event— contraction and mechanical activity.

The contraction of heart muscle depends on action potential triggering myocytes. 'Action potentials (APs), via the transverse axial tubular system (TATS), synchronously trigger uniform Ca^{2+} release throughout the cardiomyocyte'. (Cocini, C., Coppini, R., et al. 2014).

Triggering depends on transverse or T-tubule functioning: myocyte cell membranes have T-tubule sarcolemma invaginations that activate contraction through calcium ion (Ca^{2+}) release (Fig. 21). These sarcolemma invaginations have various channels for: Na^+, Ca^{2+}, Na^+ - Ca^{2+} exchange, Na^+ - K^+ ATPase, and plasma membrane Ca^{2+} ATPase. In heart failure, T-tubules become reduced in number, become dysfunctional and re-model. Spontaneous T-tubule calcium releases can cause an electrical depolarisations (sparks) that might be pro-arrhythmic; perhaps explaining cardiac electrical excitability.

Try this Q & A

Q1. Who was the first to have an ECG recorded? **A:** Thomas Goswell; **B:** Augustus Waller; **C:** Thomas Lewis; **D:** William Harvey.

Q2. Who has received credit for introducing the term 'electrocardiogram'?: **A:** Thomas Goswell; **B:** A Waller; **C:** Thomas Lewis; **D:** William Harvey.

Q3. The His bundle is nearest to which valve?: **A:** Tricuspid valve; **B:** Mitral valve.; **C:** The pulmonary valve; **D:** The aortic valve.

Q4. The left bundle crosses into the LV at which level? **A:** Near the coronary sinus; **B:** Within the membranous part of iv septum; **C:** Within the muscular part of intraventricular septum; **D:** Near the foramen ovale.

Q5. A bundle of Kent is specifically associated with: **A:** WPW syndrome; **B:** LLG syndrome; **C:** Long QT syndrome; **D:** Atrial fibrillation.

Q6. The T-wave directly relates to which myocardial cell activity?: **A:** Depolarisation; **B:** Repolarisation; **C:** A stable cell membrane; **D:** A damaged cell membrane.

Correct Answers: Q1: A; **Q2:** B; **Q3:** A; **Q4:** B; **Q5:** A; **Q6:** B.

PART 3: CARDIAC CONDITIONS

Chapter Eight
Rhythm Problems

The main considerations are:

- The heart rate (fast or slow).
- Regularity (regular, irregular or sometimes irregular).
- Continuity (intermittent or continuous). Paroxysmal refers to intermittent.
- Morbidity risk (embolism in AF).
- Mortality risk (asystole, VT and VF).
- Treatment.

Slow Heart Rates

The primary concern must always be the condition of the patient. Are they breathless, concerned about palpitation, feeling faint or unconscious? The next consideration is the pulse.

A normal heart rate at rest varies widely between 60 and 80 beats per minute (bpm), however, many 'normal' athletic people, have heart rates at rest between 50 and 60 bpm. Cardiologists will usually regard only heart rates below 50bpm as worthy of further concern; investigating those with sinus rates constantly between 50 and 60 bpm will only rarely reveal an abnormality. Consistently slow heart rates (sinus bradycardia, defined as < 60bpm) can be inherited, or the result of athletic training, sinoatrial dysfunction (sinoatrial block: a failure to generate P-waves), heart block, heart transplantation, an underactive thyroid or the effects of drugs.

Many who lack medical training and experience, become alarmed when a patient's pulse rate drops below 60 bpm. Below 50-bpm., alarm may be justified, but usually only if there is a history of syncope, light-headedness, a drop in blood pressure or poor peripheral perfusion. To comment on a patient's slow pulse, a doctor must first enquire about any disturbances of consciousness (faintness or light-headedness), the patients BP (if low, how low?) and any defect in peripheral perfusion (coldness of hands, feet and nos e).

With reduced peripheral perfusion and the patient feeling light-headed (with a low BP), concern is appropriate. The questions asked should aim to reveal the cause. Has the patient experienced the rapid onset of feeling unwell, associated with pallor, a low BP and a slow, weak pulse (vasomotor syncope)? Is complete heart block present? Has the patient had a heart attack or a CVA, or are they medicated with a drug that causes bradycardia (a β-blocker, ivabradine or verapamil for instance).

I have often found that nurses and some doctors are more concerned about the patient's pulse rate and blood pressure than about their cardiac output, and perfusion. The primary job of the heart and circulation is to perfuse all the organs with blood and oxygen, so the latter should command our attention prior to the pulse.

Learning Point: The primary function of the heart is to deliver approximately five litres of oxygenated blood to the body, each minute at rest. If it can do this adequately with a slow pulse and a low BP, there is little need for immediate concern.

The sinus and AV nodes are both connected to autonomic nerve fibres: the parasympathetic and sympathetic nervous systems. Acetylcholine

(Ach) produced at parasympathetic nerve endings, slows the gradual rate of depolarisation of pacemaker cells between the beats (Fig. 15); norepinephrine (NE), released from the sympathetic fibres, increases the rate.

It has been long thought NE and thyroid hormones inter-react: that too little thyroid hormone (T4 and T3) in myxoedema will reduce the effect of NE, and too much thyroxine T4 and T3, will increase its effectiveness and cause tachycardia. Various investigations have failed to confirm this relationship (Levey, G.S. (1971).*Catecholamine Sensitivity, Thyroid Hormone and the Heart A Reevaluation.* American J. of Medicine; 50: 413-420). It now appears (in mice at least) that NE can stimulate or inhibit the production of thyroid hormone: 'it stimulates basal but inhibits TSH-induced thyroid hormone secretion.' (Ahrén, B., Bengstsson, H.I., Hedner, P. 1986).

In the early 1970s, I used intravenous isoprenaline to test atrial pacemaker function and any sensitivity to catecholamines in patients with bradycardia. I noticed how sensitive anxious patients were to isoprenaline (similar in effect to NE). The reverse was true for those who were calm throughout the procedure. The mechanism remains to be explored.

Increased parasympathetic activity occurs in vasomotor syncope, as part of a reflex response. Such responses are stimulated by many and varied situations. They can be as diverse as the sight of blood, pain, stressful and frightening circumstances, and overheating. It also occurs in association with early febrile illness and pregnancy, and can be one of the earliest signs of both. It commonly occurs with alcohol intoxication (a toxic effect causing peripheral blood vessel dilatation, and a low BP). The sinus rate can remain slow because of parasympathetic stimulation, or because synaptic Ach remains longer with absent or ineffective cholinesterase.

Several features occur together in fainting:

- Skin pallor (as blood gets diverted to internal organs).
- Low BP.
- Slow pulse.

- Altered or loss of consciousness (in varying degrees).

- Slow recovery during which tiredness, faintness and nausea usually persist (contrasting with Adam Stokes syncope as described below).

It occurs in many situations, many of them clinical (minor ops, the sight of blood, etc.). The management is specific, and must aim to increase cerebral blood flow:

- Lay the patient flat (preferable on a head-down slope).

- Ask them to pump their calf muscles (this can improve circulation and help focus the patient's mind away from fainting).

- If they are likely to vomit, put the patient on their left side with their chin raised.

- In clinical situations, atropine or hyoscine will help to inhibit vagal influence, but will make recovery slower (visual disturbance caused by atropine and sedation with hyoscine). When fear and pain are present, analgesia and sedation will benefit patients greatly.

Pacemaker and Conduction Defects

Slow pulses are sometimes caused by sino-atrial block. Defective pacemaker function can cause occasional P-waves to be dropped, slowing the pulse rate. AV block is commoner, but both are seen in routine medical practice. In AV block, conduction through the node is reduced.

The types of AV block (Fig 18) are:

Prolongation of the PR interval alone is called **first degree AV block**; it does not affect heart rate.

Second degree AV block causes dropped ventricular beats: either regularly, as in 2 : 1 AV block (where the effective heart rate is halved), or variably, when the PR interval progressively lengthens until no beat is conducted (Wenckebach). Here, beats are dropped intermittently.

In **complete heart block,** P waves arise normally, but are not conducted to the ventricles; P waves and QRS complexes are dissociated. The ventricular rate is slow because it is driven by the intrinsic rate of the AV node (always slower than the sinus node). Such patients are liable to Adam Stokes syncope, the features of which are:

- A history of syncope without warning.

- Pallor during syncope.

- Slow pulse (40 bpm or less).

- Low BP.

- Facial flushing on recovery.

- No Nausea.

- Rapid return to feeling normal (once sinus rhythm resumes).

Learning Point: A slow recovery from syncope is typical of both vasomotor syncope and epilepsy, but not Adam-Stokes syncope.

Fast Heart Rates

We classify fast heart rates (> 90bpm at rest) by their origin. They arise either from the atria (**supraventricular tachycardia**) or from the ventricles (**ventricular tachycardia and fibrillation**).

Supraventricular tachycardia can originate in the sinus node (as in febrile illness, anxiety and thyrotoxicosis), or be caused by **ectopic activity** (from several competing pacemakers). They can result from **re-entry**, in the presence of a conduction circuit that allows repeated circular reverberation (

reciprocating wave front propagation). Something similar happens at rock concerts. A loud screaming feedback sound can occur, reverberating between the loudspeakers and the microphones. It stops only when the circuit breaks (by switching off either the loudspeakers or the microphones).

The rate of transmission of re-entry through cardiac electrical circuits, depends on the length of the refractory period of conducting cells (during phases 2 and 3 of the action potential, at which time further electrical stimulation is blocked). Longer refractory periods favour the cessation of re-entry (KCNH2 and KCNQ1 genes encode them). I will return to these to explain some rarer forms of tachycardia.

Supraventricular Tachycardias

Introducing twenty-four-hour ECG recording helped with tachycardia diagnosis.

Most patients who complain of palpitation have either atrial or ventricular (extra) **ectopic beats,** followed by a compensatory pause. Both types of ectopic activity can occur in rapid succession, giving an unpleasant feeling in the chest and throat. Sometimes atrial ectopic beats are a precursor to sustained supraventricular tachycardia. Periods of stress and alcohol consumption can induce them: occurring more often in unfit individuals. Supraventricular ectopic beats are mostly innocent and often catecholamine induced, rather than associated with ischaemia, cardiomyopathy or myocarditis. Ventricular ectopic beats are mostly innocent, but occasionally indicate cardiac ischaemia, valve defects or cardiomyopathy. They need to be investigated with this in mind, especially if they occur during exercise.

Sinus Tachycardia

Like a raised ESR, the reason for sinus tachycardia is not always obvious, but often related to fever, anxiety or a drug effect. Occasionally, it is associated with thyrotoxicosis or a toxic state (alcohol, stimulant drug taking or sepsis). Any pronounced infective or inflammatory condition, including cardiac infection (pericarditis or myocarditis), is sometimes the cause. Pulmonary embolism and cardiac infarction are other causes. Depending on the adequacy of ventricular function, breathlessness can be an associated. The poorer the ventricular function, the more readily will tachycardia cause breathlessness at lower heart rates.

Learning Point: It is mostly non-sinus tachycardias like atrial fibrillation or flutter that cause breathlessness, especially if ventricular function is impaired.

Sick patients often have sinus tachycardia. Like temperature, a resolving tachycardia has for centuries been a useful measure of clinical improvement.

Atrial Fibrillation and Flutter

Atrial fibrillation and flutter can produce fast heart rates, but can present with slow heart rates if AV nodal conduction is defective. For both reasons, the sudden onset of AF can cause syncope and/or breathlessness. Idiopathic atrial fibrosis is the usual pathological process involved, but ischaemia and atrial dilatation are also causes.

The key facts supporting active management are:

- AF increases all-cause mortality, and cardiovascular mortality (1.5 to 2-fold).

- 15% of those older than 85-years have AF.

- AF increases stroke risk, 5 to 6-fold.

AF is associated with advancing age, but also with premature atrial endocardial fibrosis. It is associated with rheumatic heart disease, heart failure and ischaemic heart disease. Other less frequent associations include thyrotoxicosis, cardiomyopathy, neoplastic disease and diabetes.

Lone AF specifically refers to cases with no pathological associations. Alcohol, cocaine and caffeine can induce it.

For patients in AF, the most important clinical objectives are to reduce the severity of any heart failure with rate control and stroke prevention.

The AFFIRM Study(2002), confirmed that reverting AF to sinus rhythm (rhythm control), was not superior to rate control alone, when considering quality of life, all-cause mortality and an extensive list of secondary end-points. John Camm summarises the evidence in: *'Why is rhythm control for atrial fibrillation becoming more popular?'* European Soc. Cardiol. (2203);1(2). The factors that should favour rhythm control in AF are: a younger age, a short history, cardiomyopathy, normal or a slightly increased LA size (left atrial volume index – LAVI), especially when there are no co-morbidities. Other reasons to favour rhythm control are when rate control is difficult or when AF occurred with an acute event. Patient choice can also become a factor. (AFFIRM study: Wyse, D.G., Waldo, J.P and AFFIRM writing group (2002). *A Comparison of Rate Control and RhythmControl in Patients with Atrial Fibrillation. NEJM; 347:1825-1833).*

The following are key facts about AF:

The risk of stroke is greater in females.
Stroke risk increases in:

- those with a previous stroke.

- those older than > 75-years.

- those with dilated atria.

- those with LV dysfunction (heart failure).

- those with rheumatic heart disease.

- those with prosthetic heart valves.

- those with diabetes, hypertension, a calcified mitral annulus and thyrotoxicosis.

One can attempt to calculate the stroke risk using the CHA_2DS_2-VASc score. The acronym stands for specific risk factors, namely: **C**ongestive Heart Failure, **H**ypertension, **A**ge (75yrs and older), **D**iabetes, **S**troke (or TIA and cerebral ischaemia), **V**ascular Disease, **A**ge: 65 to 74yrs, **S**ex category (see clincalc.com).

CHA_2DS_2-VASc scores range from 0 to 9. A score of 9 is associated with a maximal annual stroke risk of 15.2%.

For further consideration see: Teppo, K., Lip, G.Y.H., et al. *Comparing CHA_2DS_2-VA and CHA_2DS_2-VASc scores for stroke risk stratification in patients with atrial fibrillation: a temporal trends analysis from the retrospective Finnish AntiCoagulation in Atrial Fibrillation (FinACAF) cohort.* Lancet, Regional Health Europe. 2024; 43: 100967.

Important clinical conclusions are:

- Whatever the score, those with rheumatic heart valve abnormalities, thyrotoxicosis and cardiomyopathy need anticoagulation.

- Aspirin and warfarin (vitamin K antagonists) are not now regarded as the most efficacious forms of anticoagulation (stroke reduction benefit versus GI bleeding risk). Factor Xa inhibitors (factor Xa is the enzyme serine endopeptidase) and the thrombin inhibitor dabigatran have taken over the role except in special cases. Apixaban has been found superior to others because of fewer bleeding episodes. (AVERROES and ARISTOTLE trials: see bibliography).

- Controlling the ventricular rate to strictly below 80bpm, rather than more leniently below 110bpm, does not improve mortality or morbidity (RACE II Trial:Isabelle C. Van Gelder, M.D., Hessel F. Groenveld, M.D. et al. N Engl J Med 2010; 362:1363-1373).

We can attempt to calculate the risk of bleeding from prophylactic anticoagulation by using the ORBIT score. The score is produced as follows:

- Those older than 75-years score 1;

- Those with a haematological factor (Hb. below 13g/dl in men and < 12g/dl in women; a low haematocrit or history of anaemia, score 2.

- A history of GI bleeding, intracranial bleeding, or haemorrhagic stroke scores 2;

- An eGFR < 60, scores 1;

- Treatment with an antiplatelet agent scores 1.

An ORBIT score of < 3 represents a LOW risk of a major bleed (24/1000 per annum on a DOAC). A score of 3 carries a MODERATE risk (47/1000 per annum). A score between 4 and 7 has a high risk of major bleeding (81/1000 per annum). The major bleeding referred to is a fatal or symptomatic bleed in a critical area (intracranial, intraspinal, intraocular, retroperitoneal, intra-articular or pericardial, or intramuscular with compartment syndrome. Also, bleeding causing a fall in haemoglobin level that requires transfusion of at least two units of blood.

See: O'Brien EC, Simon, D.N., et al. *The ORBIT bleeding score: a simple bedside score to assess bleeding risk in atrial fibrillation.* Eur. Heart J. 2015 Dec 7; 36(46): 3258-3264.

Other Considerations

The patho-electrophysiology of atrial fibrillation (AF) still holds some mystery. The rhythm results from the complex combination of many localised atrial re-entry circuits. Atrial flutter represents rapid wavelets arising from one, regularly repeating atrial re-entry circuit. In both flutter and fibrillation, only occasional wavelets will propagate through the AV node. The adequacy of AV conduction determines whether a fast ventricular rhythm or bradycardia, is seen in AF.

If we take the beat of a metronome as 'regular', the pulse in atrial fibrillation will never synchronise with it. Most times, AF is obviously irregular; at other times seeming almost regular. Without the pulse being felt for at least 20-30 seconds, one can miss AF. It can help to view the JVP: one might see 'f' waves' superimposed.

Allessie, et al., recorded electrograms from 192 points within the atrial myocardium. They showed AF being maintained by 3–6 wavelets, rotating simultaneously (Allessie M. A. L. W., Bonke F. I. M., Hollen S. 1985. J . *Experimental Evaluation of Moe's Multiple Wavelet Hypothesis of Atrial Fibrillation.* New York, NY, USA: Grune & Stratton).

Regaining sinus rhythm can follow ablation of endocardial areas around the pulmonary veins. The idea is to block re-entry. Pulmonary vein ablation often works, demonstrating that left atrial conductive fibres (associated with autonomic ganglia) are important for AF maintenance.

Any drug that reduces the refractory period of atrial action potentials (like digoxin) will sustain AF. Vagal stimulation (producing Ach) and epinephrine from sympathetic nerve fibres, also reduce the refractory period of action potentials (they are thus pro-arrhythmic).Flecainide, amiodarone, bretylium and sotalol, lengthen the refractory period. In the past, I occasionally prescribed flecainide to end AF. Class III antiarrhythmic drugs (like amiodarone and d-sotalol) slow the rapid repolarising component of the potassium channel current (I_{kr}), lengthening action potentials and their refractory period.

Learning Points:

Atrial size (diameter and volume on echocardiography) strongly determines the success of DC reversion (conversion is a religious process). Cardiac conditions which dilate the atria, like mitral and tricuspid valve leakage, are risk factors for AF, and its maintenance.

In atrial fibrillation, the ventricles receive varying amounts of blood during diastole. The ventricles respond by remodelling muscle. This risks later heart failure, and is one good reason to attempt the reversion of AF to sinus rhythm.

The sudden onset of AF, can make patients with valve problems and slightly weak ventricular muscle, acutely breathless (inducing heart failure). It is then expedient to slow their heart rate with a drug that will not diminish myocardial function. There is often a trade-off to be made: cardiac output is better at slower ventricular rates (Frank-Starling curve), but some drugs that slow the rate, lessen ventricular contractility. Some β-blockers have a negative effect on contraction (negative ionotropic effect), yet have a net beneficial effect on cardiac output, so breathlessness improves. Bisoprolol and sotalol are more effective than older β-blockers like propranolol and atenolol. Remember that propranolol can cause breathlessness by inducing asthma.

In older patients with dilated atria (on echocardiogram), one often has to accept AF, and the potential failure of DC reversion. It is then traditional to prescribe a digitalis alkaloid (digoxin or digitoxin) to slow AV conduction, and ventricular rate. Digitalis alkaloids take time to become effective and have variable efficacy. It is easy to exceed therapeutic blood and tissue levels, resulting in anorexia and nausea (often with associated typically cupped T-waves on ECG). Given their unpredictable responses, beneficially using digitalis alkaloids remains an art to be learned. The same variability of action, and similar artistry, is common to warfarin management (often given concurrently with digoxin to prevent thromboembolism). Digoxin has lost favour over recent decades, being replaced mostly by cardio-specific $β_1$ receptor blockers like bisopralol (propranolol is now inappropriate because it also blocks $β_2$ receptors and can affect lung function).

When first introduced, digitalis was an extract of purple foxglove (*digitalis purpurea*). A wealthy Scottish doctor, William Withering (who became rich enough to lease Edgbaston Hall in 1786), introduced it in 1776. He was working in Edgbaston when he observed a heart patient improved by a herbal medicine, given to her by a gypsy. As Withering was dying, one of his friends (noted for his black humour) said, 'the flower of Physic is withering'!

Learning Point: Remember that the kidneys excrete digoxin, and digitoxin mostly passes through the liver. Since it is common for older adults with heart failure to have some renal impairment, digitoxin might be the better choice. (Digitoxin is more commonly used in the EU than in the UK).

Atrial Tachycardia

Atrial tachycardias mostly arises from either atrial micro re-entry circuits or from competition between atrial pacemakers (multifocal atrial tachycardia or M.A.T), with different P-waveforms being visible. MAT appears not only in otherwise healthy subjects, but in those with chronic pulmonary disease (hypoxia), digitalis toxicity, electrolyte imbalance, mitral valve disease and cardiac ischaemia.

Re-entry Tachycardia

The re-entry phenomenon depends on there being several circuits of endocardial conducting tissue. The circuits comprise the sinus and AV nodes, preferential pathways between them, and inherited extra pathways (Fig. 9). When re-entry involves the ventricles, the AV node and extra pathways, a fast ventricular rhythm might remain until some electrical event blocks it. To stop constant reverberation (feedback loop), an increase in the refractory period, somewhere in the circuit must occur. This happens after ectopic beats, after a stimulus from an implanted artificial pacemaker, or in response to an infused drug. Cutting, freezing, and microwaving part of the circuit can create an open, disconnected circuit, which can no longer re verberate.

Now consider the main reverberating (re-entry) circuits seen. These are:

- AV Nodal Re-entrant Tachycardia, and

- AV Reciprocating Tachycardia.

AV Nodal Re-entrant Tachycardia

Like nerve fibres that conduct fast or slow, the AV node has slow antegrade fibres (from atrium to ventricle), and fast retrograde fibres (returning current from the ventricle to the atrium). The current can also reverberate within the AV node. An atrial premature beat can start it, giving rise thereafter to retrograde P-waves (from retrograde atrial activation).

Because both sympathetic and parasympathetic nerves innervate the AV node, autonomic manoeuvres can sometimes terminate a re-entrant rhythm (see carotid sinus pressure and Valsalva manoeuvre described below).

A **Valsalva manoeuvre** increases lung pressure with forced, blocked expiration. This happens when trying to blow into a balloon that expands with difficulty. I often used a sphygmomanometer. I would get patients to blow down the tube connected to the pressure gauge (having first wiped it clean), after which I would ask the patient to maintain pressure (that associated with hard constant blowing). The heart rate rises during this blowing phase, driven by sympathetic stimulation. Once blowing stops and lung pressure drops, AV nodal parasympathetic stimulation occurs, and the SA and AV nodes become flooded with acetylcholine (Ach). This slows the atrial rate and sometimes stops AV nodal re-entrant tachycardia.

AV Reciprocating Tachycardias

These are associated with several accessory pathways. One condition that has received much attention is the Wolff-Parkinson-White (**WPW**) syndrome, involving the Bundle of Kent. It was first described by Wolff L, Parkinson J, and White P.D. in 1930. The typical pattern on ECG is of bundle-branch block with a short P-R interval. This appears in healthy young adults prone to paroxysmal tachycardia, occurring in 0.1 to 0.3% of the population. The Bundle of Kent connects the left atrium to the left ventricle, bypassing the AV node. There are other variants which connect the right atrium to the right ventricle.

In WPW, the PR interval is short because the depolarising wave front more quickly reaches the left ventricle. Because the wave front from the sinus node takes an abnormal path to the left ventricle (not via the AV node), the initial phase of the QRS can appear slurred (referred to as a delta [δ] wave. See Fig. 19). The odd thing about WPW is the infrequent occurrence of SVT in all but a minority.

As the wave front spreads throughout the left and right atria, it can take one of two paths. Either normally through the AV node (designed to delay transmission so that the ventricles have time to fill), or through the Bundle of Kent to reach the left ventricle directly. The wave front can pass back up through the AV node (retrograde transmission), but only while the AV is responsive, not refractory. This retrograde path is faster than the normal antegrade pathway. While the circuit is open to transmission, wave fronts can reverberate around the atria and ventricles causing re-entrant SVT.

Although the extra conduction pathway described, can conduct in both directions, the commonest is antegrade AV nodal conduction with retrograde Bundle of Kent conduction. When activated, we refer to this as **orthodromic reciprocating SVT.** With the ventricles normally activated through the AV node (unless LBBB occurs), it presents with narrow QRS complexes.

When the Bundle of Kent is the first to conduct wave fronts, conducting in an antegrade fashion, the SVT is called **antidromic reciprocating tachycardia** (5-10% of WPW cases). The ECG then displays wide QRS complexes because the ventricles are being abnormally activated. The latter is easily mistaken for ventricular tachycardia.

Learning Point: Exercise tests and 24-hour ECGs, can reveal if the WPW circuit is rate limited, or able to conduct at very fast rates. With the latter capability, one should assume a risk of sudden death. Ablation of the extra pathway then becomes essential.

Electrophysiology Studies

Sometimes the direct measurement of electric wave fronts and their propagation are necessary. The discovery of extra pathways, and the definition of where heart block is occurring, are reasons for this investigation.

The investigation involves inserting multiple electrode catheters intravenously, then guiding them under fluoroscopic control to various parts of the atria and right ventricle. While working in the first pacing unit in the UK at St George's, London, I became responsible for measuring the conduction intervals before and after His bundle activation.

His bundlegrams can be obtained with a multiple electrode catheter placed across the superior aspect of the tricuspid valve, in a position close to the AV node (Fig. 13). The catheter is first advanced into the right ventricle then slowly pulled back until a His bundlegram appears. Measuring the time taken through the AV node (P to His), and thereafter (His to ventricular activation; normally 40msecs.) is then possible and the location of heart block determined. Atrial pacing at different rates, will demonstrate the reliability of AV conduction. Sinus node recovery time is a measure of SA nodal function. One can calculate it as the average recovery time after fast atrial pacing.

The detection of extra pathways requires employs catheters with simultaneous electrogram readings taken from many and various places. These recordings can now be analysed using computer software.

Ventricular Tachycardias

Ventricular Fibrillation

The ECG pattern is a variable sine-wave (Fig 17), indicting no regular beating of the heart at all. Left untreated it is incompatible with life; seen often as the last stage of life in monitored dying people. It is mostly associated with serious structural heart disease (unless induced by an external electrical shock). It can revert spontaneously when, during cardiac catheterisation, a coronary artery is blocked, or the catheter stimulates an excitable area in the one of the ventricles. Otherwise the patient has to be defibrillated to stay alive.

There is a vulnerable period ½ to ¾ of the way through the T-wave, when electrical stimulation can spark off VT or VF. The period immediately follow the refractory phase. During intracardiac pacing studies and external defibrillation it is best to avoid any stimulus synchronising with this short period.

VF is associated with long QT syndrome (LQTS), occurring in one in 2500 people. The normal interval is inversely related to heart rate so it has to be corrected (QTc) for clinical interpretation. QTc intervals > 440msecs are borderline; those > 500msecs are abnormal. Long QTs express abnormalities of myocyte repolarisation. Prolongation can be caused by certain drugs like Class 1 cardiac suppressants; macrolide and quinolone antibiotics; anti-psychotics like chlorpromazine and haloperidol, but is also low levels of potassium, calcium and magnesium.

VF is thought caused by a re-entry rotor. Otherwise multiple-origin wavelets could be responsible.

Together with VT, the main associations are age, bradycardia (an escape rhythm), recent defibrillation, aortic valve disease, sleep apnoea, sepsis and CVA amongst many others. Although rare, one must be aware of Brugada syndrome, characterised by right bundle branch block and persistent ST-segment elevation. Because of its association with VF, it can warn of sudden cardiac death.

The treatment is defibrillation, with emergency administered drugs that prolong the refractory period. Lidocaine and amiodarone are often used. Epinephrine must be avoid since it makes VF much more likely.

For those with recurrent VF and VT and implanted defibrillator is now advised.

Ventricular Tachycardia

In some patients who are syncopal, or experiencing palpitation or a sudden reduction of consciousness, VT can be the cause. Others may hardly notice it happening, and diagnosed only on 24h ECG monitoring. It is often seen on ECG as mono-morphic as opposed to multi-morphic VF. Its differentiation from wide complex SVT can be difficult.

Some important features are:

- Originating in the ventricles (mostly outflow tracts, in bundle-branches), the heart rate is usually > 100 bpm.

- An irregular (broad complex ECG) rhythm does not exclude it.

- Most broad complex tachycardias will be VT (80%), but SVT with BBB can be the cause.

- Most (90%) have underlying heart disease.

Depending on the risk of sudden death, one should consider the following options:

- Ablation.

- Implantation of a defibrillator.

- Sotalol.

- Class 1 anti-arrhythmic drugs.

- A more recent treatment addition is stereotactic radiation.

Chapter Nine

Valvular Heart Disease

When I first worked in a cardiac unit in the late 1960s, patients with conduction problems, rheumatic heart disease and coronary artery disease, occupied most of our time.

Heart valves can become stenosed, leaky (incompetent) or both, becoming deranged by age, the effects of rheumatic fever or infection (bacterial endocarditis, for instance). Other diseases affect the heart valves such as SLE, rheumatoid arthritis, carcinoid and Whipple disease, but these are rarely seen.

We regularly diagnosed rheumatic valve problems in the 1960s; mostly affecting the mitral and aortic valves. Many students then found it challenging to recognise the murmurs produced by each diseased valve, and to discern the degree of stenosis and /or leakage (incompetence) present. The effects of rheumatic fever usually took between twenty and forty years to affect the valve structure.

Mitral stenosis has been amenable to straightforward valvotomy surgery since 1927, while the first aortic valve replacement took place in 1962. The clinical focus for cardiologists dealing with valve problems remains the assessment of mechanical severity and the suitability for surgery. In this respect, three considerations have changed little:

- Will surgery improve symptoms?

- Will surgery improve life expectancy?

- What are the surgical risks?

Mitral Stenosis

Patients with significant mitral stenosis (MS) can present at a young age with symptoms, the same as those of left heart failure. They may have experienced progressive shortness of breath, orthopnoea (breathless when lying flat), and have pulmonary oedema, but this is not left heart (ventricular) failure. The onset of shortness can be sudden for a patient with MS who develops fast AF (they are prone to AF because accompanying atrial muscle pathology). These are symptoms more usual in older adults, but when they occur in the second and third decade of life, one must always think of MS. Whatever the patient's age, left ventricle function will usually be normal, although rheumatic nodules and myocardial fibrosis can impair it.

The specific auscultatory signs of MS are an opening snap, followed by a mid-diastolic murmur. The P_2 component of the second sound will be loud if pulmonary hypertension is present.

The opening snap occurs only when the valve leaflets are pliable. (A measure of pliability, 'splitability index' or the Wilkin score, can be derived from 3D ultrasound scanning). The stiffer the mitral cusps, the lower will be the audible frequency of the opening sound.

The typical cadence of MS murmurs, can be difficult to decipher. In sinus rhythm it is:

rrrRUB - de – **drrrrrrrrrrr**   **rrrRUB** - de – **drrrrrrrrrrr**   **rrrRUB** - de – **drrrrrrrrrrr**

'**RUB**' represents the first heart sound, then – 'de' – the second sound; '**drrrrrrrr**' stands for the opening snap and mid-diastolic murmur, caused by blood passing through the mitral valve. In sinus rhythm there can be presystolic accentuation of the later component of the mid-diastolic murmur, caused by atrial systole. This causes an initial '**rrr**' to make the first sound seem like **rrrRUB.**

If the patient is in AF, presystolic accentuation is lost. The opening snap 'd', is followed by a mid-diastolic murmur 'rrrrrrr'. Together they make the sound—'drrrrrrrrr'.

Practical Point: The timing of the murmur can be confusing. It is made easier by having the patient lying in the left lateral position with a hand feeling for LV systole (the outward impulse times with first heart sound (or the 'RUB' sound in the diagram). The snap is best heard in this position with the patient's breath held in expiration. Never forget to instruct the patient to breathe again!

When the patient is not in AF, the pre-systolic accentuation occurring with atrial systole can be prominent and confusing. The cadence is then:

rrrr**RUB** - de – drrrrrrrrrrrrrrrr**RUB** - de – drrrrrrrrrrrrrrrr**RUB** - de – drrrrrrrrrrrrrrrr**RUB**

In Fig. 22, there is a stylised heart sound recording (phonocardiogram) from a patient with mitral stenosis, sinus rhythm and a pliable valve.

Learning Point. The length of the mid-diastolic murmur equates to the severity of the mitral stenosis. A problem exists, however. When the murmur is short, the stenosis can be minimal or severe. When long, assume the valve to be moderately stenosed and clinically significant.

Simple investigations may show bifid P-waves on an ECG (P-mitrale) (see Fig. 16), and a prominent left atrial shadow on a CXR. Sometimes mitral calcification is visible on CXR, together with prominent pulmonary arteries and horizontal, Kerley B lines in the lung fields.

The mitral valve structure differs from the other valves. It has two cusps (anterior and posterior) that resemble the opening of a conch shell (the tricuspid valve by comparison is thin and diaphanous)(Fig. 7). The normal aperture area is usually 4.0 to 5.0 cm^2, with symptoms occurring when it is less than 1.5 cm^2. We can assess the severity of stenosis by measuring the pressure gradient across the valve, and estimating the antegrade velocity and flow. All such measures are heart rate sensitive.

In mitral stenosis the leaflets can fuse, sometimes with calcified rheumatic nodules present. When found in the mitral valve leaflets and valve ring, calcification can be visible on plain X-ray, X-ray fluoroscopy and echocardiography.

In 1923, Cutler and Levine performed the first mitral valvotomy at the Peter Bent Bringham Hospital in Boston Massachusetts, USA. (Cutler E.C, Levine S.A. *Cardiotomy and valvulotomy for mitral stenosis; experimental observations and clinical notes concerning an operated case with recovery.* Bost. Med. Surg. J. 1923;188(26):1023–7).

Balloon catheters can split pliable valves. Those that are fibrotic, calcified or incompetent, will need replacing rather than splitting. Following surgery for isolated mitral stenosis, the rapid restitution of exercise tolerance is often remarkable.

Mitral Stenosis Management

To prevent clotting and embolism, all patients with mitral stenosis should be anticoagulated. With significant stenosis, all will need a slow heart rate, achieved by using drugs or DC reversion in AF. The aim is to prolong diastole and increase the time available for ventricular filling. Thirty to forty percent of MS patients will experience AF. When the atrial component of ventricular filling is lost with AF, and there is a faster heart rate than normal, symptoms of shortness of breath will arise.

Patients presenting with pulmonary hypertension and a mitral valve area < 1.00 cm^2, must be considered for surgical intervention. There are important considerations to be made before any valve repair or replacement:

The presence of significant mitral incompetence contraindicates balloon valvotomy.

Use transoesophageal echocardiography (TOE) to exclude a left atrial appendage thrombus.

Consider mechanical valve replacement for young patients, and a bio-prosthesis for those over fifty years.

All valve abnormalities, other than MS (and rare tricuspid stenosis), can affect ventricular function. Ventricular function is crucial to the assessment and management of all valve abnormalities. If the ventricular function is poor (ejection fraction < 40%), changing a valve may not benefit the patient's symptoms or prognosis.

Learning Point: Assessing Ventricular Function need not rely on quantitative measures alone. One can reliably visualise it. Whether looking at the ventricle using ultrasound or on an angiogram, there are key factors to observe:

What is the heart rate? No normal heart beats efficiently with a pulse rate > 140 beats per minute. For failing hearts, the rate will need to much lower (see the Frank Starling relationship curve. Fig.11).

What is the LV cavity size? The more dilated the ventricle, the less efficient it will be as a pump.

Is contraction reduced or sluggish? Both a 'normal' and 'sluggish' LV are easily distinguishable on an echocardiogram or angiogram. Sluggishness is often the result of heart muscle weakness (ischaemia, amyloid, fibrosis, degenerative thinning), but sometimes the result of beta-blockade (one reason for my lack of keenness to use them in heart failure).

Mitral Valve Prolapse (MVP)

As the left ventricle contracts and the mitral valve closes, papillary muscles hold the valve leaflets in place. Closure will not be complete if one leaflet (or part of it) is unrestrained; at the peak of LV pressure, it could enter or bulge into the left atrium. This can be associated with a sharp clicking sound in mid-systole, occasionally followed by a late systolic murmur. One can usually confirm these features with echocardiography. A prolapse > 5mm beyond the valve annulus is significant.

Malformed or weak papillary muscles or leaflets, stretching more than usual cause MVP (some have Marfan's syndrome or Ehlers-Danlos syndrome). It is sometimes associated with supraventricular ectopics, perhaps because an eccentric jet of blood from the left ventricle, hits the left atrial wall. In 1963, Barlow described a 'click-murmur' syndrome (Barlow JB, Pocock WA et al., *The significance of late systolic murmurs.* Am Heart J 1963; 66:443-452). Floppy mitral valve leaflets, or elongated papillary muscles are the usual cause. Although I have regularly seen patients with palpitations and mitral valve prolapse, large-scale studies have not confirmed what I have always referred to as Barlow's syndrome. Some clinical exceptions are bound to escape statistical verification.

Having a prolapsing mitral leaflet was once thought to attract circulating bacteria (after dental extraction, etc.) and a risk of (subacute) endocarditis. This is now thought too rare for prophylactic penicillin to be advisable. When discussed with patients, some disagree. Their argument can be: 'why take any risk when the cost of prevention is only a few tablets of penicillin?'

Key Point. The Prophylaxis Issue: Do patients with mitral valve prolapse need prophylactic antibiotics while having general or dental surgery? The latest guidelines no longer recommend it, although it is recommended for those who have had endocarditis, those with prosthetic valves and those who have had congenital heart disease repaired. See Wilson, W. et al. *Prevention of infective endocarditis.* Circulation (2007); 116: 1736-1754.

Only 2% of the adult population have MVP. Although mostly thought innocent, some have associated chest pain, palpitations and syncope. In my experience, these are often anxiety related symptoms caused by the finding of MVP. 'Mitral Valve Prolapse Syndrome' is likely to arise more in those who have been made anxious by the diagnosis. Because a few will proceed to develop significant mitral incompetence, heart failure and AF later in life, some follow-up at five to ten year intervals is appropriate. Although this risks making people into patients, some will enjoy the attention.

Mitral Regurgitation (MR)

Although rarely an acute condition, apart from acute cardiac infarction affecting the mitral papillary muscles, it can still result from acute rheumatic fever and endocarditis. Most cases are now caused by chronic myxomatous valvular changes, with progressive MR occurring over decades. Occasionally, a papillary muscle being tethered to an ischaemic papillary muscle causes chronic MR.

Cardiac infarction is the commonest cause of acute MR and pulmonary oedema. The murmur can fill systole (pan-systolic murmur), starting with the onset of left ventricular systole, and ending with the aortic second sound. The murmur varies in intensity, being loudest with slight to moderate regurgitation, and much less audible when severe. It is not possible to grade the severity of MR accurately from the loudness of the murmur alone. Also, where best to hear mitral murmurs is often depicted incorrectly in some textbooks. One can often hear MR murmurs over wide areas—in underweight patients, everywhere to the left of the lower sternum; in obese patients they might best be heard near to the LV apex.

When grading MR, consider the amount of LV and LA dilatation, and left ventricular function, evaluated using echocardiography. If found difficult, one can use cardiac MRI. A stress echocardiogram is also useful for evaluating LV function.

Other factors in MR are:

- The extra passive filling of the ventricle in early diastole, can produce a prominent 3rd heart sound.

- Dilatation of the left atrium can lead to AF, and associated with risks of clotting and arterial embolism.

- Tracheal carina widening, from left atrial appendage enlargement, may be visible on CXR (the LA sits beneath the tracheal bifurcation).

During cardiac catheterisation, the following are important:

- A prominent 'v' wave in the pulmonary wedge pressure.

- Pulmonary hypertension.

- A raised right end-diastolic pressures (right heart failure).

- When performing a left ventricular angiogram, severe MR (grade 4/4) causes the left atrium to fill completely during ventricular systole. Observing how much of the atrium fills with MR, will help decide its severity. One out of four is minimal.

- Left ventricular functioning is important. In compensated MR without symptoms, the ventricle can be hypertrophied.

- The LV may later dilate (decompensated MR with symptoms of right and left heart failure).

Apart from rheumatic heart disease, MR has other causes. Myxomatous valve changes now predominate, but do not forget antiphospholipid syndrome and some rheumatic conditions as causes. Increased connective tissue elasticity occurs in Marfan's syndrome and Ehlers-Danlos syndrome; these can affect the papillary muscles and mitral annulus.

In those considered for surgery, it is important to first visualise the abnormal parts of the valve leaflets.

MR Management

The considerations are:

There are no controlled trials comparing medical treatment (digoxin or beta-blockers for rate control and diuretics) with surgical intervention.
Try to correct MR before LV dysfunction occurs.
Always consider surgery in specialised centres for severe MR; when the LV diameter is greater than 4cm, and the patient is asymptomatic.
Repair is preferable to replacement.
Transcatheter repair with MitraClip, etc., allows edge-to-edge repair, whereas others employ annuloplasty.
Replacement is preferable for rheumatic valves, where the valve is unusually thickened, or there is a loss of tissue substance. Of those repaired, 0.5 – 1% will return for further consideration.
One should not consider surgery for those with obvious LV dysfunction (LV dilatation and reduced contractility).

Aortic Stenosis (AS)

Aortic valve stenosis causes an ejection systolic murmur, heard maximally in the upper right chest (to the right of the upper sternum), although sometimes audible in the neck on the right side. If severe it will be associated with LVH. Accompanying cystic changes in the media of the ascending aorta, can sometimes cause accompanying, proximal aortic dilatation.

Aortic valves are usually tricuspid. Those that are bicuspid (1–2% of the population) are prone to stenosis and leakage in middle age.

Aortic stenosis is now the commonest valve problem encountered by cardiologists working in western countries.

The pressure overload (obstruction) of stenosis causes concentric LV muscle hypertrophy. When the hypertrophy is severe, the LV chamber dimensions can be reduced to smaller than normal. A smaller LV will generate more force, although, severe muscle hypertrophy risks ischaemia. Laplace related the pressure generated within a cavity to the inverse of its radius (Pierre-Simon Laplace, 1806). The relationship predicts that a dilated ventricle (as in AI and MR), will generate less force than one with normal or smaller dimensions. Diastolic LV failure, occurs when the normal diastolic expansion becomes limited.

The commonest cause of AS in those over 60-years of age, is a degenerative process and inflammatory response (macrophages, T-cells and fibroblasts become activated). Now common only in undeveloped countries, rheumatic heart disease occurs in younger people. Therapeutic radiation of the chest can also cause it. More rarely, the obstruction is sub-valvular (obstruction with normal valve leaflets), or supra-valvular; some with associated congenital abnormalities (Williams syndrome and the Shone complex).

Because the ventricular muscle compensates for the obstruction, symptoms occur late in the natural history of AS. Dyspnoea, dizziness, syncope and angina can then occur. Untreated patients with AS, who have angina or syncope, have a poor prognosis (average survival no greater than five years).

The physical signs of AS are straightforward. The obstruction causes the arterial pulse to be slow rising (*parvus*) with a delayed peak pressure (*tardus*), hence the classical description of the pulse in AS: *pulsus parvus et tardus*. Some experience is necessary to detect these features. They are more readily felt by palpating a large artery, like the carotid or femoral. Apart from the ejection systolic murmur (crescendo – decrescendo), a fourth heart sound (associated with LVH) is sometimes audible. This soft pre-systolic sound (before the first sound), occurs with atrial systole. Its low frequency can make it difficult to hear.

Palpation of the LV apex, with the patient lying in the left lateral position, will find it pushed outward, towards and beyond the anterior axillary line when hypertrophied. The hypertrophied LV causes the apex pulse to feel sustained and forceful (especially in lean patients).

Learning Point: There is a risk of mistaking the murmur of AS for that of HCM in younger patients; hypertrophic obstructive cardiomyopathy (HCM) is easily missed. The murmur is similar, but in HCM, the arterial pulse is occasionally bifid (because of mid-LV obstruction). Also, the murmur of HCM can be louder during a Valsalva manoeuvre, while the murmur of valvular AS is usually softer. None of these factors are reliable enough for definitive diagnosis; instead, echocardiography has replaced the m.

Investigating Aortic Stenosis (AS)

On ECG one can expect to see the following in LVH:

- An S wave voltage in V1, plus the R wave voltage in V5 or V6, are together greater than 35 mm (3.5mVs)(Sokolow-Lyon Criterion).

- T-wave inversion. This usually relates to increased muscle mass (or ischaemia).

Using a sum > 60mm (> 6mVs), the ECG voltage specificity for LVH is 83% to 97% (Romhilt-Estes LVH Point Score System).

It is important to note that:

Because the AV node sits next to the aortic valve annulus, one may see all degrees of associated AV block. AS is a cause of complete heart block (CHB).

In slim people, the sum of V1 and V5 (or V6 voltages), can exceed 4mVs without hypertrophy; obesity can make the voltages smaller, even with LVH present.

One important question arises for all those with syncope and AS. Is CHB or valve obstruction, the cause of their syncope?

A posterior-anterior (PA) CXR, may reveal a boot-shaped heart (Fig. 4), in most cases of LVH. An echocardiogram may show thickened, calcified valve leaflets and annulus, increased LV wall thickness and increased LV contractile function. Attempted calculation of the gradient across the valve, using the Bernoulli equation, can overestimate the severity of stenosis sometimes.

Cardiac catheterisation in AS has three objectives:

1. To measure the aortic valve gradient accurately (if the valve can be crossed with a catheter from the aorta; not always easy for novices), and

2. To calculate the aperture area of the valve. This requires a measure of cardiac output (cardiac output in litres/min., divided by the square root of the peak gradient in mm. Hg.). The normal valve aperture area is 2-3 cm^2.

3. To obtain coronary angiography information prior to surgery.

Learning Point. Pressure gradients are flow-dependent, and it is important to assess any valve gradient in relation to the flow rate. Low flow will lower the gradient; high flow will cause it to be higher.

The pressure drop across the valve is one measure of severity:

If less than 36mm. Hg., it is **'mild' stenosis.**
Moderate stenosis yields a gradient between 36 and 63 mm. Hg.
In **severe stenosis,** the gradient will be greater than 63mm. Hg.

The Management of AS

The management of AS relates to the severity of stenosis and its rate of progression (faster when risk factors for coronary disease are present).

The main considerations are:

- Severe symptomatic AS needs valve replacement.

- Severe asymptomatic AS with an ejection fraction < 50%, needs valve replacement..

- Severe asymptomatic AS with an abnormal exercise test, needs valve replacement.

- Prophylaxis for endocarditis is not recommended, unless the patient has had rheumatic fever within 10-years (up to 21-years of age) or previous endocarditis.

- **Anti-hypertensive drugs must only be used with consider-**

able caution.

- Mild to moderate AS should be observed repeatedly every 3-5-years.

Percutaneous balloon valvuloplasty and transcatheter valve replacement are now possible. The benefits of balloon valvuloplasty can be short-lived, most redeveloping stenosis within one year. **Transcatheter aortic valve implantation (TAVI)** and surgical valve replacement, one can are recommend equally for those over 65 years-of-age. For those > 80-years of age, transfemoral TAVI is the intervention of first choice.

Aortic Regurgitation (AI)

Aortic incompetence causes an early diastolic murmur (EDM), starting with A_2.

Learning Point: The murmur of AI is best heard at the lower left sternal edge, with the patient sitting forward, and breath held in expiration (never forget to allow them to breathe occasionally!).

As with MR, a short murmur will indicate either minor or severe regurgitation. A long EDM, is found in mild-to-moderate AI. The severity will relate to the amount of LV hypertrophy or dilatation. With the backward flow into the LV, the anterior mitral valve leaflet may vibrate (seen on echocardiography). This can gives rise to a diastolic murmur, specifically heard in the mitral area: an Austin Flint murmur. The mitral valve leaflet must be pliable for it to happen.

Unlike AS, which results in LV pressure overload only, AI can cause both pressure and volume overload. With each stroke, the LV has extra blood to pump (normal volume + regurgitant volume). The initial response of the ventricular muscle is concentric hypertrophy. Later on, the ventricle dilates and the end-diastolic pressure rises. This rise in pressure, restricts blood flow into the ventricle from the LA. If the LA pressure rises, pulmonary oedema, pulmonary hypertension and right heart failure can re sult.

There are a few, not-to-be-missed causes of AI: endocarditis and aortic dissection. Both need urgent intervention. Other causes can be evaluated at leisure, namely: rheumatic disease, a biscuspid aortic valve, Marfan disease and various inflammatory collagen diseases (ankylosing spondylitis, Takayasu disease, etc.).

The anorectic (weight loss) drugs, fenfluramine and phentamine, have been reported as a cause of AI. I am sceptical. I worked alongside two doctors who used these drugs for decades. None of their many patients developed AI. As a side-effect, this is either wrong or of very rare occurrence (less than one case in a lifetime).These drugs are amphetamine-based and addictive, and not looked upon favourably.

Clinical Presentation

An important historical feature of AI is that it can be tolerated well for decades, and then deteriorate rapidly over months, with symptoms and signs of LV failure soon becoming apparent. LV dimensions can dilate towards a critical point, beyond which the force generated by ventricular contraction, will be inadequate (Laplace relationship). It often becomes urgent to replace the valve before this point (even though the patient is asymptomatic). Beyond this point, replacing the valve is unlikely to help symptoms or improve the poor prognosis associated with a dilated LV.

With severe AI, the width of the arterial pulse waveform shortens, and the downstroke gets faster. This is like the action of a water-hammer (Corrigan pulse), and sometimes associated with visible pulsation in fingernail beds.

Palpation of the LV apex, with the patient lying in the left lateral position, may find it pushed outward to the left, sometimes reaching beyond the anterior axillary line. Adverse changes in LV dynamics make the apex pulse feel sustained and forceful (especially in lean patients).

The Investigation of AI

The most useful investigations assess regurgitant flow. One can use echocardiographic Doppler imaging to assess regurgitant flow, LVH, LV dilatation and LV dysfunction. One can also verify a bicuspid aortic valve, and valve vegetations (seen better using TOE).

Indicators of severe AI are:

- Over 65% of LV forward blood flow, returns to the LV during diastole.
- A reversal of the aortic flow in diastole > 25 cm/sec.
- A fast collapse in the post-peak arterial pulse pressure.
- An aortic valve regurgitant orifice > 0.3 cm^2.

Although less specific, a PA, CXR, may reveal significant aortic dilatation and LVH, with LVH confirmed on ECG voltage criteria.

The Management of AI

Acute AI is rare, but when it occurs, lowering the BP and increasing the pulse rate with a vasodilator, can help prior to urgent surgery.

For chronic AI cases with LV dysfunction, the sudden death mortality changes from < 0.2% per annum when asymptomatic, to 10% per annum when symptomatic.

Vasodilators (calcium channel blockers), ACE inhibitors and beta-blockers, have their place in medical management, but asymptomatic patients with progressive aortic dilatation and LV dysfunction (ejection fraction around 50%), will benefit from surgery. One should review all patients annually, and evaluate those with symptoms urgently. To benefit, they will usually have had only recent symptoms, and an LV cavity diameter of < 5cm

I never waited until my patients significantly dilated their LV. I monitored them annually and based my management on the rate of progression of LV dilatation. My aim—always to refer them for surgery before symptoms arose.

There are two surgical options: valve repair or replacement: homograft or TAVI (Transcatheter Aortic Valve Implantation).

Tricuspid Regurgitation (TR)

This occurs mostly as a secondary feature of other heart problems (left heart failure and mitral stenosis) or pulmonary disease that results in pulmonary hypertension (functional TR is commonest). TR is only rarely rheumatic in origin, but sometimes found with the Ebstein anomaly, carcinoid syndrome and connective tissue disease.

Patients have symptoms and signs that relate to increased venous pressure, namely dependent oedema, hepatic congestion and ascites. GI symptoms will often occur.

The JVP will show prominent 'v' waves. With one hand on the LV or RV apex (beneath the xiphisternum), one can observe 'v' waves concurrent with the apex beat (occurring in ventricular systole).

Near to the lower right end of the sternum, one might hear a soft pan-systolic murmur and a right sided S_3. Both are of lower frequency and intensity than those generated by the left side of the heart. If TR is secondary to pulmonary hypertension, P_2 should be loud. The right apex (below the xiphisternum or next to the lower left sternum) can feel sustained with RVH present. Liver pulsation may be palpable, and synchronous with the 'v' wave of the JVP. In time, cardiac cirrhosis can develop and lead to jaundice.

Echocardiography can show a dilated tricuspid annulus, and/or a flail valve leaflet. The tricuspid valve, which usually has three leaflets, is more diaphanous than the mitral valve. Both a dilated RA and RV are the consequences of TR. Doppler measurement is used to grade TR.

A PA, CXR, may show a dilated heart, and an ECG in sinus rhythm with bifid 'P' waves.

When other valve conditions are being managed surgically, contemporaneous surgical repair (annuloplasty) of TR is a worthy consideration. Because intervention so improves quality of life, sole repair or replacement of the valve is now considered. Annuloplasty for isolated severe TR, is now amenable to percutaneous approaches. (Nickenig, G., Weber, M. et al. *Transcatheter edge-to-edge repair for reduction of tricuspid regurgitation: 6-month outcomes of the TRILUMINATE single-arm study.* The Lancet 2019; 394:2002–11). One year after the procedure, both repair and replacement (TriClip) improve the patient's quality of life. No improvement in mortality from heart failure occurred in the short-term. (Sorojja, P. et al. *Transcatheter Repair for Patients with Tricuspid Incompetence.* N Engl J Med 2023; 388:1833-1842 DOI: 10.1056/NEJMoa2300525).

Valve replacement patients need full anticoagulation, but the effect of any liver damage needs consideration. Also, pacemaker implantation is often required, given the proximity of the AV node and bundle of His, to the valve.

Few adult cardiologists, see other valve defects.

Tricuspid Stenosis (TS)

I have never seen a case, but one cause is rheumatic. I did, however, once see tricuspid flow severely reduced by a pedunculated right atrial myxoma.

Treatment of TS using percutaneous intervention is now possible; surgical intervention being advised, only when the repair or replacement of other valves is necessary.

Pulmonary Stenosis (PS)

As an isolated defect, it is uncommon and usually of congenital origin. I saw only one or two, post-rheumatic cases. Sometimes found in association with a VSD, it can be part of Noonan syndrome (autosomal inheritance or mutation, short stature, low-set ears, a webbed neck and musculoskeletal defects).

On examination, look for prominent 'a' waves in the JVP waveform. Unless the gradient across the valve is > 50mm Hg., no action is needed.

See the next section for congenital defects.

Pulmonary Regurgitation (PI)

It is common to see minor degrees of PI on routine echocardiography. Although usually ignored, its cause cannot. It is an important consideration in cases of Fallot's tetralogy.

Chapter Ten

Congenital Heart Disease

Because of advances in paediatric cardiology (diagnosis and treatment), adult cardiologists now see more adult congenital heart disease cases. There are three types to consider:

- Shunt defects,
- Obstructive lesions, and
- Complex malformations.

For a fuller understanding, review cardiac embryology.

Shunt Defects

There are three types of **atrial septal defect (ASD)**:

1. Primum,

2. Secundum, and

3. Sinus venous shunts.

Ventricular septal defects (VSD), are either:

In the upper membranous septum, or
In the septal muscle which usually close in childhood.

Another common shunt is the **patent ductus arteriosus**.

ASD

In every foetus *in utero*, the right and left atria connect through a hole in the atrial septum: the *foramen ovale*. This allows umbilical, oxygenated blood (venous) arriving from the placenta, to circulate in the foetus systemically. Oxygenated blood passing through the foetal right atrium, crosses into the foetal systemic circulation via the *foramen ovale*.

The adult atrial septum forms from the fusion of two septae: the *primum* septum and the *secundum* septum. At birth, the *foramen ovale* is normally closed by a flap in the *primum* septum. As breathing starts, it will close as pulmonary vein flow increases and left atrial pressure rises.

Secundum ASD defects are the commonest shunts. A failure of the *primum* septum to seal the *foramen ovale* causes them. The central hole can vary in size, but does not affect the mitral or tricuspid valve.

Primum defects at the lower atrial septum, can involve the mitral valve, the tricuspid valve and the upper ventricular septum (AV canal defect). Sometimes found in Down's Syndrome, the failure of both the *primum* and *secundum* parts to close the *foramen ovale*, causes a *primum* defect. An intact atrial septum will not then form.

In the much rarer unroofed coronary sinus type of ASD, and more complicated sinus venosus defects, the pulmonary veins and both venae cavae can be anomalous.

The finding of an enlarged right heart sometimes prompts clinical identification in adults. Because the ASD equalises atrial pressures, there is usually no variance in the splitting of the second heart sound (**fixed splitting** of the second heart sound occurs). The $A_2 - P_2$ interval is unchanged by respiration. There may also be a pulmonary ejection murmur, caused by increased pulmonary blood flow, from the left-to-right atrium.

An ECG may show RVH or RBBB with P-pulmonale (the initial peak of the P-wave is prominent). When the RA becomes stretched, AF can occur. A CXR may show right-sided enlargement, with prominent pulmonary arteries from increased blood flow into the lungs.

An **Eisenmenger complex** occurs when the PA and RA pressures rise with increasing pulmonary artery resistance. This can then reverse the shunt. Blood will then flow from right to left, causing cyanosis and the possibility of **paradoxical embolism** (clot emboli from peripheral veins passing from the right atrium to the left atrium and then to the systemic circulation, causing arterial blockages in the brain and elsewhere). The Eisenmenger complex is more common in cases with a VSD.

Considerations of ASD repair should start before the patient is 40-years old, especially if the left- to-right shunt is 1.5 times greater than the normal flow in the right atrium and pulmonary artery. An important proviso is that the PA pressure is less than half the systemic blood pressure (or pulmonary artery resistance is $< 1/3$ of the arterial vascular resistance). Early intervention can prevent pulmonary hypertension.

VSD

This is the commonest shunt found in children. Important points are:

- Eighty percent of which are found in the membranous part of the ventricular septum. The have the highest rate of spontaneous closure.

- 10% of all ventricular shunts are muscular VSDs.

- Supra-cristal shunts involve the right and left outflow tracts. They form 5% of the total. Associated aortic incompetence can be caused by the prolapse of a coronary cusp.

- A few shunts are A-V canal defects.

In all cases, larger defects produce low velocity jets; small defects cause high velocity flows.

Sometimes a VSD is one part of a complex presentation, transposition of the great vessels, for instance. The complications of VSDs are the same as those for ASDs, except for endocarditis. Endocarditis is more likely in VSD, associated with a right aortic valve cusp prolapse and AI.

A VSD can result from anterior cardiac infarction. It is has a poor prognosis, but much improved by surgery or percutaneous intervention.

Small VSDs produce loud pan-systolic murmurs, audible across the chest, but especially at the lower end of the sternum, left apex and neck. The loudness and widespread nature of the murmur, provides the diagnostic clue. A hand placed over the lower left sternum can sometimes detect a palpable thrill from the associated jet of blood.

An ECG can show both right and left ventricular hypertrophy. With small VSDs, the CXR will usually reveal a normal heart size; larger VSDs have all the same changes seen with significant ASDs. We used to use cardiac catheterisation to assess the clinical significance of the shunt; now Doppler echocardiography will often prove sufficient.

In later life, patients who develop hypertension or IHD, will sometimes get increased blood across their VSD. This can occur as the LV wall becomes stiffer, and the LV end-diastolic pressure rises.

Suggested reasons for surgical VSD closure are:

- Those with right-sided flow > 1.5 left-sided flow,

- Those with low pulmonary artery pressure and resistance (see ASD criteria);

- Those with a history of endocarditis.

- Those with aortic incompetence.

Both surgical or percutaneous closure is possible. At present (2025), surgery is more often the preferred approach for VSD closure; percutaneous methods for ASD closure.

Patent Ductus Arteriosus

While the foetus is *in utero*, the ductus arteriosus remains open. It will close within weeks of birth. It usually presents in adults as an isolated defect (sometimes associated with rubella in childhood). Since the ductus connects the aorta to the left pulmonary artery, it is a cause of cardiomegaly and pulmonary venous engorgement (seen on CXR). Significant flow can lead to an Eisenmenger situation, but PDAs only occasionally need closure in adults (ligation or placement of a percutaneous coil or other device).

Auscultation usually reveals a continuous (machinery) murmur, heard anteriorly and in the back; lateral to the spine, and medial to the left scapula.

Stenotic Defects

The only significant lesions are pulmonary valve stenosis and coarctation of the aorta.

Pulmonary Valve Stenosis (PS)

Common only in congenital heart disease, PS occurs in approximately one of two thousand births. It is also associated with Fallot's tetralogy and Noonan syndrome and sometimes found during routine echocardiography.

Pulmonary valve gradients of < 25mm Hg. need no intervention (now using the percutaneous approach). More severe lesions will need correction if irreversible RVH is to be avoided. Patients rarely get to see adult cardiologists since most are diagnosed and dealt with early in life. Those who do, will have ECG evidence of RVH, an ejection systolic murmur and wide splitting of the second heart sound (between A_2 and P_2), heard most easily to the left of the upper sternum. One may feel a right ventricular heave below the xiphisternum, or to the right of the lower sternum.

Coarctation of the Aorta

This is a stenotic lesion found in the area of the ductus arteriosus, thought caused by the contraction of associated tissues after birth. It has important associations like Turner syndrome, bicuspid aortic valve (resulting in an ejection click), VSD, mitral valve abnormalities and aortic aneurysm. It is sometimes associated with aneurysms of the circle of Willis (< 10% patients).

We often find coarctation by chance while investigating hypertension. In severe cases, the BP will be higher in the upper limbs than in the lower limbs. It is important to discover and correct severe cases in order to prevent claudication, aortic dissection, stroke and accelerated coronary disease. The development of collaterals will lessen the future risk.

An ejection (or continuous) murmur is best heard in the back.

A CXR can reveal a dilated aorta, LVH, and rib notching caused by collateral artery formation. An MRI scan best displays the position and structure of the coarctation. Cardiac CT scanning (coronary calcium score), and sometimes coronary arteriography, are important when assessing accelerated atherosclerosis.

One should advise intervention if the gradient across the defect is > 20 mm. Hg. Percutaneous stenting is now favoured since surgical correction and Dacron patching are associated with the later occurrence of aneurysmal aortic dilatation.

Rarer Congenital Abnormalities

Because of the complexity of some congenital cardiac defects, I provide only brief descriptions here. Congenital heart disease in children is a separate study topic.

The following overview includes Fallot's tetralogy, the Ebstein anomaly, transposition of the great arteries, and single ventricles. Every adult cardiologist needs to know something about them. Their definition and assessment depend on adequate investigation, and their management will often involve collaboration with cardiac surgeons.

Fallot's Tetralogy

The component features are:

- A VSD,

- Right ventricular outflow obstruction, and

- RVH (all are present in Fallot's tetralogy).

- An aorta that overrides the ventricular septum.

Sometimes an ASD also exists. With a right sided aorta, the left anterior descending coronary artery can arise anomalously from the right, rather than from the left coronary artery.

On examination, look for clubbing, cyanosis, a right ventricular heave, and a murmur from the VSD. If the right-sided obstruction does not divert venous flow across the VSD, there will be little or no cyanosis (so called 'pink' Fallot's tetralogy).

Cardiac MRI is the investigation of choice. An ECG usually reveals RBBB.

Adults risk arrhythmias and sudden death (an implanted ICD may be necessary when there is evidence of repeated VT).

In the Blalock-Taussig surgical procedure, a shunt between the subclavian and pulmonary arteries provides early palliation. Full correction entails closing the VSD and relieving the RV obstruction. Pulmonary valve incompetence and RV dilatation and dysfunction, are indications for surgical intervention.

Ebstein Anomaly

Consider this whenever a large rounded, 'globular' heart, appears on CXR. Both significant TI and atrial arrhythmias are common features (25% have a WPW presentation). Adults may present in right heart failure. A typical feature is an abnormal tricuspid valve, the anterior leaflet being enlarged and sail like; the other two cusps and the RV may be underdeveloped. Other abnormalities like ASD and patent *foramen ovale* (50% of patients), are often associated—the reason for some patients being cyanosed on exercise.

The ECG may show tall P-waves (Himalayan), a short PR interval, and RBBB.

Because of a liability to paradoxical embolisation, the repair of any shunt is important. Repair of the tricuspid valve may also be necessary.

Transposition of the Great Arteries

The embryonic ventricles can rotate to the right (Dextro or D-version), or to the left (Levo or L-version). In the D rotation type, venous blood returning to the right ventricle, is ejected into the aorta, bypassing the pulmonary circulation. The pulmonary venous blood will enter the left ventricle, but is pumped into the pulmonary circulation. There are, therefore, two parallel, non-interconnecting circulations, incompatible with life. Life is possible only when a shunt is present; either the patent ductus must be maintained or surgical intervention must provide an ASD.

The L-rotation variation can be asymptomatic and discovered only in adulthood. The right ventricle is to the left of the left ventricle. It receives pulmonary venous blood which is pumped into the aorta. Venous blood, returned to the left ventricle, gets pumped into the pulmonary artery. Surgical intervention can avoid ventricular dysfunction; this might include cardiac transplantation.

The PR interval can be prolonged, with complete heart block occurring in 2% of cases.

Single Ventricle

The presentation is one of cyanosis, systemic venous congestion and hepatic damage. Surgical intervention (Fontan procedures), can direct systemic venous return directly to the pulmonary artery. I would advise specialised management in a cardiac centre with appropriate experience.

Chapter Eleven

Atherosclerosis

Coronary Artery Disease and Peripheral Vascular Disease

Epidemiology

Cardiovascular disease is a leading cause of death in the world, now that infectious diseases are in decline. It is the third leading cause of mortality worldwide, causing 17.8 million deaths annually (*GBD 2017 Causes of Death Collaborators. Global, regional, and national age-sex-specific mortality for 282 causes of death in 195 countries and territories, 1980-2017: a systematic analysis for the Global Burden of Disease Study 2017. Lancet. 2018 Nov 10;392(10159):1736-1788*).

Atherosclerosis and arteriosclerosis underlie most of the cardiovascular problems cardiologists see today.

According to the American Heart Association, cardiovascular disease (CVD), listed as the cause of death, accounted for 868,662 deaths in the

US in 2017. (*2021 Heart Disease and Stroke Statistics Update Fact Sheet At-a-Glance*. American Heart Association. Heart Disease and Stroke Statistics 2021 Update). They also reported that between 2015 and 2018, 126.9 million American adults had a form of CVD. Only 1% of those who died were younger than 40 years old. Approximately 800,000 US citizens had a myocardial infarction each year between 2005 and 2014.

There are (age-adjusted) differences between races. In 2017, US deaths among the African American black community were 208 per 100,000. For white (non-Hispanics) people, the number was 169; for Hispanic people 114; for Asians or Pacific Island groups, it was 85.5. The poor and ill-educated among them, suffer more. The prevalence of CAD in US women (8.8%) is less than of US men (9.4%), with women presenting ten to twenty years later in life on average than men.

In 2017 the WHO Global Burden of Disease attributed 17.8 million deaths (32% of all deaths) to cardiovascular diseases (The Cleveland Clinic Cardiology Board Review. 3rd Edition. 2022. p543). Between 1990 and 2013, there was a worldwide increase of 46% in cardiovascular related deaths (mainly among poorer communities). Growing poor populations, partly account for this. In all populations, three to five times more CVS deaths occur among the poor, compared to the richest subgroups. Referred to as 'the health divide', it also applies to cancer deaths.

Between 1990 and 2013, cardiac deaths rose worldwide. In contrast, Northern Europe saw a dramatic decrease in both CAD-related mortality and other cardiovascular events between 1990 and 2017. The event rates (morbidity) decreased by 10% in women and 14% in men; the CAD-related mortality decreased by 28% in men and by 33% in women. Therapeutic intervention and prevention strategies, could be responsible for one third and two-thirds of the reductions respectively.

Atherosclerosis

Atheromatous plaque formation is the usual pathological basis for coronary artery disease, although found throughout the arterial tree, it occurs at points were the artery is subjected to physical stress (bending and bifurcation). Atheroma exists in different forms, not all of which have critical consequences. Steady growth of intimal plaque over decades, however, can obstruct blood flow. When plaques fissure and ulcerate, they can induce clot formation, and cause complete arterial blockages responsible for distal tissue necrosis. The presence of an adequate collateral circulation will often prevent infarction.

Without genetic research, it is clear that atherosclerosis and its consequences (angina and cardiac infarction) are highly heritable. The public are mostly unaware of this, and for decades have believed that food (saturated fat) and lifestyle, are the major causes of heart attacks and strokes. For decades, starting in the 1960s, in western countries we were all urged never to eat fatty meat or eggs. The reasoning was simple but flawed: ingested fat causes hyperlipidaemia; blood lipids deposit on the walls of arteries to form plaques. This naïve explanation found acceptance because it is logical, but only partially correct. Although rabbits fed fatty meat, rather than their usual vegetarian diet, will develop more atherosclerosis, it is intimal metabolism that creates most of the fatty accumulation.

Here are some other important points:

- Epidemiological studies have revealed risk factors for populations, but their application to any individual is far less certain. They are: smoking, fatty food causing hyperlipidaemia, diabetes, age, obesity and hypertension.

- Cardiologists treat individuals, not populations. Generalised risks will not always apply to individuals. Individual family history is more critical.

- The epidemiology gives us guidance based on statistical analysis and appraisal. In individuals, however, there are often many aetiological exceptions.

- In younger patients (< 40 years), genome-wide studies have revealed 9p21, APOE, and LPL, as associated genetic variants.

- Those who have had one myocardial infarct are 1.5 to 15 times more likely to die (80% sudden deaths) than age-matched controls.

- Plaques > 4 mm are associated with thrombotic embolisation.

Evidence of coronary atheroma has been found in two Egyptian mummies, with widespread atheroma found in many more (Allam, A.H. , Thompson, R.C. et al.(2011). *Atherosclerosis in ancient Egyptian mummies: the Horus study. JACC Cardiovascular Imaging.* Apr;4(4):315-27). This study provides evidence for atherosclerosis existing 1500 years BCE, proving it is not solely a modern problem.

Key Point: From the clinical perspective, what matters most about CAD in individuals is their age, their family history and their inherited tendency to atherosclerosis and hypertension. Individual genetic polymorphism may influence the tendency to generate atherosclerosis.

The atherosclerotic process is patchy. It occurs throughout the arterial tree, but more where there is turbulent flow: at bifurcations, and where the arteries are subject to continuous bending, as in the femoral, popliteal, carotid, and coronary arteries.

Plaque microscopy, allows plaque morphology to be divided into three types: cholesterol-rich, calcified (calcium apatite), and fibrous. Many are mixed. Plaques can be 'active' or inactive; the active plaques contain inflammatory cells (macrophages, etc.) with oxidised LDL (ox-LDL) partly responsible for inciting the inflammatory process. 'Inactive', fully calcified plaques, are not much inflamed. The involvement of cytokines (IL-6), insulin (metabolic syndrome), immune responses to stress (acute phase protein and C-reactive protein), and the production of nitric oxide (NO) by the intima (endothelium), are all important. Nitric oxide (NO) is a vasodilator capable of inhibiting leukocyte chemotaxis and platelet adhesion. It also provides an anoxic tissue environment which, in theory, should reduce ox-LDL formation. The presence of plaque can inhibit endothelial NO synthetase (eNOS), and reduce NO production from its L-arginine substrate.

In the coronary arteries, fixed, stable plaque (often calcified), will cause angina whenever a greater than 80% obstruction reduces blood flow (without a collateral circulation). Lipid-rich plaques can fissure and ulcerate (vulnerable plaques), and have the potential to induce clot formation. The sudden occurrence of an obstructive clot in a coronary end artery, can cause cardiac infarction. Clot formation that is slowly progressive, or is regularly subject to lysis and reformation, can cause unstable angina.

From my own (unpublished) studies of carotid atheroma (3000 patients over twenty years), I observed that:

The presence of carotid atheroma correlates poorly with blood lipids, except for HDL cholesterol. The average blood HDL is lower in those with proven atheroma.
Over 95% of those with symptomatic coronary artery disease have carotid atheroma (it is a widespread, albeit patchy disease).
It is rare to find evidence of atheroma in those under 30 years of age, but it is more common in those with a family history of vascular disease.
Atheroma can grow measurably within five years; rarely, within two years.

I concluded that:

- One should assume that patients with chest pain and carotid atheroma, have CHD until proven otherwise.

- One should treat atheroma, in preference to dyslipidaemia, but treat both.

- In the right dose, 'statin' drugs can halt the progression of atheroma. Coronary artery atheroma progression has been shown to predict coronary artery incidents (CLAS Study. Azen, S.P. et al. 1996).

- The useful follow-up of those taking a 'statin' (at two year intervals) should verify which dose halts their atherosclerotic progression.

- The effective 'statin' dose varies between individuals.

Fig. 23, shows an image of carotid atheroma.

Coronary Artery Disease (CAD)

The term coronary artery disease (CAD) implies atherosclerosis affecting the coronary arteries, although a few other pathological processes also cause it, like systemic lupus, rheumatoid arthritis, polyarteritis and psoriasis. Accelerated atherosclerosis can be a feature of these conditions, mediated by circulating activated immune cells, and elevated inflammatory cytokines (TNF-α, IL-1β, IL-6, and IL-7). Ischemic heart disease (IHD) occurs only when coronary blood flow is insufficient.

One can better appreciate the significance of CAD with knowledge of coronary artery anatomy (Fig. 8). There are only two coronary arteries. They arise from the left and right coronary aortic sulci, just below the aortic valve. After a short **main stem**, the left artery divides into the **left anterior descending** branch (following the intraventricular septum, and sending perforating branches into it), and the **circumflex** artery which descends posteriorly. The **right coronary** follows a right lateral path, supplying oxygenated blood to the right atrium, both sinus and AV nodes, and right ventricle.

The blood flow down each artery can be equal, but one can dominate. The presence or absence of collateral vessels connecting the arteries can be crucial. With enough collateral flow, a blocked anterior descending artery, for instance, would not cause anterior cardiac infarction as long as another coronary artery donates blood to the left. The presence of such cross-flow has major implications for patient morbidity and mortality.

It is important to know which thrombosed coronary artery causes which type of myocardial infarction (and accompanying ECG changes). It is not always as simple as the left anterior descending occlusion causing anterior infarction; right coronary occlusion causing inferior infarction, and circumflex occlusion causing posterior infarction. Review some research I contributed to, while working in Amsterdam. My hard-working colleague, Dr. Pim de Feyter, did most of the emergency work that led to this publication (de Feyter, P., Van Eenige, M.J., Dighton., D.H. et al. (1 982). *Prognostic Value of Exercise Testing. Coronary Arteriography and Left Ventriculography, 6-8 Weeks after Cardiac Infarction.* Circulation 66 (3): 527-536). The results relate the occluded artery to ECG changes at rest and on exercise. The paper may now be old, but has not devalued. Unfortunately, much excellent research can lose political relevance, and get overshadowed by new publications. Even scientists can harbour bias. They might believe that whatever is new must be more reliable—the chronology bias at work. Because of it, references more than five years old, only rarely get quoted.

As a general rule, IHD becomes symptomatic, only after coronary arterial flow drops below 20% of normal. There is an important physical principle here. By pumping water through a rubber tube bearing a ligature (tourniquet), the flow of water from the end of the tube will only drop after an 80% ligature restricts the lumen. Bernoulli's equation (much simplified, is that fluid pressure + flow velocity^2 remains constant). This helps us understand the pressure / flow velocity relationship in an artery: as pressure drops across a restriction, the flow increases up to a point. Once the restriction is > 80%, the flow drops off rapidly. None of this may apply when an effective collateral circulation exists.

The Detection of CAD

At the time I qualified, patients had to wait to develop angina or cardiac infarction, before we diagnosed CAD. Pre-symptomatic discovery was unknown, and regarded by many as undue interference. The early detection in asymptomatic CAD patients came much later. Apart from indirect blood tests with insufficient individual diagnostic accuracy, like blood cholesterol, it is still not NHS policy to go further. Unfortunately, blood tests for hyperlipidaemia, inflammation and diabetes, lack diagnostic accuracy for the diagnosis of CAD in individuals.

Biomarkers are of dubious relevance to cardiologists. They are statistically relevant, but not always to individual patients. Because atherosclerosis is only partly an inflammatory process, one might expect raised blood hsCRP levels sometimes; other markers like fibrinogen, Lp-PLA$_2$, and homocysteine, also have their uses. I have measured them all, but gave them up when I found I could better predict coronary atheroma risk using carotid atheroma detection.

I started investigating asymptomatic patients in the 1970s, using exercise ECGs. Some classic ECG indicators of cardiac ischaemia on exercise, remain unchallenged, like ST changes. I would sometimes followed such ECG changes with coronary arteriography when positive. Later I added the direct detection of atheroma using carotid artery ultrasound, as an intermediary diagnostic indicator.

CT scanning (Agatston or calcium score, and CT angiography) can reveal only calcified atheroma (with calcium apatite). They cannot detect the more dangerous lipid rich plaques or the innocent fibrous ones. (Iori E, Bendinelli S, et al. (2003). *Coronary artery calcium identified by multislice CT as a marker of early coronary artery disease.* Monaldi Arch. Chest Dis. 2003; 60(1):63-72).

Coronary angiography uses radio-opaque dye injection, but visualises only a silhouette of the coronary artery lumen. Injecting radio-opaque dye into the coronary arteries, can reveal much, but not always enough information to determine the nature of the blockages. Before coronary bypass surgery was introduced, the main use of coronary angiography was to confirm the diagnosis.Once coronary bypass surgery became an option, surgeons needed to image the coronary anatomy using angiography. Now IVUS (intravascular ultrasound) has a place in some cardiac centres.

Important Point: At the moment only coronary angiography with IVUS can prove the presence of vulnerable plaque.

Statistically Relevant Risk Factors

Tobacco

Those who have smoked tobacco have increased their risk of developing CAD, two to four-fold. Not every patient with CAD has been a smoker. I have assumed it to be an accelerating factor in those genetically predisposed. Very few of my wealthy patients with CAD had ever smoked.

The cessation of smoking is relevant, even to those with CAD. The Framingham Heart study showed a 50% decrease in presentations of those with CAD, one year after they stopped smoking. Fifteen years after smoking cessation, the group average risk of CAD becomes equal to that of non-smokers.

Blood Lipids

Over 75% of patients with CAD have hyperlipidaemia (high LDL, low HDL). Many individuals with hyperlipidaemia have no evidence of atherosclerosis, but many with normal lipid profiles do. In individual patients, blood lipids are unreliable CAD predictors. They are, however, significant population-based risk factors, based on averages.

In my own unpublished studies of 3000 cases, only a low HDL correlated with the presence of carotid atheroma. A low HDL or high LDL, was found in only 60% of those with atheroma. One important question arises: should we treat both the discovery of atheroma and hyperlipidaemia as risk factors (with statins)? Those with both are obvious candidates for treatment. But what of those with atheroma and no hyperlipidaemia, and those with hyperlipidaemia and no atheroma? At present, only hyperlipidaemia is the NHS criterion for statin treatment. Many of my patients without a family history objected to taking a 'statin'. They opposed it after learning they had hyperlipidaemia, but no (carotid/coronary) atheroma. We agreed to examine them for carotid atheroma in the future. Many would accept treatment if it arose.

At present, UK GPs will not prescribe statins for those with proven atheroma without hyperlipidaemia. That needs to change.HMG-CoA reductase inhibitors have saved many lives. They reduce lipid levels in the blood, and in the arterial endothelium.

Hypertension

Eighteen percent of cardiac infarction cases have hypertension (systolic > 130mm Hg / diastolic > 80 mm Hg.) as a risk factor.

Diabetes

Over 50% of diabetics, have dyslipidaemia. Large group studies, have shown that diabetes increases the risk of stroke, sudden cardiac death, and cardiac infarction. Diabetics have worse outcomes following PCI and CABG. In my (pre-selected) wealthy patient group, diabetes and pre-diabetes were not common in those with CAD.

Obesity

Obesity is a contributing factor in 20% of those with cardiac infarction. Waist size (> 102 cm in men; > 88 cm in women), not BMI, is more pertinent. Weight reduction is difficult for most patients. A high protein type diet) is expensive, but works for those disciplined enough to continue with it. Semaglutide and liraglutide are now widely available. They are GLP-1 receptor agonists, but also stimulate insulin release. They are especially useful for obese Type 2 diabetics with a BMI > 27. Amphetamine-like appetite suppressant drugs, like phentermine and diethylpropion present problems: they suppress appetite but are addictive.

The metabolic syndrome (raised fasting blood glucose, obesity, hypertriglyceridaemia, other dyslipidaemias, and hypertension) now affects 30% of all middle-aged and older people in the UK and USA. It is associated with insulin resistance. In my experience, this is seen less in educated, wealthy groups.

For a review of all weight loss methodologies, see my book: *Who Loses Wins* (2024).

Alcohol, Food and Atheroma

In the 1960s, as a medical student, pathologists at the London Hospital often made a consistent observation. There were then many homeless, methylated spirit drinkers, sleeping on the streets outside the hospital. When one died, the pathologist undertaking the post-mortem would note the absence of atheroma in their arteries! Methyl alcohol is toxic and often lethal, but prompts the question: how beneficial is ethyl alcohol to atheroma generation? I once facetiously wrote: 'If a patient smokes, I would advise them to drink alcohol!' (Dighton, D.H. (2005). *Eat to Your Heart's Content*. Heartshield).

Patients ask if red wine is better than white wine. There is no clear answer. Flavonoids in red wine and berries, acting as anti-oxidants, might protect us from CAD, but no research study I know of, proves it. One of my patients asked me what he should eat, having just had a CABG for 3-vessel coronary disease. I had to admit that I was poorly informed on the subject. As a result, I did some research. After a thorough literature search, I reached some speculative conclusions, based on animal experiments. Using animals to study atheroma, dietary cardio-preventative dietary factors seem to include: arginine, selenium, zinc, manganese and magnesium; omega-3, 6 and 9 oils. In animal experiments, saturated fat can be shown to be atherogenic.

Another question arose. Which foods contain most of the potentially atheroprotective nutrients, and which are the least atherogenic?

I set out to calculate a ratio— of atheroprotective nutrients to atherogenic ones, using dietary composition data. I called it the Cardiac Value™ of food. Using the scale this generates, mussels are best for atheroma prevention, and white chocolate is the worst. For more information, read my book: *Eat to Your Heart's Content*. David. H. Dighton (2005). HeartShield Publications. ISBN 0-9551072-0-2. This contains some of the technical detail. For patients, I wrote, *Heart Sense. How to look after your heart*. David. H. Dighton (2006). HeartShield Publications. ISBN 0-9551072-1-0.

I never saw carotid atheroma reverse, having repeatedly screened hundreds of patients, some for over 20 years. The most one might expect from any diet, therefore, is to halt the atherosclerotic process. That this can happen, remains unproven for any dietary intervention. My guess is that those diets containing the most protective foods would need to be started in childhood, to be effective. Because the aetiology of atherosclerosis has strong genetic components, I doubt that any diet can cut atheroma morbidity or mortality, by more than 5%. From my observations of patients over fifty years, and what I have learned from research studies, nature rather than nurture would seem to take precedence in the aetiology of atheroma and CAD.

The secondary prevention of cardiac infarction associated with a Lyon diet, might suggest otherwise (de Logeril, M., et al. 1999). Atherogenesis and the reduction of secondary cardiac infarction rates are, however, somewhat different.

Aortic Disease

Atherosclerosis and hypertension often precede aortic dilatation and dissection. Cystic degeneration of the aorta occurs with age, and underlies the occurrence of aneurysms. Aortic atheroma, ulceration, intramural haematoma, aneurysm and dissection, may present similarly and share the same management criteria including urgent intervention.

Artery Wall Degeneration

Cystic medial degeneration is a common predisposing factor for aortic disease. The loss of elastic fibres and smooth muscle cells, can cause the aortic wall to weaken. This occurs with age, but hypertension, Marfan and Ehlers-Danlos syndromes are other causes. Rare inflammatory diseases also occur, like giant cell arteritis (white populations), Takayasu (in Africans and white Asians) and BehÇet disease (Middle-Eastern and Asian patients). I saw only one of each, in fifty years of practice in the UK and Holland.

Atherosclerosis can weaken the aortic wall. Atheroma in the ascending aorta and arch, sometimes create cerebral and other emboli. Trauma can split the aorta, occurring mostly at the level of the left subclavian artery or in the abdomen.

Aortic Dissection

Chest pain can have ominous causes, one of which is aortic dissection. The arterial intima can split, to produce a flap and false lumen, between the intima and muscle layer. This usually progresses in an antegrade fashion. Rupture of a vasa vasorum (an artery supplying blood to the artery wall), sometimes initiates the process.

Patients are mostly hypertensive males, over fifty years of age. They describe their pain as ripping, or having a tearing element. Ascending dissections (Type A), often produce anterior chest and neck pain; others (descending aorta—Type B) can produce posterior thoracic and even lower back pain. Types A and B, refer to the Stanford classification system.

Ascending Aorta Dissection. If dissection of the ascending aorta involves the aortic valve (40-80% of ascending dissections), it can cause AI. Pulse deficits will occur in the arms and legs, depending on which arteries are involved. The coronary ostia (right more than left) are involved in 2% of cases. The dissection can reach the pericardial space, causing tamponade. Blood can fill the space, restricting passive ventricular filling and reducing cardiac output. A pericardial rub might be heard, but only if there is minimal fluid in the pericardial space. An urgent CT scan should show it. Sometimes the renal and spinal circulations (causing paraplegia) are involved. Hypotension is a common feature of serious dissections (with or without tamponade).

One can use CT angiography, trans-oesophageal ultrasound (TOE), and MRI angiography, to confirm the diagnosis. Catheterisation is now used less. TOE is most useful when LV and aortic valve function need review.

Following the diagnosis, management rests on reducing blood pressure. In those with ascending dissections, one might prescribe pre-operative beta-blockade and alpha-blockade. Verapamil and diltiazem are alternative drugs for lowering the blood pressure. Type B dissections are now considered for intravascular repair (TEVAR: thoracic endovascular aortic repair), whereas Type A dissections need urgent surgery (minimum expected mortality: 25%).

Important Point. In patients with tamponade, one must resist percutaneous pericardiocentesis—it can cause aortic rupture. If possible, wait until the patient is lying on the operating table.

Cardiac catheterisation and aortic clamping, sometimes cause **iatrogenic dissection**. On one tragic occasion, left coronary artery dissection happened to one of my Dutch patients, while I was performing a routine pre-op coronary arteriogram. The coronary catheter dissected her left coronary artery. We rushed her to the operating theatre where she successfully underwent emergency CABG. Unfortunately, she died two weeks' later from renal failure.

Risk Alert. Undertaking risky procedures is not for every doctor. Those who perform them regularly, will be very lucky to avoid damaging a patient, or worse—a fatal complication. Regulatory authorities will never appreciate, without appropriate clinical experience, that many necessary medical practices are inherently dangerous. They rightly advocate risk-free medical practice, while being personally non-conversant with operative risk. This can lead them to punish doctors who experience serendipitous side-effects. While doctors continue to do their best for their patients, complications will never stop. While completely removed from patients and their management, bureaucrats and lawyers who work for regulators, have a natural, but impractical inclination—to removal all clinical risk. Although their duty is to keep the public safe, they have nothing to lose when punishing doctors.

Aortic Aneurysm

Patients can present with chest pain, described as having a 'ripping' or 'tearing' quality. Many diagnoses (80%) are fortuitous, with many resulting from routine CXRs. Ascending aortic aneurysms can compress surrounding structures causing dysphagia, stridor and hoarseness (laryngeal nerve compression). Compression of the *vena cava* (superior *vena cava* syndrome) also occurs. When the aortic valve is involved, acute heart failure can result. If the coronary ostia are involved, angina and chest pain become likely. Because large aneurysms can cause blood stasis, clots can form and result in embolic symptoms.

An osteopath once referred a patient to me with back pain. Having failed to help his patient's posterior thoracic pain, he correctly concluded that his pain was not of musculoskeletal origin. I found a descending aortic aneurysm to be the cause. I well remember osteopath, Mr Andrew Joseph of South Woodford, Essex. He was an astute exponent of the art of medicine.

Only 20% of patients with aortic aneurysm, present as above. Regardless of presentation, rapidity of aneurysm progression is the major concern. It is urgent to operate on those who are progressing rapidly. Removal of the patient to the most appropriate specialist centre is advisable. If there is data supporting progression and high risk, one must act fast, before one needs t o.

Aortic aneurysms are either **acute** (up to 2 weeks' duration) or **chronic** (when they have existed for longer). Two classifications are in use:

DeBakey: Type 1 originates in the ascending aorta. **Type 2** involves only the ascending aorta. **Type 3a** involves the descending aorta. **Type 3b** involves the descending and abdominal aorta.

Stanford: Type A involves the ascending aorta, and **Type B** only the descending aorta.

Important note. Some CT scanning views can overestimate the size of an aneurysm.

Management

Some important points are:

- Repeatedly evaluate those with an aortic root > 4.0 cm

- Genetic screening should be made available to the families of those with an aneurysm.

- Aneurysm rupture and dissection risk, increases with their size.

- Annual risk in individuals increases from 2% to 7% as the aneurysm grows from 4 to 6 cm.

- Surgical intervention is recommended for ascending aortic aneurysms > 5 cm, and for descending aortic aneurysms > 6 cm.

- If an aortic aneurysm grows by more than 0.5 cm per annum, or symptoms develop, surgical intervention must be considered.

- In Marfan syndrome, beta-blockade can slow progression. The same is assumed to be valid for other types of aneurysm.

- Those known to have Marfan syndrome who become pregnant, or engage in athletic pursuits, risk developing an aortic aneurysm.

Abdominal Aortic Aneurysm

The main antecedent cause is atherosclerosis. It is five times more common in men, especially in smokers and those with hypertension. Twenty-five percent have a positive family history.

The finding of a pulsatile mass during routine abdominal examination of patients older than 65 years of age, can lead to the diagnosis. It can present with abdominal or back pain, but sometimes with embolic features (look for *livedo reticularis* and blue toes). Ultrasound, CT scanning, and MRI can all confirm abdominal aneurysms.

The key management features are:

- Regularly observe those with an aneurysm greater than 4 cm in diameter;

- Observe those with an aneurysm diameter between 4 and 5 cm, every 6 months; for them, beta-blockade is the accepted treatment.

- Offer intervention to those with an aneurysm > 5 cm (the annual risk of rupture is 22%).

- Aneurysms expanding > 0.5 to 1 cm per annum mostly need intervention.

- The risk of rupture is greater in women.

- The preferred current intervention is endovascular stent-grafting.

Peripheral Vascular Disease (PVD)

Claudication is the discomforting tightness of calf muscles that occurs with walking. It usually dissipates after a short rest. Thereafter, patients will sometimes walk further than before. The discomfort is the ischaemic equivalent of angina pectoris in the calf muscles. Classic symptoms, as described, occur in only 11% of those with PVD. The rest have no pain (34%) or atypical pain (55%) (PARTNERS STUDY. Hirsch, A.T., et al. JAMA (2001); 286(11):1317-1324).

Chronic ischaemia can cause skin ulceration and skin changes. Restricted mostly to the feet and below the knee, the limbs are typically pale, cold and pulseless. Patients with severe pain at rest might have an acute arterial thrombosis; their ischaemia could lead to gangrene (CLI or critical limb ischaemia). Many find that dangling the affected leg out of bed relieves their night pain. Regard such symptoms as an emergency requiring urgent revascularisation.

When blood flow restriction occurs from stenotic iliac arteries (or anywhere more distal), the cause is nearly always atherosclerosis. It affects 30% of some older adult populations. Five percent of patients with PVD proceed to amputation.

Patients may not notice the condition unless they can walk fast or far enough. Diminished leg pulses might be felt (the dorsalis pedis is a less reliable indicator than the posterior tibial) and graded in strength from zero to three. With pulses absent, they score zero; when diminished, the score is one. A score of three indicates bounding pulses, and two indicates normal pulses.

Between the brachial and ankle arteries, a BP difference might be found. A Doppler device will allow the ankle / brachial index (API) to be calculated. It uses the highest pressures measured. The difference will be more reliable after exercise. Suspect PVD if the highest ankle pressures are less than 90% of the brachial pressures. When the ratio is less than 50%, severe PVD is likely.

Patients can have angina pectoris and claudication, but with one restricting the other. If claudication occurs early, the patient may not be able to walk far enough to get angina pectoris. Both symptoms arise more quickly in cold weather. Obesity, diabetes, unfitness, and any lung disease associated with smoking are risk factors.

All PVD patients should be checked for atherosclerosis in other arteries; it may also be found in the coronary (60 - 80%), cerebral, carotid (25%) and renal arteries. CT angiography, ultrasound, and MRI angiography scanning, can confirm the distribution of atheroma. Subtraction angiography allows detection after intravenous dye injection. In unclear cases, suspect spinal problems as a cause of leg problems and walking difficulty.

When I was a junior doctor, smoking was more common than it is today. Almost every patient with claudication was a male smoker or ex-smoker—once the most dependable associations in circulatory medicine.

Medical intervention for PVD has always disappointed since the problem is one of physical blockages to blood flow. Some drug trials (like that of cilostazol, a platelet aggregation inhibitor) exaggerated their effectiveness. Patients cannot use the drug when in heart failure and it often causes diarrhoea and headache. Some drugs claim a 50% increase in walking distance, but patients will hardly get much benefit from being able to walk 150 meters, rather than 100 meters!

Treatment should include smoking cessation, 'statins', and evolocumab which is known to reduce the need for surgical revascularising procedures by 20%, and to reduce the rate of decline. Rivaroxaban and aspirin together, are more effective than aspirin alone (see Rymer, J, et al. (2023) the VOYAGER PAD trial). Blood pressure control trials have shown mixed benefits for PVD patients. However, it is advisable to control BP with an angiotensin receptor blocker. Some beta-blockers reduce peripheral blood flow, so doctors must exercise caution when prescribing them for PVD patients.

Adequate diabetic control is crucial. It can reduce microvascular events and slow the progression of PVD. The HbA1c is best kept < 7%. Interval exercise regimes (frequent stops and starts) can boost walking distance by 80%, but only with moderate PVD.

External iliac endofibrosis can sometimes cause young cyclists to develop claudication. In the young, always consider aortic coarctation and vasculitis.

Percutaneous angioplasty and stenting work well if the arterial anatomy is suitable (discreet stenoses). For those with more widespread disease, only surgical bypass using autologous veins is advised. Angioplasty works best for iliac stenoses, and less so for more peripheral lesions. IVUS is sometimes useful for confirming the atheromatous lesions most suitable for angioplasty.

Chapter Twelve

Heart Failure and Cardiomyopathy

Heart failure results from many pathophysiological processes, all of which reduce cardiac output to a level below that necessary for normal human functioning. It is mostly a chronic, long-term condition, that deteriorates progressively with age. Heart failure is occasionally acute, occurring with the sudden onset of AF, cardiac ischaemia (without angina), cardiac infarction, papillary muscle rupture and aortic dissection causing severe AI. Just occasionally, a normal heart will fail to meet the increased demand imposed by medical conditions like thyrotoxicosis, septicaemia or morbid obesity.

I once had a morbidly obese, middle-aged patient, with a small but normally functioning heart. Her heart lacked the capacity to supply enough blood (and O_2) to her vital organs. This resulted in her obvious shortness of breath on minimal exercise. With a normal heart failing her needs, reducing the demand by losing weight was her only option. I likened her cardiovascular situation to a lawnmower engine, trying to power a Challenger 3 tank!

Single ventricle heart failure can occur, but it mostly involves both ventricles.

Clinical Presentation

Right heart failure can occur as pulmonary resistance rises secondary to emphysema, lung fibrosis and multiple emboli. The first response is for the right ventricle to hypertrophy. With advancing age, the right ventricular muscle weakens, and chamber dilatation results. The pressure in the relaxed ventricular phase (the end-diastolic pressure) will rise, reducing ventricular filling and causing increased back pressure. This back pressure will cause the vena cavae to fill at a higher pressure, sometimes resulting in an engorged liver, ascites, bowel dysfunction and oedematous feet. Increased superior vena caval pressure and distension, cause the jugular venous pressure to rise. Both the head and neck can become oedematous and cyanotic; mostly seen after patients have laid supine for long periods.

Any shortness of breath that occurs in right heat failure, results from the pulmonary condition and the extra effort required by weight gain from fluid retention (from oedema and from the consequences of acid/base changes resulting from pulmonary pathology).

In left heart failure, shortness of breath and tiredness are the key features. As the left ventricle weakens, and the end-diastolic pressure rises, LA pressure and then pulmonary venous pressure will rise. As the pulmonary venous pressure rises, pulmonary oedema will occur, with fewer open alveoli then available for gas exchange. This results in shortness of breath. It is usual for breathlessness to occur on exercise, but also while lying flat (orthopnoea); the latter being a pathognomonic symptom of left heart failure. In the classic scenario, the patient wakes breathless during sleep, sits upright, and opens a nearby window for more air (paroxysmal nocturnal dyspnoea).

Learning Point: In the presence of cardiac ischaemia (especially with anterior descending coronary stenoses), shortness of breath (from left heart failure) is a common symptom, often preceding the occurrence of angina.

If left heart failure is severe enough, forward peripheral flow diminishes and perfusion of the hands, feet and nose, will become insufficient; the result is cold and often cyanosed limbs. The cyanosis occurs because of slowed perfusion, leading to greater blood oxygen extraction. Eventually, right heart failure follows, and we then refer to it as **congestive heart failure.**

The Frank and Starling curve, relates heart rate to cardiac output (CO). In a normal heart, CO will peak with a heart rate of approximately 140 bpm. As heart failure progresses, this peak rate for maximum CO reduces; the heart becomes less compliant and needs more time for filling. The peak CO in heart failure can be as low as 50 bpm. The sudden onset of fast AF, can cause sudden shortness of breath, even in those with minimal heart failure.

Digitalis Alkaloids.

> *The English botanist and physician William Withering was first to use digitalis to treat heart failure, in 1785.*

(*An Account of the Foxglove and some of its Medical Uses With Practical Remarks on Dropsy and Other Diseases*). The dried leaves of *Digitalis lanata* were originally used to produce digitalis, now made available in two synthetic forms: digoxin and digitoxin. They can slow the sinus rate and delay AV conduction (allowing fewer impulses through the AV node). We now know that digitalis reduces the refractory period of atrial Purkinje fibres (mimicking acetylcholine), perhaps sustaining AF, with the added potential of inducing other forms of tachycardia (drugs that lengthen the refractory period, like flecainide, amiodarone and quinidine, can prevent or stop AF). By partially inhibiting myocardial cellular Na^+ - K^+ ATPase, digitalis alkaloids allow intracellular sodium and calcium to increase, possibly improving myocardial contractility. Whether digitalis alkaloids clinically improves myocardial contractility, is contentious. The main benefit in prescribing them is to reduce the heart rate and slow AV conduction.

Causes of Ventricular Dysfunction

Two fundamental physiological processes account for ventricular dysfunction:

- Reduced contraction (ionotropy) and,
- Reduced ventricular relaxation (lusitropy) and filling.

They are modified by:

- Pre-load (venous return),
- After-load (forward resistance, as in hypertension), and
- Heart rate (as indicated by the Frank Starling relationship).

The four phases of ventricular activity are:

1. Isovolumetric contraction: **phase 1**. Before the aortic and pulmonary valves open.

2. Contraction and ejection: **phase 2**;

3. Isovolumetric relaxation: **phase 3**, after the aortic and pulmonary valves close and before the mitral and tricuspid open;

4. Ventricular filling: **phase 4**. Passive at first, later supplemented by atrial contraction.

Reduction of these phases will lead to reduced stroke volume and CO. The stroke volume (about 70mls) is the difference between the end-diastolic volume (end phase 4) and the end-systolic volume of each ventricle (end phase 2).

The Laplace equation allows an understanding of cardiac output. Laplace related the pressure generated within a cavity to the inverse of its radius (Pierre-Simon Laplace, 1806). The greater the end-diastolic volume, the lower the stroke volume (ejection fraction) generated by ventricular contraction. The pumping action of large ventricles is less efficient than small ones. Try hugging a large tree trunk. Now squeeze a broom handle. Which generates the most force? The radius of the object squeezed, predicts the force generated. The larger the radius, the less force possible. For the heart, this squeezing force (contractility), also depends on the thickness (and health) of the ventricular wall. The force needed, will determine myocardial oxygen consumption.

A few other factors are of dynamic importance. In heart failure, fluid retention adds to pre-load with an initial increase in contractility and lusitropy to compensate. In diastolic heart failure, the ventricular walls may be stiff, but the ejection fraction is preserved (HFpEF) initially. As heart failure proceeds, the ventricles remodel and dilate (with increased compliance).

Cardiogenic shock occurs when the stroke volume drops too low (because of reduced contractility) and there is hypotension. This will exacerbate ischaemia and initiate a downward vicious spiral of deteriorating ventricular function. Organ failure can then follow.

As heart failure proceeds, other mechanisms become incurred. There will be:

- Increased autonomic sympathetic activity.

- More circulating catecholamines.

- Increased renin-angiotensin-aldosterone (RAAS), and

- Increased inflammatory cytokine activity.

These are adaptive at first, but can then become maladaptive, especially if increased demand gets prolonged, with catecholamines driving contractility. Loss of myocyte sarcolemma T-tubules occurs together with myocyte necrosis. My former boss, cardiologist Dr. Peter Nixon, sometimes treated cardiogenic shock with intravenous adrenalin (he referred to it as 'God's own ionotrope'). Unfortunately, any beneficial effects were short-lived.

There are several important drug trials showing benefit from reduced catecholamine stimulation in heart failure, using beta-blockade (CIBIS-II: bisopralol; MERIT-HF: metopralol; COPERNICUS: carvedilol). Lowering the heart rate alone with ivabradine (SHIFT), can help (as predicted by the Frank Starling relationship).

Dysfunction of the RAAS cascade, affects renal perfusion and diuretic use. RAAS activation leads to increased circulating blood volume, and increased pre-load, raising atrial volume and pressure. These will subsequent increase stroke volume and CO.

The angiotensin converting enzyme (ACE), cleaves angiotensin-I to produce angiotensin-II. In heart failure, angiotensin-II has many negative effects. Some of them are:

- Fibroblast activation.

- Increased collagen production.

- Nor-epinephrine release.

- Blood volume expansion.

- Myocardial toxicity

- Cardiomyocyte growth.

- Ventricular remodelling.

It is no surprise, therefore, that RAAS blockade can be beneficial. There are relevant trials: for ACE inhibition, CONSENSUS and SOLVD for enalapril; for ARB's, Val-HeFT for valsartan, and CHARM for candesartan.

Three other drugs are worthy of consideration: spironolactone (the RALES trial), eplerenone (EMPHASID-HF trial), and neprilysin inhibitors (ARNI: angiotensin-receptor neprilysin inhibitors. PARADIGM-HF trial).

Ventricular Remodelling

Increased cardiomyocyte size is the initial response to a failing ventricle. The concurrent increase of myocyte apoptosis, allows their replacement with collagen.

These changes cause chamber enlargement, increasing wall stress and stiffness; they also reduce the oxygenated blood delivered to cardiomyocytes through capillaries. This, in turn, reduces myocardial contractility. As the Laplace formula predicts, chamber enlargement reduces pumping efficiency.

Developing Issues:

Endothelial neuroregulin (NRG) has protective effects. Its inhibition using trastuzamab and anthracycline (or genetic deletion, as in experimental animals), could enhance the deleterious remodelling process.

NRG limits apoptosis, regulates calcium exchange, and may limit atherosclerosis. Similar polypeptides may shortly be available for therapeutic use.

Resynchronising pacing can limit remodelling in some heart failure cases. If that fails, cardiac contractility modulation might help.

Non-stimulatory pacing of the right ventricle during the refractory period has been shown to increase contractility (Patel, P. A.(2021). *Heart Failure Review*. 26(2):217-226).

Pathophysiology of Heart Failure

Natriuretic peptides, the role of inflammation, oxidative stress, renal function and their effects on the peripheral circulation, need consideration.

In heart failure, both B-type natriuretic peptide (BNP) and amino terminal NT-proBNP blood levels, correlate with ventricular pressure load and volume. BNP production in cardiac tissue, occurs in response to ventricular wall tension and hypertrophy. It has vasodilatory and natriuretic effects. Both levels in the blood have been used (sometimes in the absence of cardiac clinical acumen) as indicators of heart failure and its progress. Before either became accepted, the diagnosis of heart failure presented diagnostic problems for some physicians, but hopefully not for cardiologists. Being able to observe and measure the JVP, is the required skill.

Many inflammatory markers, like C-reactive protein and TNF-α, are raised in heart failure. IL-6 is raised in response to hypertrophy and TNF-α in response to ventricular dilatation. Endothelial dysfunction (related to reduced endothelial NO synthesis) could allow myocardial damage from oxidative stress. Oxidised LDL and myeloperoxidase (MPO), can both be raised in heart failure; the latter having prognostic value.

Renal tubular function underlies the salt and water retention in heart failure. In response to lower perfusion, the kidneys retain fluid. Initially, angiotensin II release stimulates aldosterone production by the adrenal cortex. As heart failure progresses, however, the GFR usually falls. Sympathetic nervous system activation and vasopressin release, cause vasoconstriction and hyponatraemia respectively. Hyponatraemia serves as a marker for both heart failure severity, and prognosis in later stage cases.

As renal perfusion diminishes, autonomic nervous system changes prioritise perfusion of the skeletal muscles and the brain. Hypoperfusion and the reduction in size and number of mitochondria cause skeletal atrophy and weakness.

In heart failure, it is typical for the heart rate not to increase normally with exercise. Together with increased ventricular filling pressures, and increased pulmonary resistance, exercise ability can become progressively limited. Sleep apnoea and orthopnoea exacerbate fatigue. It is typical for exercise oxygen consumption to be low due to increased lung dead space and reduced ventilation/perfusion. Both reduce tissue oxygen delivery, while concomitant changes in CO_2 exchange, can cause blood acid-base changes.

The Pathology of Heart Failure

The pathological causes of heart failure are **ischaemic** and **non-ischaemic** (infection, infiltration, genetic, and valvular). Among the more important causes are:

- Myocarditis,
- Familial cardiomyopathy,
- Amyloidosis,

- Haemochromatosis,

- Sarcoidosis.

- Chemotherapeutic effects.

- Diabetes and hypertension both contribute pathological changes to heart failure.

Cardiomyopathy presents as one of two types: **dilated cardiomyopathy** (the commonest form), and **hypertrophic**.

Hypertrophic obstructive cardiomyopathy (HOCM or HCM) can be difficult to diagnose clinically, but is not to be missed. In early life it is associated with dysrhythmia and sudden death. Because it is inherited, a family history of early deaths should raise suspicion.

When patients in heart failure present with a dilated heart (on CXR or echocardiogram), finding the cause is often a challenge. When no cause is forthcoming, it may be attributed to age or to small vessel coronary atherosclerosis that is difficult to prove clinically.

Age is always an important consideration. Young people with large hearts, might have myocarditis, a pericardial effusion or Ebstein's anomaly. Some will have severe valvular heart disease, although this has become much rarer in the western world. Older adult patients may have a degenerative myocardium, or one that has sustained several infarctions; especially anterior infarction. By the 1980s, most post-rheumatic patients had been surgically treated; few new cases presented in the UK thereafter. In some third world countries, rheumatic fever is common and continues as a cause of heart failure.

HOCM (HCM is now the acceptable WHO acronym)

Family surveys often discover cases of hypertrophic (obstructive) cardiomyopathy. Sometimes atypical chest pain, syncope, palpitation and breathlessness will lead to the diagnosis in a young, otherwise healthy, athletically fit individual. In older adults, cardiac amyloidosis can present similarly.

It is human nature to relish the rare, rather than the commonplace. This applies to HCM, occurring as it does, in approximately one in 500 people. Most cardiologists will diagnose a number during their working life, but it will not be a daily event. The diagnosis remains a challenge because without symptoms, it may only get diagnosed on routine investigation.

Patients with sustained hypertension, and some athletes, have hearts similar to those of HCM Because patients are mostly young, some will be athletes. Left ventricular hypertrophy, associated with primary hypertension, gets confused with HCM; they sometimes occur together, although at the histological level, only HCM will have asymmetric hypertrophy and sarcomere disruption. Non-sarcomeric forms, like amyloidosis (two types), Fabry, sarcoidosis, iron storage, and lipid storage diseases also exist. When histological definition is necessary for management, cardiac biopsy may be necessary. This is especially important when the myocardium is constrictive and possibly mistaken for myocarditis. Other reasons for biopsy are to detect tumours and diagnose drug effects.

A widespread belief is that sudden death and syncope are the commonest presenting features of HCM In fact, AF, stroke and heart failure are more common.

Consider the findings on investigation. Typically, the ECG will look odd, but subject to many variants. Many will suggest LVH with deeply inverted T-waves.

The echocardiogram in HCM can show:

- Asymmetric septal hypertrophy, mainly affecting the LV outflow tract.

- An outflow gradient (with severe hypertrophy),

- Sub-aortic stenosis (with an associated murmur),

- Systolic opening of the anterior leaflet of the mitral valve.

- A reduced diastolic LV volume and diastolic dysfunction.

The investigations of choice is MRI, T_1 mapping.

The annual rate of death from HCM is 1%, but many achieve longevity and will lead an asymptomatic life. Others develop arrhythmias and progressive heart failure. Sudden death is the most fearsome risk for the young. News of a young athlete dying suddenly, will cause many others to worry.

The primary objects of treatment are to reduce septal thickness and to control dysrhythmias. Although the former can be achieved surgically, myosin inhibitors may soon make this procedure redundant.

Individual and family genetic profiling for HCM is now on offer, but must include counselling.

Reliable genetic testing for cardiac connexins (40 and 43, as in AF) could indicate the risk of arrhythmia.

Cardiac Amyloidosis

The commonest clinical presentations of cardiac amyloid are dyspnoea on exertion, palpitation, chest pain, pre-syncope, and syncope. Otherwise, patients have all the symptoms and signs of heart failure. Amyloid can occasionally affect cardiac arterioles, causing angina and infarction.

As the mean age of a population rises, and diagnostic techniques improve, the number of older adult patients diagnosed with cardiac amyloidosis will rise. Ten percent of heart failure cases (over 65-years of age) with preserved ejection fraction (HFpEF), have cardiac amyloid.

Amyloid infiltration causes heart muscle to become stiff; a known cause of restrictive diastolic heart failure. Both sarcoidosis and iron storage can do the same. Primary amyloidosis (short chain type), and the senile type (transthyretin type) of amyloid, both affect the heart. Secondary amyloidosis occurs with chronic inflammatory diseases (amyloid A). Inherited amyloid usually affects only the atria.

With cardiac amyloid present, the following investigation features may be present:

- Low ECG voltages and various degrees of heart block on ECG.
- Increased ventricular wall thickness, typically with a speckled appearance, on echocardiography.
- An incongruity may occur, like thickened ventricular walls with unexpected low ECG voltages.
- Dilatation of the atria is usual in the inherited form.

Both MRI and nuclear scanning can reveal the diagnostic features, although cardiac biopsy remains the gold standard for definitive diagnosis.

Clinicians must consider whether obtaining biopsy data will benefit the patient.

Over 120 gene variants have been identified in cardiac amyloidosis; Thr60Ala and Val122Ile are the commonest. There is an extracellular accumulation of proteoglycans, glycosaminoglycans, collagen, and laminin. (Shams, P Ahmed, I. (2023). Cardiac Amyloidosis. StatPearls (Internet). National Library of Medicine).

The mainstay of treatment for amyloid induced heart failure often proves to be a diuretic. Beware of patients becoming hypotensive with an ACE, ARB or calcium channel blocker. Beta-blockers can induce syncope and should be avoided. For those with serious arrhythmias, treatment with amiodarone can be effective.

The treatment for amyloidosis is often that for myeloma. For those young and fit enough, cardiac transplantation may need to be considered. A significant concern is that the transplanted heart will also develop amyloid, although the use of stem-cells has improved the prognosis for some.

The average survival time of untreated primary amyloid cases, is nine to twenty-four months. It is seven to ten years in the familial type; five to seven years for those with senile amyloidosis. The prognosis of those with secondary amyloidosis is usually over ten years.

CHAPTER THIRTEEN

Myocarditis and Systemic Disease

Although these conditions are all important clinical entities needing consideration from cardiologists, they occur infrequently. For this reason, I have included them all in one chapter.

Systemic diseases affect the heart, only when they share the same pathological process: infection, infiltration, autoimmunity or neoplastic disease.

Myocarditis

Myocarditis and pericarditis share causes. General infections can affect the heart and many other organs. Myocarditis and pericarditis often have a viral origin (65% of those with myocarditis have detectable viral genome on biopsies).Myocardial lymphocyte infiltration and anti-CD3/CD4/20 and 28 cytokines, typify viral myocarditis. Like many other infections, it will follow an acute or chronic time course. The commonest viruses involved are enteroviruses, like Coxsackie. Parvovirus 19 (PVB19), herpes 6, SARS-CoV-2 and COVID-19, can also cause acute myocarditis, chronic infection and heart failure.

In viral myocarditis, several mechanisms account for myocardial damage: the cytotoxic effect of the virus itself, macrophage activation mediated auto-immune damage, and cytokine expression.

With each virus, there is a different associated risk. Coxsackie B3 has a high mortality; Coxsackie A9 is usually benign and short-lasting. Parvovirus PVB19 can target endothelial cells, and with sufficient damage, cause cardiac ischaemia and heart failure.

The clinical presentation is usually a younger person with a febrile illness and persistent mild chest pain. There may have a recent history of an URTI (tonsillitis, etc.) and arthralgia. Few will have signs of heart failure, but those who do, have the poorest prognosis. Some will have tachycardia at rest, others will be syncopal (Adams Stokes syncope) caused by heart block. When rheumatic fever is the cause, the patient may have a rash—erythema marginatum; others may have subcutaneous nodules and/or joint swelling. Those with sarcoidosis will usually have lymphadenopathy. In South America, those with Chagas' disease (CHAH-gus), are infected with Trypanosoma cruzi; they often have cardiac conduction abnormalities.

Hypersensitive patients can get eosinophilic myocarditis. They may have an allergic, maculopapular rash.

The investigation findings usual reflect the inflammatory process (raised CRP and ESR and leucocytosis); some have raised troponin and cardiac enzymes levels. A raised troponin I, is associated with a risk of heart failure. Raised markers like TNF-alpha and IL10, are also associated with a poor prognosis. With autoimmune diseases, various anti-nuclear antibodies may be present.

An **ECG**, apart from sinus tachycardia, might reveal various forms of conduction block. Occasionally a 'pseudo-infarct' pattern is seen, although actual infarction is possible (parvovirus PVB19 can disrupt the endothelium of coronary arteries and cause actual infarction). On occasions, this could prompt the need for a coronary angiogram.

An **echocardiogram** may show no abnormality, or every feature of a dilated and poorly contractile heart. Intracardiac emboli are a major concern in all fulminant types and in COVID-19 infections.

MRI can show intracellular and interstitial oedema, capillary leakage, cell necrosis and fibrosis.

Anti-myosin scintigraphy is sensitive to myocardial inflammation, but has poor specificity.

Because of many false negatives, **myocardial biopsy** may not provide definitive answers. Use it only to confirm the need for specific treatment.

The acute phase of myocarditis usually lasts three days. From day fourteen onwards, 20 to 50% of acute cases will progress to the chronic phase. In this phase, autoimmune mechanisms and fibrosis can become prevalent.

Treat all patients as if they had chronic heart failure: taxing exercise is ill-advised, and the standard protocol for heart failure should be started. This should include anticoagulation (especially for those with AF, and those with ventricular aneurysms). A pacemaker may become necessary for those with heart block and syncope. Mechanical assist devices and transplantation, are sometimes required.

Routine immunosuppression is not now recommended (including steroids), although the results of trials suggest that the sooner treatment starts, the better the outcome. The TIMIC study, in virus negative myocarditis, showed improvement on prednisone and azathioprine. (Frustaci A, Russo MA, Chimenti C. (2009) *Randomized study on the efficacy of immunosuppressive therapy in patients with virus-negative inflammatory cardiomyopathy: the TIMIC study*. Eur. Heart J.: 1995-2002). Over 85% had improved LV function, six months after treatment. At the moment, we reserve this treatment for refractory patients and those with giant cell myocarditis, although some with eosinophilic and sarcoid myocarditis could benefit.

Several monoclonal antibodies (immune checkpoint inhibitors or ICIs), are now available: ipilimumab, nivolumab and others.

The prognosis of myocarditis varies from full recovery in a few weeks, to rapid fulminant heart failure and death. The young and aged, have the worst prognoses. The average one year mortality is 20%; the four-year mortality is 60%. In secondary myocarditis, the prognosis will be that of the disease process. Late occurrence heart failure (dilated cardiomyopathy) following myocarditis, varies in different studies between 12% and 50%.

Endocarditis

The commonest form of endocarditis is subacute, rather than acute.

Rheumatic heart disease was once the foremost cause of acute endocarditis. Now, endocarditis presents mostly as a chronic degenerative valve condition. Introducing cardiac catheterisation, prosthetic valves, haemodialysis, pacemakers and other devices, has promoted endocarditis. Following any of these procedures, one should expect that endocarditis might arise between 72-hours and eight weeks.

SBE patients were once mostly young; now they are mostly older adults. The commonest bacterium involved was once *streptococcus viridans*; *staphylococcus aureus* is now commoner. The latter more often presents as acute sepsis, rather than a sub-acute condition. Many other bacteria and fungi can be involved.

As an example of rarity, and unpredictability, I once saw a young patient die from *trichophyton* valve infection, secondary to his athlete's foot. Other fungi like Candida species, are more common.

Because of the change in bacterial prevalence, what were once the classic features of SBE (splinter haemorrhages, Osler's nodes, and Roth spots) are no longer common.

Other facts of importance are:

- Left-sided valve lesions are more common than right sided ones.

- Right-sided endocarditis occurs in those with indwelling venous catheters, in congenital heart disease, and in the immunosuppressed.

- Mortality is high (10% to 30%).

- MRSA and gram-negative bacteria have the worst prognoses.

- The risk is higher for mechanical prosthetic valves than for bio-mechanical valves.

- A bicuspid aorta valve remains a risk factor.

- Injected drugs of abuse now present a significant risk.

Valve infection occurs on roughened or damaged endocardial surfaces. Bacteria, platelets and fibrin, can adhere and form vegetations. Valve leakage is the most common outcome of valve infection.

It is now thought (from animal models) that chronic bacteraemia more often causes endocarditis than one-off episodes of bacteraemia, caused by surgical procedures (like tooth extraction). Daily chewing and toothbrushing, can be associated with chronic bacteraemia. This may explain why a history of previous dental extraction is uncommon, and why prophylactic penicillin is no longer thought necessary for low-risk, valvular heart conditions (prolapsing mitral valve, and bicuspid aortic valve).

We use prophylaxis for:

- Those who have had prior endocarditis.
- Those with prosthetic valves.
- Those with congenital heart disease.
- Post-transplantation cases with unrepaired valve incompetence.

High-risk procedures are those that involve the gums, and those involving incisions into infected tissue. The drugs used mostly are amoxicillin and cephalexin. For those with a penicillin allergy, one must consider clindamycin and azithromycin.

Because prophylactic penicillin use is low risk, and endocarditis has serious consequences, patients who request prophylaxis may need counselling. To enable an early diagnosis, they should all know about the early adverse symptoms.

Many of my patients disagreed with the policy of not giving penicillin prophylaxis. They asked, 'why take any unnecessary risks, even when low risk, if the antibiotics used are effective and harmless?' To practice the art of medicine well, requires respect for every patient—their opinion, knowledge, intelligence and ability to assess risk. Fundamental to practicing the art of medicine is some personal knowledge of each patient, allowing individual assessments and personalised clinical judgements to be made, while respecting guidelines.

The diagnosis and management of endocarditis rely on positive blood cultures. When the culture is negative, it could be because of previous antibiotic treatment. Both long-term culture and bacterial genome detection using PCR amplification, now have roles.

The clinical presentation of endocarditis can have the following features:

- It usually presents as a fever of unknown origin.
- Those in heart failure, may not have a pronounced fever.
- Some present with pneumonia or septic arthritis.
- On examination, Osler's nodes (tender nodules in the fingers), splinter haemorrhages (in the nails), and Janeway lesions on the palms and soles (painless macules indicative of vasculitis) may be visible. These are all infrequent findings.
- Most patients will have a heart murmur; one that gets worse in the short term (7 to 10 days).
- A few will have petechial haemorrhages and splenomegaly.
- Infected emboli derived from vegetations, can cause toes and fingers to become painful and ischaemic.
- Embolism can also cause renal and splenic infarcts (some with left shoulder pain), and stroke.
- Infected emboli can lodge in arteries and cause mycotic aneurysms (liable to burst). Skin biopsy of such lesions can be diagnostic. In the brain, they can cause cerebral abscess and haemorrhage. When they are multiple, they can cause confusion and sometimes psychosis.

In long-term cases, some patients will present with unexplained progressive weight loss. Their debilitation can be like that of advanced cancer.

Investigations: Undertaking blood culture samples requires strict sterility. Take several 20 ml samples at intervals; ideally, take them before giving antibiotics.

Echocardiography has proven to be a major advance for diagnosing endocarditis, although it is mostly transoesophageal ultrasound (TOE) that detects vegetations, abscess formation and prosthetic valve abnormalities. If a transthoracic echocardiogram (TEE) does not detect vegetations in a case of suspected endocarditis, repeat it after a few days or perform a TOE. Echocardiograms are used to monitor vegetations during treatment; for this, three-dimensional echocardiography has proven valuable.

Complications of Endocarditis.

Partial destruction of valves can occur, and fistulae can form that connect heart chambers. Mitral chordae can rupture, and vegetations can embolise to coronary and other arteries. Although infected emboli can seed abscesses in any organ, those in the spleen and brain occur most frequently.

Functional problems occur most when aortic and mitral incompetence occur. Infection involving any valve annulus has serious implications. When an abscess affects the aortic valve annulus, it can affect the nearby bundle of His and cause heart block. The surgical evacuation of abscesses, together with valve replacement, is sometimes necessary.

Neurological complications range from transient ischaemic attacks, to meningitis and cerebral haemorrhage from mycotic aneurysms. These occur at the bifurcations of distal arteries, with middle cerebral artery aneurysms occurring in one to 5% of cases. In order to detect early changes, it is reasonable to get an early brain CT or MRI scan, in those with left-sided endocarditis.

Left-sided, systemic embolisation, can cause renal, splenic, hepatic, mesenteric and bone abscesses. Embolisation to an arm or leg artery, can cause a tender pulsatile mass. Haematuria should suggest renal embolisation, while melaena and haematemesis, suggest the rupture of a bowel aneurysm. With right-sided endocarditis, infected pulmonary emboli can o ccur.

Therapy

Therapy with appropriate bactericidal antibiotics should follow the specific bacteriological findings. After ten days of parenteral therapy, oral antibiotics are sufficient (Iversen, K., Ihlemann, N., Gill S.U. et al.(2019). *Partial oral versus intravenous antibiotic treatment of endocarditis.* N. Engl. J. Med., 380: 415-424).

For implanted devices, staphylococcus is the commonest infection.

For those with vegetations > 1.0 cm, and those with multiple embolic episodes, cardiac surgery needs consideration. The severity of remnant valve regurgitation or ventricular dysfunction, will help decide any need for surgery.

Systemic Diseases and the Heart

As a medical student in the 1960s, I was told— never forget syphilis. This still holds true, although in my whole career, I only saw one case of tertiary syphilis with aortic leakage. On CXR, the whole of his aneurysmal aortic arch showed partial calcification. I was faced with a dilemma: to leave him untreated, or to treat him with penicillin and risk a Herxheimer reaction (rapid swelling of his coronary arteries leading to occlusion and sudden death). Treated or not, he risked aortic aneurysm rupture (80% likely perhaps). My dilemma was resolved in a tragic way. Following my diagnosis, his girlfriend abandoned him, and soon after he committed suicide.

The incidence of syphilis varies between countries. According to the World Health Organisation, 7.1 million adults acquired syphilis in 2020.

Other general infections like diphtheria and Chagas' disease, can still be found in Latin and South America. Both can cause infection-related conduction disorders.

Endocrine Disorders

Hypothyroidism is common enough for most physicians to have seen many cases; some with a history of previous hyperthyroidism. I was a student at the London Hospital when Beall, Roitt and Deborah Doniach, first described autoimmune thyroiditis and its relationship to hyperthyroidism (LATS, or the long-acting thyroid stimulator is causative). Chronic damage to thyroid tissue causes hypothyroidism in later years. They published their work after I qualified (Beall, G., Doniach, D. et al. 1969).

Hypothyroid patients typically feel tired, cold and slowed, with hair loss and dry skin. Some will wear warm clothes, even in hot weather.

They can have non-pitting swelling of their feet (myxoedema), easily confused with the pitting oedema of heart failure. They are prone to heart failure and bradycardia. Their ECG may show a low-voltage pattern, sometimes with bundle branch block. Treatment must start slowly since as the dose increases, some will develop angina (they have a known proneness to atheroma and coronary artery disease).

Hyperthyroidism is an uncommon cause of palpitation, nervousness and hypertension. Some are wrongly diagnosed as anxious. One reason for the clinical features is that thyroid hormone seems to exaggerate the effect of catecholamines. Questioning patients about their temperature control is critical to making the diagnosis. Those with hyperthyroidism feel hotter than others; they typically wear light clothes when others are wearing heavy overcoats to protect themselves from the cold. This is never a feature of lone anxiety.

They often have sinus and supraventricular tachycardia, but are also prone to AF (for which they need anticoagulation). Beta-blockers help. Some proceed to heart failure, although this is now a rare occurrence with improved early detection. For younger women, the use of radioactive iodine will raise fertility as a potential problem; for men and menopausal women, this is not an issue.

Acromegaly causes the heart to enlarge under the influence of growth hormone and hypertension. Many patients get accelerated atherosclerosis. Heart failure is common. Some will have paid large sums of money for 'rejuvenating', growth hormone injections. I had one such patient who developed the typical facial appearance of acromegaly, as well as hypertension. Both reversed after stopping his ill-advised treatment.

Much rarer than the latter conditions, are Cushing's syndrome, Conn's syndrome and Addison's disease (adrenal insufficiency). In fifty years, I only saw one of each.

Cushing's and Conn's syndrome can both cause hypertension, hypokalaemia and dysrhythmias. Hypokalaemia can cause flat ECG T-waves. Addison's disease is a cause of hypotension associated with a profound lack of energy and buccal pigmentation. In the single case I diagnosed and treated, the patient discovered a remarkable relationship between the energy he had, and the replacement dose of cortisol he took. I always remember how lucky he felt. He said, 'How many people do you know, who have total control over their energy'? Inadvisably, he used to increase his cortisone dose, if he needed to work long hours or desired to dance through the night!

Granulomatous Diseases

Sarcoidosis is an uncommon granulomatous disease. I encountered only once. Although it mostly causes lymphadenopathy and interstitial lung disease, it can cause pericarditis and a restrictive form of cardiomyopathy. Any lesion close to the AV conducting system is liable to cause heart block, as well as incite ventricular arrhythmias.

In the heart, granulomata are most often found in the distal ventricular septum. The best imaging modalities are MRI or PET scanning, using flu-deoxyglucose (18F).

Takayashu Arteritis is also a granulomatous condition. It occurs most frequently in the Japanese (ten times more common), most often affecting the aortic root, but also the coronary arteries. It is a cause of superficial artery tenderness (carotids, etc.) and is associated with arthritis.

BP readings between arms can become different. It may present as a general illness and weight loss, with pain and tenderness in many parts of the body. It is also a cause of claudication in the arms.

Takayashu arteritis can be confused with giant cell arteritis, sarcoidosis, Behçet disease and polymyalgia rheumatica.

Churg-Strauss syndrome is an eosinophic granulomatous condition, causing asthma, and occasional myocarditis. Necrotising arteritis is one feature, affecting small and medium-sized arteries. It is a rare cause of restrictive cardiomyopathy.

Wegener's granulomatosis is a polyangiitis affecting the nasal passages and respiratory tract. It can also cause myositis and pericarditis.

Connective Tissue Disorders

Systemic Lupus Erythematosus (SLE) is an autoimmune disease that can cause pericarditis, pericardial effusion, and fatal coronary arteritis. The only case I saw during my career was in 1984. My patient Max, was an advertiser involved with the Los Angeles Olympics. While in California, I arranged a visit for him, with a leading expert on SLE at Stanford University. Anti-malarial drugs were prescribed. The benefits were soon apparent, but unfortunately, they were only transient. He died two years later from coronary arteritis and cardiac infarction.

The aortic and mitral valves are most commonly involved. SLE can cause cardiac infarction from the embolisation of non-infected valve vegetations. Enhanced coagulation can occur when antiphospholipid antibody (APLA) is present. Those with APLA need anticoagulation. Although rarely spontaneous, haemorrhage can occur.

Pregnant women with SLE can give birth to babies with complete heart block. The foetus may develop conduction system fibrosis while *in utero*. Intrauterine steroids have been used to reduce the risk of heart block.

Some drugscause SLE-like conditions. Among them are atenolol, procainamide, statins, hydralazine, and the ACE inhibitors enalapril and captopril.

Rheumatoid arthritis is fairly common, but a rare cause of pericarditis and effusion. It can enhance coronary artery disease.

Ankylosing spondylitis (associated with HLA-B27 antigen) and Reiter's disease, can be associated with AI, secondary to aortic dilatation.

In **Polymyositis**, older adult patients usually respond rapidly to steroids; this is a pathognomonic feature of polymyalgia rheumatica. It is occasionally associated with myocarditis and pericarditis.

CREST Syndrome (*calcinosis cutis*, Raynaud's phenomenon, oesophageal dysfunction, sclerodactyly and telangiectasia) occurs in scleroderma. It is associated with pulmonary hypertension, and a rare cause of systemic hypertensive crisis, related to renal involvement.

Polyarteritis Nodosa can cause necrotising arteritis. It can affect the kidneys, causing renal failure and hypertension. It occasionally causes small coronary artery aneurysms. Coronary atherosclerosis may become exaggerated (as with most of the inflammatory conditions described).

Fibromuscular Dysplasia is a non-inflammatory condition, found mainly in females. It affects the arterial walls; mainly those in the kidneys and the carotids. It rarely affects the coronary arteries where it can cause spontaneous dissection. Give thought to it, if a young female patient presents with an acute cardiac infarction.

Chapter Fourteen

Heart Disease in Pregnancy

Cardiologists are often asked to review pregnant women, especially those with heart murmurs, and hypertension. Few obstetricians will now want to take any responsibility for the cardiac aspects of pregnant patients, and will refer pregnant women with minor issues.

Coincident with the decline in rheumatic heart disease, are more pregnant women being found with adult congenital heart problems. The increased demand put on the circulation by pregnancy, will make some pregnant patients symptomatic for the first time. Managing the mother's state of health takes priority over that of the foetus.

The physiological changes responsible for cardiac problems in pregnancy are:

- An increased plasma volume (40-50%).

- An increased red cell mass (20-30%).

- A fall in peripheral resistance.

- Increased pulse rate.

- Limited diaphragmatic excursion, offset by an increase in tidal volume.

The increased stroke volume associated with a faster pulse, will elevate the cardiac output by 30-50%. The fall in systemic resistance will cause blood pressure to drop by 10-20 mm Hg in the 2nd and 3rd trimesters.

During labour, uterine blood gets transferred to the circulation. Normal circulatory physiology is usually restored within one month of delivery, although a raised cardiac output can persist for months.

During pregnancy, aerobic exercise reduces uterine blood flow. This could account for the lower average birth weights of babies, born to mothers who exercise regularly during pregnancy.

Many metabolic changes occur. There is an increase in both insulin resistance and insulin production. Some pregnant women become overtly diabetic for the first time, although some will have been pre-diabetic before. Diabetic control during pregnancy can become unsatisfactory. The babies of diabetic women often have an increased birth weight; the offspring, being more likely to develop type 2 diabetes and obesity in later life.

During the Second World War, many pregnant women in Holland had to withstand imposed famine. A follow-up study of the babies born to them showed an
increase in obesity, type-2 diabetes, renal disease, coronary artery disease, and mental health problems (Dutch Family Cohort, 2021).

Some symptoms of normal pregnancy mimic heart disease. Tiredness and fatigued, shortness of breath and ankle oedema, mimic heart failure. Some develop physiological murmurs with a pronounced 3rd heart sound, and more obvious splitting of the second heart sound (S2). These all reflect an increased blood flow and a greater circulating blood volume.

The symptoms and signs of heart disease in pregnancy are:

- Paroxysmal nocturnal dyspnoea.
- Orthopnoea.
- The presence of pulmonary oedema.
- Fixed S_2 splitting.
- A 4th heart sound.

All are reasons for further investigation.

The Management of Pregnant Patients with Heart Disease

Few are aware of the increased risk pregnancy brings to those with known heart disease. Ideally, they would be counselled before considering pregnancy.

Mitral stenosis was once a common cardiac condition in pregnancy; it remains one of the commonest. It is tolerated well once when the mitral valve aperture area is $< 1.5 \text{ cm}^2$; pulmonary oedema is then to be expected. Heart rate control is crucial and diuretics often needed. Defibrillation for AF is considered safe in pregnancy, although low blood pressure may then become a problem. In those who deteriorate during pregnancy, mitral balloon valvuloplasty in the second trimester should be considered.

Aortic stenosis with a valve gradient of $< 50\text{mmHg}$ is usually well tolerated. When more severe, balloon valvuloplasty is considered.

Hypertension in pregnancy is a common cause for cardiological referral. Pre-eclampsia (hypertension, oedema and proteinuria), as well as pre-existing chronic hypertension, need to be distinguished. Labetalol, methyldopa and calcium channel blockers, are the mainstays of treatment. Nitrates and hydralazine will be rarely needed. ACE inhibitors and ARBs are contraindicated.

Other Cardiac Risks of Pregnancy

The WHO classifies the cardiac risks of pregnancy (Canobbio, MM., Aboulhosn, J. et al. *The management of pregnancy in patients with complex congenital heart disease.* Circulation (2017); 135(8): 85-87). The following have the highest risk:

- Ventricular dysfunction (heart failure).
- Pulmonary hypertension.
- Coarctation.
- Previous peripartum cardiomyopathy.
- Eisenmenger syndrome (with pulmonary hypertension and reversed shunting).

Pregnancy is a justifiably contraindication for those with these prior conditions.

Mechanical heart valves, HCM, and mild aortic dilatation in Marfan syndrome, all pose an increased risk. ASD and VSD both present a small risk.

No detectable risk is associated with a repaired ASD, VSD or PDA, or mild PS.

Both atrial and ventricular ectopics are common, innocent findings in pregnancy. They do not present any tangible risk.

Five percent of women with congenital heart disease, give birth to a child with a cardiac defect. One must consider Caesarian section for those needing anticoagulants, and those with more significant cardiac lesions. It helps to avoid foetal brain haemorrhage during labour. Regardless of the severity of their cardiac condition, maternal mortality remains high for the week following delivery.

Prophylaxis for endocarditis needs to be considered for some. This is especially the case for those with mechanical valves, those with valve stenosis, those who have had endocarditis in the past, and those with unrepaired cardiac lesions.

Paradoxical embolism is a potential risk for those with an unrepaired ASD. Closure prior to pregnancy is advisable.

Mitral valve prolapse does not present a significant risk, and established **mitral incompetence** and **aortic valve leakage**, are often well tolerated. **Acute papillary muscle rupture** can lead to pulmonary oedema.

HCM is usually well tolerated when the septum measures < 30mm. The judicious use of beta-blockade needs consideration.

When pregnant patients have **coronary artery disease**, they risk cardiac infarction. Although rare (< 1 in 10,000 pregnancies), it usually occurs in the third trimester. The foetus must be shielded from radiation if angioplasty is undertaken.

Statins used in pregnancy are not known to cause congenital foetal abnormalities.

All those with known CAD should have their coagulation status assessed (including antiphospholipid antibody). Because of insufficient data, the safety of artery stenting and clopidogrel, have yet to be assessed in pregnancy.

Arrhythmias Occurring in Pregnancy

The commonest arrhythmias in pregnancy, match those of the non-pregnant state; they are mostly atrial and ventricular ectopics. Atrial fibrillation, when it occurs, can be treated with beta-blockade, digitalis, anticoagulation and defibrillation. When refractory, catheter ablation could be appropriate.

When VT occurs, it may be associated with long QT syndrome, cardiomyopathy, thyrotoxicosis and hyperemesis gravidarum (intractable vomiting). Sotalol is the only acceptable drug; amiodarone is not acceptable.

Coagulation, Anticoagulation and Pregnancy

Pregnancy is a hypercoagulable state; present from the start, and progressing throughout pregnancy. Venous thromboembolism is approximately five to ten times higher in pregnancy; the postpartum risk is fifteen to twenty times higher than in the non-pregnant state.

During pregnancy, women experience progesterone-induced venous dilatation. This promotes venous stasis, venous compression by the uterus, and compression of the left iliac vein by the right iliac artery. In addition, pregnancy causes a hypercoagulable state, that includes decreased protein S activity, increased protein C resistance, and other factors that increase thrombin production (higher levels of factor VIII, factor IX, and fibrinogen, for instance). Dr.Martina Murphy, MD. *Anticoagulation During Pregnancy*. Ash Clinical News (Feb 2017).

When anticoagulation is necessary, heparin and warfarin are both acceptable. Warfarin crosses the placenta and is a known teratogen; it is best avoided during embryological development. The foetal safety and efficacy of rivaroxaban or dabigatran in pregnancy has yet to be established.

For those with mechanical valves, anticoagulation should be given throughout pregnancy, although, given during the first trimester, warfarin can cause foetal bone and cartilage abnormalities (< 10%). It is safer in the 2nd and 3rd trimesters, although foetal CNS abnormalities can occur (optic atrophy, mental retardation, spasticity). Only dosages below 5mgs/day are advisable. Because the foetus is more sensitive to warfarin than the mother, foetuses can have intracranial haemorrhages resulting in spontaneous abortion and stillbirth.

The heparin of choice is low molecular weight heparin (LMWH). It does not cross the placental barrier, but can cause maternal osteoporosis and thrombocytopenia. In all cases, the aPTT should be kept between two to three times the control level. If a caesarean section is to be undertaken, heparin should stop four hours before. Neither LMWH nor warfarin get into breast milk.

For cardiac patients who become pregnant on warfarin, a switch to LMWH during the first 12 weeks is advisable; thereafter, patients can revert to warfarin.

For further review see: Marino, T., Lange, R.A., et al. (2022). Anticoagulants and Thrombolytics in Pregnancy. https://emedicine.medscape.com/article/164069

Chapter Fifteen
Cardiac Tumours

Cardiac tumours are rare but not to be forgotten. The only cardiac neoplasm I ever saw, was a left atrial myxoma. I once recorded a pedunculated tumour protruding through the mitral valve with atrial systole. This was in the early 1970s, when St. George's Hospital at Hyde Park Corner, London, had the only echocardiograph machine in the UK.

Myxomas are the commonest of benign cardiac tumours (25% of all). Typically, they occur between the 3rd and 6th decade of life. They variously cause syncope, a PUO, weight loss, and symptoms related to embolism. On auscultation, the sound produced is a mid-diastolic 'plop', followed by a mid-diastolic murmur.

Some are familial (LAMB and NAME syndromes). LAMB (Lentigines, Atrial Myxomas, and Blue naevi); NAME (Naevi, Atrial Myxoma, Myxoid neurofibromas, and Ephelides or skin freckles). Most myxomas are solitary and in the left atrium. They are gelatinous and pedunculated, varying in size between 4 and 18 cm. They can attract thrombosis and are a potent source of emboli. Surgical excision is required.

Papillary fibroelastomas can form fringe-like structures, arising from the non-contact areas of the leaflets of the mitral and aortic valves. Although benign in themselves, there are major risks from clot formation and embolisation. Surgical excision is recommended.

Lipomas and fibromas form in both sub-epicardial and sub-endocardial locations. When large, they can cause obstruction to blood flow and interfere with conduction; they are otherwise harmless. On echocardiography they appear as dense rounded structures, within the ventricular walls or septum. Surgical removal is required only if they become symptomatic or secondary problems are likely. Unlike fibromas and lipomas, **haemangiomas** are likely to regress if left.

Rhabdomyoma is a paediatric condition, usually occurring in the first year of life. These tumours are hamartomas and not neoplastic. They can also regress if left.

Malignant Primary Cardiac Tumours

Twenty-five percent of all cardiac tumours are malignant. Secondary tumours are more common. They are often present with pericardial effusion.

Most primary malignant tumours are **angiosarcomas** (usually right-sided, and associated with RV failure). **Rhabdomyosarcomas** occur in children (causing obstruction and eosinophilic syndrome). **Leiomyosarcomas** (usually in the LA, can cause obstructive symptoms). **B-cell lymphoma** in the immunocompromised, can cause blood flow obstruction and arrhythmia. **Mesothelioma**, usually arising from the pericardium, is associated with past asbestos exposure.

Malignant cardiac tumours can proliferate rapidly and cause death within 6 – 12 months. Removal before they metastasise may be possible in the early stages. Radiation and transplantation are the treatment options.

Secondary malignant tumours are 20 – 30 times more common than primary tumours. Most are in the pericardium and are secondaries from melanoma, lung, breast or oesophageal tumours.

Carcinoid tumours cause flushing and GI hypermobility from serotonin release. In the heart, they are associated with right-sided valvular fibrosis. They can cause thickening of the pulmonary and tricuspid valves, with incompetence and right heart failure. A raised 24-hour HIAA urinary secretion supports the diagnosis. Treatment with octreotide may help flushing and diarrhoea. Surgical removal of lesions and valve repair may become necessary.

Long-term use of fenfluramine as an appetite suppressant has occasionally resulted in similar features.

PART 4: CARDIAC TREATMENT

Chapter Sixteen

Pharmacology and Therapeutics

Treating cardiac conditions can often require active management. The nature of cardiac disease and its attendant risks, together with the relative toxicity of the drugs used, make regular observation and adjustment of doses essential. 'Prescribe and leave' styles of intervention, are not always appropriate for cardiac patient management. The over-dosing and under-dosing of cardiac patients occurs often, caused by varying patient responsiveness.

Specific expertise is required to treat the following:

- Heart failure.

- Thrombosis and anticoagulants.

- Rhythm management.

- Hypertension (systemic and pulmonary).

- Ischaemic heart disease (angina, cardiac infarction);

- Atheroma prevention and its control.

Individual entities can arise for management, or occur together in various combinations. One may have to manage a patient with atrial fibrillation and heart failure, caused by ischaemic heart disease. Each condition requires separate consideration.

Cardiac drugs can have a varied effect on the cardiac pathophysiology. An acquaintance with these, and the many variations of patient response to the drugs we prescribe, is essential. With practice one can get to know the usual responses of each drug. Although most cardiologists (following the BNF) will give lower doses to older adults and young patients, and vary doses according to the patient's BMI. This strategy will not always yield a predicted outcome.

Cardiac drug responses are notably individual, and not as predictable as presumed. It is helpful to ask patients about previous adverse drug responses. Just how misguiding guidelines could be often surprised me. Some low body weight, older patients, will tolerate high doses, while some young overweight people can be unduly sensitive. Among my patient group, I encountered a three-fold individual difference in the responses to drugs like digitalis and warfarin. These drugs existed, before the introduction of stricter qualifying pharmacodynamic criteria. Many of the factors influencing drug responsiveness are genetic. Gender, drug – drug interactions, pregnancy, non-pharmaceutical substance taking, diet, defective absorption (gut microbiome), target organ bio-availability, metabolism, and the effects of associated diseases (deactivation in the liver and excretion) can all be important. They can all contribute to adverse reactions and treatment failure.

Enquire about responsiveness to alcohol, sedatives and anaesthesia. They can provide clues to a patient's drug sensitivity. Any allergic reactions must also be noted.

Some pharmacogenomic associations with enzymes help to establish personalised treatment. For warfarin, they include VKORC1, CYP2C9, and CYP4F2. For clopidogrel, CYP2C19, and for simvastatin, SLCO1B1. There are more to discover.

Heart Failure Drugs

Digitalis Alkaloids

Of all the anti-arrhythmic drugs available, constituents of digitalis extract are perhaps the oldest. A physician at the Stafford Infirmary (UK), Dr. William Withering, discovered it in 1775 as a constituent of herbal tea. He published his *'Account of the Foxglove'* in 1785. (D. R. Laurence, 1963. *Clinical Pharmacology*. J.&A. Churchill Ltd). He observed that it improved a patient with dropsy (swollen feet). He recognised the active ingredient to be in foxglove extract, from which several active alkaloids were later isolated. In the UK, digoxin is the usual choice; in the EU, digitoxin is more often used and has a much longer half-life. Digoxin is mostly excreted through the kidneys, whereas the liver metabolises digitoxin. This is important for patients with defective renal and hepatic function.

The efficacy of digitalis is variable, but still useful for those with a need for heart rate control. Whether it increases ventricular contractility is in dispute. If therapeutic doses increase cardiac contractility, it is only be by a small amount. The problem with digitalis alkaloids is their accumulation. Once started, the maintenance dose will need to be lowered and often changed later. It was wise, in retrospect, for Victorians to have left off all their drugs on Sundays.

Before ECG monitoring and blood level evaluation, the method for administering digitalis was to start with a low dose, and steadily increase it until the patient experienced anorexia and/or nausea. At that point, the dose was halved.

Digitalis causes cardiac myocyte potassium loss. This is associated with typically cupped, ECG T-waves. Because diuretics can add to cellular potassium loss, they can potentiate the effects of digoxin. Oral potassium or foods containing lots of potassium (bananas and citrus fruit), help to reduce these adverse effects.

Because digoxin reduces the refractory period of atrial tissue, it could maintain AF, at the same time slowing AV conduction and heart rate. A reduction in the refractoriness of cardiac conducting tissue can sometimes convert flutter to fibrillation. Because high doses of digitalis alkaloids can increase myocardial excitability and shorten the Q-T interval, extrasystoles and VT can occur. Procainemide, and other arrhythmic drugs, are to be used with caution in those with digoxin toxicity. The combination can produce heart block.

Other Drugs Used in Heart Failure

These are ivabradine, phosphodiesterase inhibitors and diuretics. Others include beta-blockers, ACE inhibitors and ARBs. These I describe later.

Ivabradine

Pacemaker cells gradually depolarise and at threshold, create an action potential. The slow diastolic depolarisation (the I_f current) is controlled by cyclic adenosine monophosphate (cAMP), and autonomic influence. Ivabradine reduces the rate of pacemaker cell depolarisation (blocking the I_f current channel), slowing the cardiac pacemaker rate.

By slowing the heart rate to under 70bpm., those with a reduced left ventricular ejection fraction (of 35% or less), will be helped by ivabradine. Unfortunately, it can induce atrial fibrillation, and should then be withdrawn.

Ivabradine can cause sinus bradycardia, sinus node dysfunction, atrioventricular block, and bundle branch block. Because it is metabolised by enzyme CYP3A4, one must avoid the concomitant use of verapamil and diltiazem.

Lowering the heart rate in heart failure has been thought to affect its outcome. Two trials of ivabradine showed no difference in the rate of death or cardiac infarction (BEAUTIFUL and SIGNIFY trials). The SHIFT trial, however, achieved its primary end-point and showed a reduction inhospital admissions from heart failure. An increased occurrence of bradycardia and optical phosphenes (visual eye problems) were noted.

Phosphodiesterase (1, 3, 4 and 5) (PDE) Inhibitors

Phosphodiesterase inhibitors can improve cardiac function in heart failure by increasing the intracellular content of cAMP and/or cGMP in cells.

PDE-1Cis present at high levels in human myocardium and is an alternative target for inotropic and cardioprotective actions.Inhibitors of PDE-3 can raise intracellular cAMP content in cardiac muscle, with inotropic effects.

PDE5 inhibitors have been used to raise cGMP content in cardiac muscle cells in animal models with pressure overload, and in those with chronic β-adrenergic receptor stimulation and ischemic injury. They are not so effective in humans and there is a major problem with their long-term use. Clinical trials have shown an increase in cardiovascular mortality.

PDE-5 inhibitors are useful for treating erectile dysfunction. The first was **sildenafil** (Viagra). It can cause hypotension, vasodilatation, palpitation, facial flushing and sometimes indigestion. It is useful in treating pulmonary hypertension. Tadalafil has similar uses. In the heart, inhibition of PDE5 signalling can be protective. Because lung PDE is lower than in the myocardium, it has little influence on lung function.

Two PDE-3 inhibitors are in use: **enoximone and milrinone**. Enoximone has myocardial ionotropic effects and causes vasodilatation without increasing oxygen consumption. It can help if used short-term in heart failure, by helping to lower end-diastolic pressures. Once a 17% increase in mortality rate was found, trials in the US were stopped (Phosphodiesterase III inhibitors for heart failure (Cochrane Database Syst. Rev. (2005) 24. doi: 10.1002/14651858.CD002230.pub2).

Milrinone is mostly used for acute heart failure with pulmonary hypertension, especially post-infarction or in association with cardiac surgery.

Roflumilast is a PDE-4 inhibitor with anti-inflammatory effects. One can use it in severe chronic obstructive airways disease as an adjunct to bronchodilator use.

Vasoconstriction is one hallmark of congestive heart failure (CHF), resulting in an increased impedance to ventricular ejection in both systemic and the pulmonary circulations. Defective nitric oxide release is one cause for this vasoconstriction.

Diuretics

Venous obstruction and inflammatory processes cause the accumulation of fluid in heart failure, and diuretics can help to remove the excess. The question is, from which physiological compartment will it come: from the intracellular fluid, the interstitial fluid or the circulating fluid? We would prefer the interstitial fluid compartment to be targeted alone, but diuretic action is not that specific.

In right heart failure, dependent oedema collects in the legs; in left heart failure and mitral stenosis, fluid collects in the interstitial lung tissue. Leg oedema is most commonly gravitational, and not caused by heart failure. A common mistake is to assume that every case presenting with ankle oedema has right heart failure. Being able to detect a raised JVP at rest, or after exercise, is key to making the correct clinical diagnosis. The absence of ankle oedema does not exclude right heart failure, especially after treatment with diuretics.

One important therapeutic objective, is the removal of interstitial fluid; ideally, diuretics will not reduce the circulating blood volume or the intracellular fluid. Because this ideal is not achievable, adverse cardiac effects often ensue. In left heart failure, it is most important to reduce interstitial lung fluid. Patients will then exchange more oxygen and have lungs that are not so stiff; both will help to reduce dyspnoea. If, however, diuretics reduce blood volume and intracardiac pressures, cardiac output may diminish. Sometimes, one has to accept some ankle oedema in order to maintain an adequate systemic circulation.

It is common enough to see patients, given high doses of diuretic for too long. They can feel weak and have cold cyanosed hands, feet and nose, as evidence of their poor peripheral tissue perfusion (caused by reduced left ventricular output). In this state they may also be sodium and potassium depleted, with significant acid-base changes (either in the blood, or cellular and intercellular spaces).

Because the latter complications of diuretic use are common, and their optimal use difficult to achieve, the use of diuretics in cardiovascular medicine is an art; a balancing act between removing fluid and maintaining CO and tissue perfusion.

While helping to remove excess fluid, one must aim:

- To maintain peripheral perfusion.
- To keep the JVP at a normal level.
- To avoid dehydration,
- To avoid overt sodium and potassium depletion (the latter can cause fatal dysrhythmias).

To achieve these aims, regular review is essential, especially in older adults. Regular review must continue until a stable equilibrium is attained. If you prescribe diuretics, you must allocate enough time to manage them.

IMPORTANT NOTE: *Whenever fluids and vital salts are being removed, there will be a large, metaphorical elephant in the room. Heart failure is a cardiac pathological problem that no diuretic or other medication can cure. The object is expedient maintenance, created by improving heart function with diuresis, exercise, weight loss, rate control, and the removal of ischaemia (CABG, coronary artery stenting or heart transplantation).*

Which Diuretic(s)?

Diuretics are not only used to remove oedema; they are useful in some hypertensive patients (thiazides), those with renal failure (bumetanide), ascites (chlortalidone) and those with cerebral oedema (mannitol). There are also endocrine disorders where they are helpful.

Important note. *Hypertensive patients are sodium sensitive or not. A diuretic may help some, but not others. Occasionally, one will encounter a moderately hypertensive patient, with blood pressures adequately normalised by a thiazide diuretic alone. A trial of a diuretic alone is a worthwhile test. There is not much to be gained by using a diuretic in combination with other anti-hypertensive drugs, if such a test proves negative.*

Diuretics either work quickly, and are short acting (furosemide), or they act slowly over longer periods (thiazides). Combining these features can work well; later on, one or the other will usually suffice. It is important to get a feel for just how fast each diuretic works (minutes or hours) before an effect occurs; how effective they are (the increase of urine volume for instance), and for how long they remain effective. Often, tablets rather than injections, work fast enough; only rarely, as in acute left heart failure, is intravenous furosemide necessary.

As an accompaniment to diuretic use, most physicians will recommend a low sodium salt / high potassium diet (it is sodium salt that retains water in tissues). Beware of sodium and potassium depletion, and be prepared to compromise since low salt foods can be tasteless and bring further misery to those already suffering.

In the UK, the British National Formulary is a primary source of diuretic information (uses, cautions and contraindications and administration). I will mention here, only the key, practical points, of those diuretics listed for use.

Slow Acting Diuretics

Thiazides take an hour or more to work, and their effect can last 12 to 24 hours. They inhibit renal sodium reabsorption from the distal convoluted tubules.

Bendroflumethazide 5mgs daily, is an effective slow acting diuretic. It will suffice for many hypertensive patients by reducing arteriolar sodium, especially for those prone to sodium retention (hypertensive menopausal women, for instance). It can help to maintain the reduction of oedema in heart failure.

Amiloride (5-20mgs daily)is an alternative drug, with the advantage of potassium sparing. In the distal renal tubule it inhibits sodium/potassium exchange.

Indapamide (2.5mgs daily)is a thiazide-like drug useful in hypertension. It causes vasodilatation with low doses, and diuresis with higher doses.

Chlortalidone (up to 50mgs daily) is a thiazide-like diuretic, used mostly for ascites and hypertension. It is longer acting than most thiazides.

Hydrochlorothiazide became commercially available in 1959; its benefits have remained, despite changing therapeutic fashions. It works well in combination with ACE inhibitors and ARB's.

Fast Acting Diuretics

These are usually loop diuretics, blocking the renal reabsorption of sodium from the ascending loop of Henlé.

Furosemide. This drug has been in use for many decades, used in doses that must be varied according to need. A 10mg dose is regarded as small; 80mgs is large; 20 – 40mgs is the usual daily dosage range. One danger is potassium depletion. Patients need to be prepared for quick onset, noticeable, and repeated diuresis. They might prefer to take them in the morning; they can then do their shopping in the afternoon and sleep at night!

Bumetanide is a powerful loop diuretic, useful for those in renal failure. Use it cautiously, starting with 1mg or less. It is not for use in hypertensive pregnant women.

Other Diuretics

Mannitol can reduce cerebral oedema. It is an osmotic diuretic given intravenously.

Spironolactone has been in and out of fashion for five decades. It is an important drug for several reasons:

- Spironolactone is an aldosterone blocker useful in hypertension, heart failure, nephrotic syndrome and liver failure. Aldosterone is a component of the renin-angiotensin-aldosterone system, binding to receptors in the distal renal tubules and collecting ducts.

- It is a mineralocorticoid receptor antagonist—receptors that cause sodium reabsorption and potassium secretion.

- It thus causes renal sodium loss, and potassium retention.

In cardiology it is useful for:

- Those in heart failure who have a reduced ejection fraction (HFrEF). The Randomized Aldactone Evaluation Study (RALES: New Engl. J. Med. 1999; 341:709-717) was a landmark trial. The results indicated a positive advantage to those with an ejection fraction of 35% or less.

- Resistant hypertension.

- Reduction of myocardial fibrosis and cardiac muscle weakness. Important evidence not only suggests that spironolactone reduces myocardial fibrosis; it also reduces left ventricular mass and decreases extracellular volume expansion in HFpEF patients (those with preserved myocardial function)(McDiarmid AK, Swoboda PP, et al. *Myocardial Effects of Aldosterone Antagonism in Heart Failure With Preserved Ejection Fraction*. J. Am. Heart Assoc. 2020 Jan 07; 9(1):e011521).

Spironolactone is the main option for treating ascites in cirrhosis, if dietary salt restriction proves ineffective.

Hyperkalaemia is a contraindication for its use, especially when renal failure is present.Men may experience gynecomastia, loss of libido, and general feminisation. Its androgen blocking effect will help those with acne.

Review the BNF for diuretic combinations, and other drug alternatives. I have only rarely found others necessary.

Thrombosis, Anticoagulants and Fibrinolysis

Warfarin was first approved in the 1950s. It is one of three Vitamin K antagonists in use. It reduces the amount of circulating clotting factors, specifically inhibiting the C1 subunit of vitamin K epoxide reductase. With warfarin, the liver becomes unable to produce factors: II, VII, IX, X, and proteins C and S.

I used warfarin for 40-years and saw no significant bleeding or clotting episodes in any patient. Expert titration is key to its safe and successful use. Because this is partly an art, regulators will not promote it: they want medicine to be totally rule-based, and practiced algorithmically for compliance enforcement purposes. Because the management of diuretics and warfarin, both require the resolution of multiple factors, both have art-form components. Scientific measurements (in this case the prothrombin time) are but a guide to safe doses for individual patients. General rules of thumb while useful, will not always suffice. Here are some of the key management points:

- Note the patients stature, health or frailty. Beware of frail older people with any evidence of spontaneous skin bruising.

- Ask about reactions to other drugs.

- Have good reasons for prescribing an anticoagulant.

- Start treatment with 5mgs of warfarin daily.

- After three days test the prothrombin time (PT) as a measure of its effectiveness. Thereafter, test again in 3 days. Adjust with dosage increments or decrements of 1-2 mgs. And test at lengthening periods. The most stable patients need testing every one to three months; some other patients may need to be tested every few days or every week.

- Instruct patients to report any instance of bleeding or clotting as a matter of urgency (calf tenderness or acute, unilateral ankle swelling).

- Note instances of bruising: nose bleeding or melaena (explain the need for patients to report incidents).

- Assume that even 1mg of warfarin has an anticoagulant effect. Warfarin is a potent poison.

- In older adults with idiopathic AF and previous skin bruising, I would occasionally stabilise a patient with a maximum PT ratio of 1.5. The reason is: THE PT RATIO IS ONLY AN APPROXIMATE GUIDE TO THE ANTICOGULANT EFFECT IN INDIVIDUALS.

- Only those with rheumatic valvular disease and prosthetic valves, need PT ratios between 2.0 and 4.00.

- PTs found higher than 3.5, may need an immediate warfarin dosage reduction with repeated reviews of the PT ratio over 2-3 days.

- Stop any drug (aspirin and vitamin E preparations) that might enhance warfarin anticoagulation.

- Be prepared to administer vitamin K to reverse the effect of warfarin.

Following cardiac catheterisation, I observed that aspirin and warfarin users had an obvious prolongation of arterial bleeding. Such consistent anecdotal observations are of clinical relevance to patient dosing. Every laboratory test result like the PT, needs interpretation in the light of clinical considerations. Always give priority to clinical observations rather than laboratory test results. When in doubt, repeat the laboratory test. This led me not to allow older patients with bruising, to have a PT ratio > 1.5. One arrogant CQC inspector, a general practice lecturer, once criticised my use of low-dose warfarin. He had no significant experience, but attributed my lack of patient side-effects to luck. My strategy, however, had served many patients well for four decades. Such conceit is what many UK doctors have to suffer from ill-informed regulators on a daily basis.

Aspirin

Aspirin can cause stomach erosions and bleeding, nose bleeding, and spontaneous bruising, especially in older adults. It can prevent arterial clotting and reduce morbidity in patients with coronary and cerebral artery disease. It is inexpensive in comparison to the latest anticoagulants, and for economic reasons, risks being derided by big Pharma.

A Short History of Aspirin Use

Bayer named 'Aspirin' in 1899: 'A' stood for 'a'cetyl, and 'spir' for the *Spirea ulmaria* (meadowsweet), from which salicin was extracted. As a derivative of willow bark (and its leaves), both ancient Egyptians and Greeks had used it as an analgesic and antipyretic. In 1853, French chemist Charles Gerhardt showed the active ingredient salicin to be salicylic acid

Gerhardt then synthesised acetylsalicylic acid (thought less irritant in the stomach). In 1873, Thomas MacLagan undertook a successful trial on patients with joint pains and rheumatism (*Lancet 1:383*-84). By 1950, the Guinness Book of Records described it as the most commonly used analgesic in the world. In 1971, Prof. John Vane discovered it caused dose-dependent inhibition of prostaglandin synthesis (Nature New Biology 1971; 231: 232). In 1974, P.C. Elwood and A.L. Cochrane et al. (B.M.J 1974; 1:436), showed that it did not prevent primary cardiac infarction, but reduced the further mortality of post-infarction cases (by 12.5% in the first six months, and 25% in the year that followed). The CAST study in 1997, showed that it was beneficial in acute stroke if given early enough (Lancet, 1997; 349:1641).

The efficacy of aspirin in primary cardiovascular prevention is doubtful, but secondary prevention remains unquestioned. When given for second heart attack prevention, an increase in the number of non-fatal strokes was noted. (Baigent, C., Blackwell, L. et al (2009) *Aspirin in the primary and secondary prevention of vascular disease.* Lancet 373; 1849-60). One vital clinical question often arises: how to minimise two major risks—clotting and bleeding.

Aspirin will cause more all-cause deaths in those over 70-years of age, if 100mgs is taken daily. (McNeil, J.J., Nelson, M.R., et al (2018). *Effect of Aspirin on All-Cause Mortality in the Healthy Elderly.* New Engl. J. Med. 379:1519-1528. It does, however, reduce bowel, stomach and oesophageal cancer incidence, when given prophylactically (Cuzick, J., Thorat, M.A., et al. *Estimates of benefits and harms of prophylactic use of aspirin in the general population.* Annals of Oncology 28: 47-57).

Clopidogrel

Clopidogrel is a useful pro-drug and antiplatelet agent. It gets metabolised to carboxyl esterase-1 in the liver, with all the metabolites becoming bound to plasma albumen. Over 90% gets metabolised to an inactive carboxyl compound; only 2% is converted to 2-oxoclopidogrel, and converted to the active metabolite. The active metabolite irreversibly stops ADP binding to $P2Y_{12}$ receptors on platelets (for their lifetime of 8-11 days). It prevents the activation of GPIIb/IIIa complex and reduces platelet aggregation.

Taking it after a meal reduces its distribution by 60%. Unlike aspirin, it can be taken without food. It is maximally effective between 30 and 60 minutes and eliminated equally in faeces and urine.

It is a useful prophylactic in angina, and can be used to treat acute coronary syndrome, myocardial infarction and non-haemorrhagic stroke. In these conditions, clopidogrel is thought superior to aspirin.

Thrombin and Factor Xa inhibitors CHADS and ORBIT scores (see NICE)

Dabigatran (introduced in 2010) as an anticoagulant thrombin inhibitor. **Rivaroxaban** (2011), **apixaban** and **edoxaban** are factor Xa inhibitors. Some refer to them as NOACs (Novel Oral AntiCoagulants, or Non-Vitamin K antagonist Oral AntiCoagulants); also, DOACs (Direct Acting Oral Anticoagulants). They are no longer 'novel'. Other terms used to describe them have been: TSOAC (target-specific oral anticoagulant), ODI (oral direct inhibitor), and SODA (specific oral direct anticoagulant).

I had two problems with their use originally. First, there was no method equivalent to a prothrombin time, to check their anticoagulant efficacy. Second, no reversing agent was available initially. Now, the anti–factor Xa assay (a measure of antithrombin) is used to measure plasma heparin (UH and LMWH) levels, and NOAC anticoagulant therapy. Both heparin and factor Xa inhibitors increase the natural anticoagulant antithrombin.

Andexanet alfa is a very expensive recombinant form of human factor Xa protein that binds specifically to apixaban and rivaroxaban, reversing their anticoagulant effect. It costs £15,000 for a usual course (2020). Measures of antithrombin are unreliable after using it. Also, heparin is ineffective after its use. **Idarucizumab** (approximately £3000 for an effective dose) is used to reverse dabigatran. Neither compare to the low cost of vitamin K used to reverse warfarin action.

Factor Xa inhibitors can improve outcomes for patients over 75-years-old, with non-valvular AF. One meta-analysis showed fewer cerebral emboli, intracranial bleeds, and haemorrhagic strokes, when compared to warfarin. Major cerebral bleeds are no different in occurrence to those on warfarin. In addition, they offer no advantage to myocardial infarction, ischaemic stroke or all-cause mortality (Zahoor, M.M., Mazhar, S., et al. (2024) *Factor Xa inhibitors versus warfarin in patients with non-valvular atrial fibrillation and diabetes mellitus: a systematic review and meta-analysis of randomized controlled trials.* Annals of Medicine & Surgery 86(2): 986-993). For further information research the RE-LY, ROCKET AF, ARISTOTLE, and ENGAGE AF-TIMI trials.

For those unable to tolerate a NOAC, warfarin remains an option, even though some art of medicine is required to perfect its use. The annual cost of monitoring warfarin in the NHS lies between £120 and £220 (much greater privately). The cost of warfarin is £10 per year on average; the cost of apixaban 2.5mgs (approximately £1.00 each), is £365 per annum. NOACs are not monitored. The financial difference is in favour of warfarin, but to use it safely, much time will be consumed. The attention warfarin demands, promotes doctor patient engagement and trust; neither will receive priority over accountancy considerations in a corporation like the NHS.

In conclusion, NOACs are easy to use and well tolerated, but much more expensive.

Fibrinolysis

My first encounter with effective fibrinolysis was in 1982, while working with Pim de Feyter at the Vrije Universiteit in Amsterdam. Pim's study involved injecting streptokinase directly into the blocked coronary artery causing cardiac infarction (within three hours of onset, but often much quicker). (de Feyter, P.J., van Eenige, M.J., et al. Eur. Heart Journal (1982): 3(5); 441-448). The immediate effects of any therapeutic intervention have rarely amazed me. As a co-author of Pim de Feyter's work, the use of intracoronary streptokinase was one of those times. As the clot quickly dissolved, the patients cardiac performance improved immediately. They quickly felt better, and became less breathless as their pallor waned as their peripheral perfusion and BP returned to normal. *'Uitstekent'* (outstanding), in any language!

Pim's work was successful, partly because of the excellent home to hospital transit times in Amsterdam (a small city with excellent ambulance services). Dr. Frank Partridge, who introduced CPR in Northern Ireland in 1957, soon learned that survival from VF depended on delivering treatment as soon as possible. Survival (as I once told an interview committee in 1967), is inversely proportional to the time taken to get to the patient. Although obvious to most junior doctors at the time, this basic fact was unknown to many senior doctors. One year previously, Partridge had introduced the mobile coronary care unit (1966): an ambulance with specialist equipment and staff, to provide pre-hospital care.

Fibrinolysis has moved on since 1982. Fibrinolysis needed to be available universally; for patients in the community and for those in hospitals without cardiac catheterisation facilities. Several fibrinolytics activate plasminogen to form plasmin, fibrin is degraded and thrombi dispersed. We can use them in acute cardiac infarction, ischaemic stroke, DVT and pulmonary embolism.

Alteplase is perhaps the most popular recombinant plasminogen activator. It is effective given intravenously, within six hours of the first symptom from cardiac infarction, pulmonary embolism and acute ischaemic stroke (confirmed under neurological care). The risk is one of cerebral and other bleeding. **Reteplase and Tenecteplase** are similar. Intravenous **Streptokinase** is not the drug of choice if previous allergic reactions have taken place. Antibody persistence will reduce its effect on subsequent occasions.

Rhythm Management

The Control of Heart Rate and Contractility

The autonomic system influences heart rate by affecting the depolarising rate of the sino-atrial node and other atrial pacemakers (including the A-V node). Acetylcholine, released from parasympathetic endings, reduces the rate of pacemaker cell depolarisation; nor-epinephrine, release from sympathetic fibres, increases it. Both the contractility of the heart and heart rate, increase when circulating adrenaline, noradrenaline (NE) from nerve endings, and the thyroid hormones (T3 and T4) are produced. Other substances can cause tachycardia (cocaine, caffeine, alcohol, nicotine). Conditions such as carcinoid syndrome and the menopause, can also influence heart rate (often with associated facial flushing).

In the short term, arteriolar resistance controls blood pressure; in the long term arteriolar medial hypertrophy permanently affects it. Sympathetic stimulation and NE, cause vasoconstriction. Many stimuli, including parasympathetic stimulation and acetylcholine release, ATP, adenosine (released by muscular activity), bradykinin, histamine, and shear stress, will activate eNOS and COX pathways and form nitric oxide (NO) to dilate arterioles and reduce resistance to flow.

These substances decrease intracellular calcium and increase myosin light chain (MLC) phosphatase activity. In smooth muscle cells, active MLC phosphatase dephosphorylates the contracted actin and MLC complex, causing MLC to relax. During arteriolar relaxation, receptor and voltage-gated calcium channels, inhibit calcium entry into the smooth muscle cell. The overall effect is to relax smooth muscle and cause vasodilation. Calcium channel blockers have an anti-hypertensive effect.

Adrenaline and NE production are crucial to heart rate and contractility control, their action being reduced by blocking sympathetic ß-receptors (ß-blockade). There are three types of beta receptor, named 1, 2 and 3. Beta-1 receptors are in the heart; beta-2 receptors control smooth muscle relaxation and glycogenolysis; beta-3 receptors induce the breakdown of fat cells.

Following the discovery of ß-blockers (James Black synthesised propranolol in 1964), it became expedient to produce pharmacological drugs that could reduce the synthesis of catecholamines (ACE and ARB drugs), rather than block adrenergic transmitters once produced (beta-blockade). These three classes of drug, have brought outstanding benefits for many patients with hypertension and heart failure.

Beta-1 Receptor Blockers

Beta-blockers have many uses that include the treatment of hypertension, congestive heart failure, cardiac arrhythmias, long-QT syndrome, HCM, coronary artery disease, hyperthyroidism, aortic dissection, and the secondary prevention of myocardial infarction. Other uses are for migraine, portal hypertension, glaucoma and essential tremor. Besides blocking beta-receptors, they some block the release of renin.

The first beta blocker, propranolol, can reduce the physical effects anxiety (tachycardia, sweating and tremor) and remains beneficial for anxious hypertensive cases. Because it reduces ventricular contractility, as well as heart rate, it needs careful monitoring. The latest beta-blockers have improved heart rate specificity, and influence ventricular contractility less. The latter is not an assumption I have been happy to make, given many direct observations of reduced ventricular contractility I made while undertaking echocardiography. Ventricular contractility can appear sluggish in those taking beta-blockers. Because of the variance in the anatomical distribution of beta receptors, my policy was to make an individual assessment of each patient's reaction.

Because beta-blockade reduces the pro-arrhythmic effects of catecholamines, some reduction in the incidence of AF., VT and VF, was to be expected. A meta-analysis of heart failure cases concluded that, 'for every heart rate reduction of 5 beats/min with beta-blocker treatment, a commensurate 18% reduction in the risk for death occurred (CI, 6% to 29%). No significant relationship between all-cause mortality and beta-blocker dosing was observed (risk ratio for death, 0.74 [CI, 0.64 to 0.86])(Mc Alister, F. A., Wiebe, N., et al.,(2009). *Meta-analysis: beta-blocker dose, heart rate reduction, and death in patients with heart failure.* Ann. Int. Med.150(11):784-94).

Congestive heart failure patients can be given beta-blockers, but only if they are in a compensated state. The selective cardiac beta-blockers bisoprolol, carvedilol, and metoprolol succinate are best. Metoprolol tartrate is not used in heart failure, but can be used for atrial fibrillation. Non-selective beta-blockers bind both beta-1 and beta-2 receptors (propranolol, carvedilol, sotalol, and labetalol) in varying amounts; those most used today (atenolol, bisoprolol, metoprolol, and esmolol) are beta-1 selective.

Because it decreases oxygen demand, beta-1 blockade can improve angina. The effect is often partial, and the benefit not comparable to surgical revascularisation. Because they can interfere with melatonin release, they often affect sleep.

Alpha-1 receptors induce vasoconstriction and increase heart rate. Some beta-blockers (carvedilol and labetalol) have additional alpha-1 receptor blockage activity. Beta-blockers that also block the alpha-1 receptor, can be advantageous for treating hypertension.

Sotalol

This is a non-selective, beta-adreno-receptor blocker, with important uses:

- For patients with sustained ventricular tachycardias,

- Prophylaxis of paroxysmal atrial tachycardias including AF (also following DC reversion).

- Prophylaxis and treatment of re-entrant tachycardias.

It should not be given to those with long QT syndrome and *torsade de pointes*. Because it is water soluble, it will be present in breast milk. In those with renal dysfunction, doses must be attenuated. Patients need monitoring for QT interval changes (it can be pro-arrhythmic), and electrolyte status (to avoid low potassium and magnesium blood levels).

Alpha Stimulation and Blockade

Sympathetic stimulation affects heart rate and myocardial contractility (inotropy). Some important facts are:

- Alpha-1 and beta-1 receptors found in the myocardium and blood vessels. They are activated by catecholamines.

- Alpha-1 stimulation causes arteriolar vasoconstriction.

- Alpha-2 adrenergic receptors are located on peripheral nerve endings and inhibit the release of noradrenaline.

- Blocking alpha-2 adrenergic receptors will release more noradrenaline.

Selective alpha-1 blockers work best with stress-related hypertension (if more noradrenaline is being released).Because selective alpha-1 adrenergic antagonists cause vasodilation (by preventing NE from activating the alpha-1 receptor), they can lower blood pressure in hypertension. They also relax smooth muscle in the prostate, enabling urine to flow more freely through the urethra. Given to older adult males, there is a risk of hypotension.

In cases of phaeochromocytoma, phenoxybenzamine and phentolamine can be used to reduce blood pressure.

Chronic beta-1 receptor stimulation (by catecholamines) is associated with adverse CVS outcomes, whereas apha-1 stimulation (affecting smooth muscle) is thought protective. Although they cause arteriolar vasoconstriction, their effect in the myocardium can be beneficial. Although fewer exist in the myocardium than beta-receptors, alpha-1 receptor stimulation is involved in cardiovascular regulation, muscle hypertrophy, inotropy and protection from ischaemia and cell death.

Whereas α-1 activation protects against heart failure (inducing physiological hypertrophy), non-selective α-1 blockade doubles cardiovascular risk (ALLHAT study:*Major Outcomes in High-Risk Hypertensive Patients Randomized to Angiotensin-Converting Enzyme Inhibitor or Calcium Channel Blocker vs Diuretic.* JAMA. 2002; 288(23):2981-2997. doi:1 0.1001/jama.288.23.2981).

In another arm of the same study(ALLHAT Study:JAMA. 2000; 283(15):1967-1975. 10.1001/jama.283.15.1967), the incidence of heart failure in subjects who received the nonselective α1-blocker doxazosin, was twofold higher than in those subjects who received other blockers. The doxazosin arm of the study had to be stopped prematurely.

Selective alpha-blockers were first introduced to alleviate the symptoms of benign prostatic hypertrophy, and are now also used for hypertension (alfuzosin, doxazosin, terazosin, tamsulosin, and prazosin). Unfortunately, they increase the risk of hospitalisation for heart failure.

Alpha-1 receptor agonists (causing vasoconstriction) are used for:

Vasodilatory shock, hypotension, hypoperfusion (methoxamine, phenylephrine), and glaucoma (apraclonidine), and
Nasal congestion (epinephrine, pseudoephedrine, oxymetazoline).

Antiarrhythmic Drugs

The Vaughan Williams classification divides these drugs as follows:

Class 1a: cardiac suppressants: quinidine, procainamide, disopyramide, etc.
Class 1b: cardiac suppressants: lidocaine, mexiletine, etc.
Class 1c: cardiac suppressants: flecainide, etc.
Class 2: beta-blockers.
Class 3: influence cellular potassium efflux. Amiodarone, sotalol, etc.
Class 4: affect calcium channels. Diltiazem, verapamil, etc.
Class 5: work in other ways. Digoxin and adenosine, etc.

Class 1a: Quinidine

Quinidine is now a 'third-line' anti-arrhythmic: it can be pro-arrhythmic and has too many other side effects. It is no longer listed in the BNF (an excellent way to restrict the use of a drug in the UK). From my long experience of its use, I would question the wisdom of its exclusion and the reasons for it.

Quinidine is anti-malarial, anti-epileptic and anti-arrhythmic. Originally derived from South American cinchona tree bark, it is a stereoisomer of quinine.

D.R. Lawrence, in his remarkable book, Clinical Pharmacology, provides the historical, anecdotal background of many drugs. In 1749, quinine was found to quell 'rebellious palpitation'. He notes that one of Wenckebach's patients, troubled by AF, helped himself by taking one gram of quinine. Six years later, after various derivatives of quinine had been tested, quinidine was introduced. (D. R. Laurence. (1963) *Clinical Pharmacology*. J.&A. Churchill Ltd).

I used quinidine for many decades and never saw one substantial complication or side-effect. I often used it to revert AF and to maintain long-term sinus rhythm, but never used it in heart failure cases. Amiodarone and flecainide replaced it (NICE 196). Only amiodarone is now recommended in heart failure cases, and for those with structural abnormalities. This is clearly a case of remote data analysis, superseding practise-based experience. One must respect the analysis of large numbers, but having used both drugs, there can be little doubt that quinidine was better tolerated than amiodarone. I also found flecainide well tolerated.

Individual patients respond in individual ways, so be prepared to assess their effects early, and to change unsuitable medication soon. Many patients will persevere with drugs, even though the side-effects are less than tolerable. Don't let this happen with cardiac suppressants.

Quinidine affects cardiac conducting tissue in ways now emulated by other suppressant drugs. The effects are:

Lengthening of the refractory period, making re-entry less likely.
Reduction of parasympathetic (vagal) activity (acetylcholine reduces refractory period);
Depression of conductivity.
Depression of excitability (fewer ectopics). Perhaps reducing myocyte, T-tube spontaneous Ca^{2+} influx sparking. Slowing of cellular calcium transport.
Inhibition the myocyte fast inward sodium current.
Slowing of potassium efflux.

These effects can cause QRS and QTc prolongation, with little effect on the PR interval. With it, one can revert AF and reduce the frequency of its recurrence.

I found quinidine well tolerated, except for minor GI side-effects. Over the many decades I used it, I never saw one case of allergy, or drug-induced SjÕrgrens or DLE, but these can happen. Heart failure lessens its effects, and liver impairment, hypokalaemia and hypomagnesemia exacerbate them. It was never to be used in pregnancy or during breastfeeding.

The long-term use of cardiac suppressants requires ECG monitoring, and measures of hepatic and platelet function at regular intervals. Regularly look for various forms of heart block, prolonged QTc and *torsade de pointes* after prescribing them.

Like quinidine, procainamide is no longer listed in the BNF. I assume that too many side-effects became registered against both. Neither were 'prescribe and leave' drugs; both needed active management over short periods; with time restraints, not always possible in the NHS. Doctors working in countries outside the UK, may wish to know that both remain effective cardiac suppressants, although my first choice would now be to use flecainide.

Disopyramide has Class 1a conducting system effects, but caution exists about its muscurinic effects. Avoid it in angle closure glaucoma, prostatic hypertrophy and myasthenia. It is as effective as quinidine, procainamide and flecainide.

Class 1b: Lidocaine and Mexiletine

Lidocaine (formerly called lignocaine) has many uses. Apart from being a class 1b cardiac suppressant, it is an anaesthetic for injection or for topical use. It is used in resuscitation for controlling ventricular tachycardias following cardiac infarction. In theory, it has the same conducting and anti-arrhythmic effects as Class 1a drugs, and sharing all the same cautions and contraindications.

Mexiletine is no longer listed in the BNF. It is a sodium channel blocker, useful for treating long QT type-3. It has had other uses: muscle pain and peripheral neuropathy, but with side-effects that have prohibited its use.

Class 1c: Flecainide

This is especially useful for preventing and halting reciprocating atrial tachycardias (WPW, etc.). I used it on occasions to 'revert' recent onset AF ('convert' implies a change of faith!). Its side-effects are worth noting, but are infrequent. Like all other cardiac suppressants, it needs active management.

Amiodarone

For many cardiologists, this Class 3 drug has replaced many former cardiac suppressant drugs for VT and VF. Unlike some of the former drugs, it is not recommended for treating supraventricular arrhythmias. I found it had frequent side-effects and was unpopular with many patients, although undoubtedly effective. Its intravenous use can replace lidocaine in resuscitation situations.

Dronedarone has a specific use: to prevent AF recurrence after DC reversion. All the same precautions apply, especially those related to heart block and a long QTc.

Adenosine

It can reverse supra-ventricular tachycardias, including those propagated by extra pathways (like WPW). One can use it in intravenous boluses (6-12mgs). High doses can cause AV block, and worsen existing AV block. Avoid it in asthmatics, in those with COPD, and those with long QT syndrome. It can cause systemic hypotension.

It is a naturally occurring, pulmonary and coronary artery vasodilator, released by exercise and cardiac ischaemia. It can attenuate contractility, ischaemia and infarct size. By inhibiting adenylyl cyclase, it slows sino-atrial and AV nodal rhythmicity and conduction (A1 receptors). A2 receptors increase coronary blood flow by activating the same enzyme in the coronary endothelium and smooth muscle,. Adenosine reduces several deleterious effects of ischaemia (Hori, M., Kitakaze, M. (1991) *Adenosine, the heart and coronary circulation.* Hypertension; 18 (5):565-74).

Electrical DC Reversion

Electrical DC reversion should be considered for patients with recent onset AF (under 48hrs) and those with deteriorating cardiac function. Those with deteriorating haemodynamic function need prior anticoagulation (for three weeks before and one month afterwards) to prevent stroke. Some will need rate control drugs. (See: B.J.C. Staff. *Direct current cardioversion and thromboprophylaxis in atrial fibrillation.* Br J Cardiol;23(suppl 2):S1–S12).

DC reversion is a component part of cardio-pulmonary resuscitation. It is a direct current shock of sufficient magnitude, delivered through the chest, rendering the heart refractory to depolarization. During DC reversion, all cardiac electrical activity stops, including re-entry. Thereafter, if successful, the dominant pacemaker, (possibly the sinoatrial node) should resume control of the heart rhythm.

For AF and tachyarrhythmias, other than ventricular fibrillation (VF) and ventricular tachycardia (VT), DC shocks must synchronise with the QRS complex (DC cardioversion). The aim is to avoid the vulnerable period (near the peak of the T wave). At this time, VF is more easily induced.

DC current gets delivered between two electrodes in a monophasic or biphasic way. In biphasic devices, the current reverses direction part way through the shock waveform. The biphasic device requires lower energy and results in higher rates of return of spontaneous circulation (ROSC). Survival outcomes are similar with both devices (Schneider, T., Martens, P.R., et al. (2000). *Multicenter, randomized, controlled trial of 150-J biphasic shocks compared with 200-J to 360-J monophasic shocks in the resuscitation of out-of-hospital cardiac arrest victims.* Circulation102:1780–1787).

Most manual and automated external defibrillators (AEDs) are now biphasic. They more efficiency restore sinus rhythm and are small (making them portable).

Various energy levels delivered by DC reversion need consideration:

Elective DC Reversion

Patients should have fasted 6 to 8 hours before, to avoid pulmonary aspiration. The procedure is frightening and painful, so brief general anaesthesia is used (one regime is: iv analgesia and sedation with fentanyl 1 mcg/kg, then midazolam 1 to 2 mgs every two minutes to a maximum of 5 mgs. Using propofol would be my preference, given its rapid recovery and lack of after-effects.

The electrodes (pads or paddles) used for cardioversion can be placed:

The first, antero-posteriorly along the left sternal border over the 3rd and 4th intercostal spaces, and the second in the left infra-scapular region.

One between the clavicle and the 2nd intercostal space along the right sternal border, and a second over the 5th and 6th intercostal spaces at the apex of the heart.

Once synchronization with the QRS complex is confirmed, a shock can be given at an appropriate energy level. As a guide to the energy delivered:

In VF, use 120 to 200 joules for biphasic devices (or according to manufacturer specification), although many doctors use the maximum device output.

The equivalent is to use 360 joules for monophasic devices.

For **synchronized cardioversion** of AF, the energy level for the first shock using a biphasic device is usually 100 to 200 joules; 200 joules if using a monophasic device. Subsequent shocks should be at the same or higher energy level for both biphasic and monophasic devices.

DC cardioversion-defibrillation can also be applied directly to the heart during a thoracotomy or through an intracardiac electrode catheter. Much lower energy levels are then required.

Indications for implanting a defibrillating device

According to American College of Cardiology/American Heart Rhythm Society (ACC/AHRS) 2008, these are:

Class I: For patients who have:

- Survived cardiac arrest due to hemodynamically instability: sustained ventricular tachycardia (VT) or VF, after evaluation and exclusion of reversible causes.

- Structural heart disease, with spontaneous sustained VT; hemodynamically stable or not.

- Unexplained syncope, with hemodynamically significant, sustained VT or ventricular fibrillation (VF), induced during an electrophysiology study (EPS).

- **Ischemic cardiomyopathy** patients with a left-ventricular ejection fraction (LVEF) < 35% caused by a prior MI; more than 40-days post-MI, and New York Heart Association (NYHA) functional class II or III.

- **Non-ischemic dilated cardiomyopathy** (DCM) with a LVEF of 35% or less, and NYHA functional class II or III.

- **LV dysfunction** because of prior MI; more than 40-days post-MI, an LVEF of < 30%, and NYHA functional class I.

- **Non-sustained VT** because of prior MI; LVEF < 40%, and sustained VT or inducible VF during EPS.

Class IIa: For patients who have:

- Unexplained syncope; non-ischemic DCM, and significant LV dysfunction.

- Sustained VT and normal, or near-normal ventricular function.

- Hypertrophic cardiomyopathy with one or more major risk factors for sudden cardiac death.

- Arrhythmogenic right ventricular dysplasia (ARVD/C) and one or more risk factors for sudden cardiac death

- Long QT syndrome, with syncope or VT.

- Brugada syndrome patients with syncope or documented VT, not resulting in cardiac arrest.

- Catecholaminergic polymorphic VT patients, with syncope or documented sustained VT, while receiving beta-blockers

- Cardiac sarcoidosis, Chagas' disease, and giant cell myocarditis.

It can also considered for non-hospitalised patients awaiting transplantation.

Class IIb: For patients who have:

- **Non-ischemic heart disease** patients with a LVEF < 35% or less, and NYHA functional class I.

- **Long-QT syndrome** patients with risk factors for sudden cardiac death.

- A **risk of sudden cardiac death**, with advanced structural heart disease, but without a defining cause defined by invasive or non-invasive methods.

- **Familial cardiomyopathy** associated with sudden death.

- **LV non-compaction**.

See: Bernier R, Al-Shehri M, et al., (2018). *A Population-Based Study of Adherence to Guideline Recommendations and Appropriate-Use*

Criteria for Implantable Cardioverter Defibrillators. Can J Cardiol.; 34(12):1677-1681. PubMed .

Specific training on various delivery devices is essential. Each has its own specific design features. Some are simple enough to be operated by untrained members of the public on football pitches and in other public arenas. The only requirement is that they can read the brief instructions. Providing them in public places has saved many lives.

All cardiac staff attending likely cases of VF and VT, must pass an annual certification course (Life Support and Advanced Life Support), since CPR involves much more than the delivery of a DC shock. An efficient team needs co-ordination enough to deliver effective cardiac massage; be able to maintain an open airway, and an ability to gain iv access for the delivery of drugs.

During the DC reversion phase, it is important to safeguard both the patient (from burns, tongue biting, etc.), and all those in attendance. They must be distanced from any risk of electric shock.

Ablation

Ablation techniques aim to interrupt re-entry pathways: those that cause AF, and extra-conduction pathways that cause tachycardia. There are three techniques in use: those using heat, cold, and pulsed-field energy. Radiofrequency ablation generates heat to damage conduction pathways; cryo-ablation uses cold, delivered directly during cardiac surgery or indirectly via a cardiac catheter.

In the 1990s, radiofrequency (RF) ablation replaced direct current. RF energy is an alternating current (350 kHz to 700 kHz; usually 500 kHz from commercially available RF generators) delivered in a continuous, unmodulated sinusoidal manner to create thermal injury.

Consider left atrial ablation if constant AF remains symptomatic, despite drug treatment. For paroxysmal AF producing unacceptable symptoms, despite drug treatment (or difficult to treat rate dependent ventricular dysfunction), one can use cardiac pacing combined with AV nodal ablation.

Certain parts of the atrium, best suit the cessation of AF and its return. A meta-analysis of surgical cases showed that left atrial ablation gave the best results, although bilateral atria and pulmonary vein ablation are other contenders. In their review of AF cases, Hanaffy, D.A., Erdianto, W.P (2023) concluded that left atrial ablation alone restored sinus rhythm in 76% of cases, and pulmonary ablation worked in 67%.

The need for re-operation and pacemaker implantation, was less for atrial ablation than for pulmonary vein ablation. Death, pulmonary vein stenosis, oesophageal perforation, stroke, phrenic nerve injury, and vascular access complications can occur, but are infrequent. It is important to know which patients will benefit best from surgical or catheter ablation, rather than continue with medical therapy.

Catheter ablation is now a leading intervention in the management of AF and other tachyarrhythmias. The procedure is complicated by needing intracardiac mapping (to detect the conducting pathways and sources of re-entry via the AV node or not), together with stimulation techniques.

The indications for its use are:

- Symptomatic supraventricular tachycardias (SVT) such as: atrioventricular re-entrant tachycardia (AVRT), atrioventricular nodal re-entrant tachycardia (AVNRT), unifocal atrial tachycardia and atrial flutter.

- AF with lifestyle-impairing symptoms, and failed antiarrhythmic agents.

- Symptomatic idiopathic ventricular tachycardia (VT).

For VT caused by structural heart disease, catheter ablation is reserved for those with failed drug-therapy, and where there have been too frequent implantable cardioverter-defibrillator (ICD) discharges.

Endocardial Mapping System

Endocardial mapping technologies provide high-resolution activation or wave front propagation maps, acquired from locating electrodes in multiple locations. Besides correlating local electrograms to three-dimensional cardiac structures, newer mapping techniques reduce the radiation exposure to both patient and cardiologist.

The most widely used electro-anatomic mapping system locates the mapping and ablation catheters in a magnetic field. Three coils beneath the patient, generate ultra-low magnetic fields that temporally and spatially, code the areas within the patient. With a magnetic field sensor in their tip, catheters can be localised in three dimensions and intracardiac electrograms recorded.

Another technology can map the electrical fields between opposing pairs of patch electrodes on the patient's chest. With six patches placed on the body to create the orthogonal axes (x,y,z), the heart is at the centre. With a transthoracic electrical field created by each pair of opposing patch electrodes, a mapping catheter can deliver a signal for processing.

Hypertension and Heart Failure Control

The Renin-Angiotensin-Aldosterone System (RAAS)

In order to appreciate the benefits of ACE inhibitors and ARBs, one must know about the Renin-Angiotensin-Aldosterone System (RAAS); a hormonal pathway, critical for regulating blood volume, electrolyte balance, and systemic vascular resistance.

The system has circulating and tissue bound components (tRAAS is important in obesity). The main features are:

Renin converts angiotensinogen to angiotensin.
Angiotensin 1 is converted to **Angiotensin II** by **Angiotensin Converting Enzyme** (ACE).
Angiotensin II, ACTH and extracellular potassium levels, regulate aldosterone synthesised in the zona glomerulosa of the adrenal cortex. Mineralocorticoid receptors (MR) mediate the re-absorption of sodium from the renal collecting tubules (they are also found in the colon, heart and brain).

The following effects of angiotensin II are of clinical importance:

The **AT_1-receptor** induces **vasoconstriction**, is anti-natriuretic, profibrotic, and inflammatory. It aids cell proliferation and induces oxidative injury.

The **AT_2-receptor** (expressed in the heart, kidney, adrenal glands, and brain) causes **vasodilatation**, is natriuretic, anti-fibrotic, anti-inflammatory, apoptotic and anti-oxidant.

Renin, produced from prorenin, increases blood pressure. While mostly found in the juxtaglomerular cells, it occurs in blood vessels, the adrenal gland, placenta, uterus, retina, testes and submandibular glands. Conditions which decrease renal perfusion and reduce tubular sodium content, cause renin release into the bloodstream.

Prorenin and renin can affect disease pathogenesis that is receptor-mediated.Both molecules can activate mitogen-activated protein (MAP) kinases (p38 and P42/44) that can alter the actin filaments of cardiomyocytes. This might explain their role in ventricular hypertrophy and cardiomyopathy. Blocking them with an ARB or ACE inhibitor, can inhibit the development of hypertrophy in hypertension (a significant risk factor for stroke)(Saris, J.J., 't Hoen, P.A., et al. *Prorenin induces intracellular signalling in cardiomyocytes independently of angiotensin II.* Hypertension. 2006; 48:564–571).

The **primitive receptor (P)RR**, found originally in nematode worms, is involved in the tissue activation of the RAAS system. The receptor plays a role in many physiological functions, including cell proliferation and differentiation, acid-base balance, autophagy, energy metabolism, blood pressure regulation, and water balance. With renin attached to (P)RR, there is a fourfold increase in the catalytic conversion of angiotensinogen to angiotensin I (Ang I).

Other, newer components of the RAAS system, are now known: **Angiotensin A** (with an affinity for the AT_2 receptors, reduces the deleterious effects of Ang II); **alantensins and alamandine** (Ala), can lead to a reduction in BP through vasodilatation.

For more detailed information see: Kanugala, A.K., Kaur, J. et al. (2023)*Renin-Angiotensin System: Updated Understanding and Role in Physiological and Pathophysiological States.* Cureus; 15(6): e40725.

ACE Inhibitors

Angiotensin-Converting Enzyme inhibitors act by:

- Inhibiting angiotensin II production. Angiotensin II normally decreases arteriolar NO production, and its inhibition will cause extra nitric oxide to drive vasodilation and the lowering of blood pressure.

- By inhibiting bradykinin breakdown. Bradykinin releases arteriolar nitric oxide (NO), allowing vasodilatation.

Many doctors now, have personal experience of prescribing lisinopril, enalapril and ramipril, for heart failure and hypertension management. There are several other ACE inhibitors to choose from, each the product of different pharmaceutical companies. All work favourably in heart failure and hypertension, but with one common side-effect—a troubling dry cough. I have also seen angioedema, and hypotension when the patient had taken long-term diuretics. Refer to the BNF for other side-effects. Pregnancy is a contraindication. In renal failure, they increase the risk of hyperkalaemia. ACE inhibitors are notably less effective in Afro-Caribbean patients.

The relative efficacy of each ACE inhibitor needs consideration. One meta-analysis conducted by Sun et al.(2016) concluded:

Enalapril might be the best option when considering factors such as increased ejection fraction, stroke volume, and decreased mean arterial pressure.
Enalapril was associated with the highest incidence of cough, gastrointestinal discomfort, and greater deterioration in renal function.
Trandolapril ranked first in reducing systolic and diastolic blood pressure.
Ramipril was associated with the lowest incidence of all-cause mortality.
Lisinopril was the least effective in lowering systolic and diastolic blood pressure and was associated with the highest incidence of all-cause mortality.

Sun, W., Zhang, H., et al (2016). *Comparison of the Efficacy and Safety of Different ACE Inhibitors in Patients With Chronic Heart Failure: A PRISMA-Compliant Network Meta-Analysis.* Medicine (Baltimore)Feb; 95(6):e2554. doi: 10.1097/MD.0000000000002554.

Angiotensin II Receptor Blockers (ARBs)

Like ACE inhibitors, these are effective in hypertension, heart failure and renal disease. Afro-Caribbean patients may not always respond well to ARBs, especially if they have hypertension related LVH (see BNF).

Angioedema and hyperkalaemia are the concerning, albeit infrequent side-effects. I had personal experience of using **candesartan, losartan** (short half-life: 2hrs), **irbesartan** (longer half-life: 11-15hrs) and valsartan. Their effect on blood pressure can be weak for hypertensive patients, but importantly, they reduce stroke rates and all-cause mortality. The LIFE trial showed fewer strokes (25% fewer) with an ARB, than with atenolol. The trial conclusion was: 'Losartan prevents more cardiovascular morbidity and death than atenolol for a similar reduction in blood pressure and is better tolerated. Losartan seems to confer benefits beyond reduction in blood pressure.' (Dahlöf, B., Devereux, R.B., et al. (2002) *Cardiovascular morbidity and mortality in the Losartan Intervention For Endpoint reduction in hypertension study (LIFE): a randomised trial against atenolol.* Lancet; 359(9311): 995-1003).

ARBs have fewer side effects than ACE inhibitors; both are useful in heart failure, hypertension and renal disease. It is no longer considered efficacious to combine ACE inhibitors with ARBs.

In treating hypertension, only modest falls in BP can be expected (7 -15 mm Hg) with an ARB. Their combination with a diuretic or amlodipine, however, can double the effect on blood pressure.

In the CLAIM studies: '**candesartan** cilexetil at doses of 16 and 32 mg/day were found to be more potent than losartan at doses of 50 and 100 mg/day, respectively. Candesartan 16mg/day also reduced clinic BP to a greater extent than losartan 100 mg/day. In a trial of **olmesartan** medoxomil 20 mg/day, ambulatory systolic BPs were lowered more than with **valsartan** 80 mg/day and losartan 50 mg/day and similarly to **irbesartan** 150 mg/day.' (Oparil S, Williams D, et al. (2001). *Comparative efficacy of* *olmesartan, losartan, valsartan, and irbesartan in the control of essential hypertension. J Clin. Hypertension (Greenwich)*; 3 :283–91.

Calcium Channel Blockers

These drugs block Ca^{2+} ion influx into vascular smooth and cardiac muscle cells.

In the 1970s, they were called calcium channel antagonists, and used to treat hypertension and stable angina. Unfortunately, they became major contributors to cardiovascular, drug-related fatalities, although over many decades, none of my patients came to suffer this fate (a cohort too small, perhaps).

Calcium channel blockers block binding to the L-type, long-acting voltage-gated calcium channels, in both the heart and vascular smooth muscle. Their metabolism is mainly hepatic.

There are two functional categories: **non-dihydropyridines and dihydropyridines**.

Non-dihydropyridines slow SA and AV node conduction, and reduce myocardial contractility. The latter helps to treat hypertension, reduce oxygen demand in coronary artery disease, and slow tachycardias. The drugs in common use are **diltiazem and verapamil**. I used both repeatedly to slow fast AF etc., but never saw them induce heart failure. They are, nevertheless, not recommended in heart failure. Diltiazem has less effect on contractility than verapamil and is more commonly used for those with angina. Because both are negatively ionotropic, their use with beta-blockade is best avoided. Use them to replace beta-blockade.

Dihydropyridines are peripheral vasodilators. This property renders them useful in hypertension and angina, especially if vasospasm contributes. **Amlodipine, felodipine, and nifedipine** are those I used most, although my preference was always to revascularise patients with angina if possible. I only persisted with nitrates or calcium blockers, if revascularisation was not possible. **Nicardipine** has a specific use in malignant hypertension and arterial dissection associated with hypertension. Use it intravenously, to lower the BP in an emergency.

Widely used amlodipine is cheap and effective in hypertension. It has one common side effect—ankle oedema. Oedema is so common, I would always ask patients with ankle swelling, one pre-emptive question. Were they taking amlodipine? Many were. Felodipine seemed more acceptable, but more expensive. Neither reduce myocardial contractility and are useful in heart failure.

For both classes of calcium blocker, doses need to be reduced when hepatic impairment is present. They are safe in pregnancy and during breastfeeding.

Angina and Nitrate Use

Nitrates are a long established, first-line treatment for angina.

Both **nitrites** (organic and inorganic) and **nitrates** (organic), donate nitric oxide to the vascular endothelium (NO is an endothelium-derived relaxing factor). They cause the vasodilatation of arterioles, venules and capillaries. They increase coronary blood flow in angina, especially if vasospasm is a factor (sometimes after angioplasty).

Nitrates were first discovered in 1847 as a treatment for chest pain. For patients with stable and predictable angina, long-acting nitrate can increase exercise tolerance. For patients with acute anginal pain, short-acting nitrates can offer symptom relief. Nitrates are used for angina pectoris and acute coronary syndrome, but are no longer considered necessary for treating arterial hypertension and heart failure.

The sequence of how nitrates affect the cardiovascular system (the NO-cGMP-cGK-I signalling pathway) is:

- Nitrates convert to NO, and NO activates soluble guanylate cyclase.

- Increased levels of cGMP and cGMP-dependent protein kinase I (cGK-I) result.

- cGK-I inhibits inositol-1,4,5-trisphosphate-dependent calcium release.

- Decreased intracellular calcium levels result, and that

- Inhibits myosin light chain kinase.

- Unphosphorylated myosin light chains causes myosin heads to detach from actin, causing smooth muscle relaxation.

See: Münzel T, Steven S, Daiber A. (2014). *Organic nitrates: update on mechanisms underlying vasodilation, tolerance and endothelial dysfunction.* Vascul. Pharmacol.; 63(3):105-13.

Whether nitrates can actually dilate atherosclerotic coronary arteries is debatable. Dilating drugs are mostly ineffective when used in peripheral vascular disease (caused by atherosclerosis). Dilating venules could be an important factor. Dilated venules lower LV preload and reduce oxygen consumption. Both should improve angina.

The NO-cGMP-cGK-I signalling pathway, not only causes arterial vasorelaxation but also platelet disaggregation and prevention of platelet adhesion. This can be important in acute coronary syndrome.

The nitrates in use are:

Nitroglycerin (GTN) – for angina pectoris (treatment/prophylaxis) and acute coronary syndrome. Many patients with angina of effort, carry GTN with them in anticipation of getting angina on effort.

Isosorbide mononitrate (ISMN);

Isosorbide dinitrate (ISDN)--both for the treatment and prophylaxis of chronic angina pectoris.

Tolerance to nitrates and nitrites can develop. Patients notice flushing, headache and palpitation with many preparations. All can cause hypotension and syncope. Do not use them for patients with HCM or those taking PDE inhibitors like sildenafil. Many patients who use glyceryl trinitrate, will store their tablets and use them only after six months when their potency has decreased.

Amyl nitrate is now a drug of abuse.

In the cardiac exercise testing environment, GTN must be available for patients who might develop angina on exercise.

Administration can take several forms, the most usual of which is sub-lingual GTN. It is also available as a skin patch, ointment, capsule, sub-lingual spray or intravenous infusion. The intravenous form is useful for hypertensive emergencies.

Pulmonary Hypertension

Iloprost is a manufactured prostaglandin, capable of relieving pulmonary artery vasoconstriction. Administered using a specially designed inhalation (I-neb) delivery system, it can cause coughing and symptoms associated with hypotension.

Others drugs in use are **endothelin receptor antagonists.** Selective inhibition of the endothelin receptor type A (ETA), causes pulmonary arterial vasodilatation by inhibiting the intracellular pathways that lead to vasoconstriction. For this purpose **Ambrisentan, Bosentan and Macitentan** are used by those specialised in the field. Anaemia and hepatic dysfunction are among the commoner side-effects.

Atheroma Prevention
'Statins' (HMG-CoA Reductase Inhibitors)

Some 'statins' occur naturally. Lovastatin, can be isolated from Aspergillus, and non-therapeutic quantities of simvastatin are present in red yeast rice. The therapeutic importance of these drugs is thought related only to the lowering of total blood cholesterol and LDL (statistically supported risk factors). They do this by affecting liver metabolism (HMG-CoA reductase inhibition). There are other aspects to their value in individuals.

'Statins' can slow or stop, the progress of atheroma. Over a twenty-year period, I observed this in the carotid arteries of hundreds of patients (unpublished observations). They also stabilise the intima of vulnerable plaque, thereby reducing the risk of plaque rupture and acute infarction. They can also reduce plaque thrombus formation. They were once reported to reduce plaque size by 1% per annum. If true, it would take one hundred years to remove occlusive lipid rich plaques. Most atheromatous plaques are fibrous or include calcium compounds; which this medication cannot affect.

In my experience, side-effect rates vary with each statin. Thirty to 50% of all patients will experience muscle pains. Here are some of my other observations:

- Atorvastatin and rosuvastatin less frequently cause muscle pains than simvastatin.

- The minimum dose needed to stop the progression of atheroma (from my own observations) is 80mgs daily for simvastatin, and 40mgs daily for atorvastatin.

- I treated carotid (and coronary) atheroma, even when the patient had normal lipid levels. In my cohort of individuals, neither total cholesterol nor LDL, correlated significantly with the presence of carotid or coronary atheroma. HDL, however, did show a significant inverse relationship.

- I treated hypercholesterolaemia with a statin, only when the patient had had a significant past cardiovascular event, when they had detectable carotid atheroma (often the first artery to develop it), and there was or a significant family history of cardiovascular events (preferably, they had both risks).

- Some of my patients opted to take a statin, only after developing observable atheroma; many refused to take them otherwise. Some patients refused to accept that the proven, general risk of CVS events associated with hypercholesterolaemia, applied to them (or at least could be proven to apply to them).

- I scanned the carotid arteries of patients with atheroma every two years and doubled their dose of 'statin' when no slowing of atheroma progression was apparent.

- My method of managing atheroma follows no current guideline. I have yet to publish my data to justify my management policy. I offer this as 'insider information' until peer reviewed.

Of my four thousand patients, managed over twenty years in the above way, only two were diagnosed with cardiac infarction. The sub-clinical ones, will have escaped detection.

In 2023, there were 44 million people over the age of 30-years in the UK. Heart attack admissions in the UK were approximately 100,000 per annum (0.23% of the over 30-year-old population). I had 4,000 under review, so I could have expected to see 9.2 cases of cardiac infarction per annum. In fact, I saw only two cases in twenty years. Further deductions would be entirely speculative, but a large reduction in morbidity might be possible if my management scheme for atheroma and CHD prevention is proven correct. The biggest confounding factor is patient selection; all my private patients elected to come and were drawn from a socio-economic group with the fewest CVS problems.

My facilities for patents were more labour-intensive and costly than any provided in the UK by the NHS. But, if we could manage atherosclerosis better, would any government authority sanction it?

NICE recommends that those with a < 10%, 10-year risk of a CVS event (heart attack or stroke, using Qrisk3) should have the choice of taking 'statin' treatment. Given a lack of diagnostic sensitivity and specificity for the Qrisk3 algorithm predicting atheroma, many patients with it, but no hyperlipidaemia or hypertension, will not be offered specific prophylactic treatment in the UK.

CHAPTER SEVENTEEN

Surgical Interventions

INVASIVE CARDIOLOGICAL PROCEDURES

A Short History

The splitting of a stenosed mitral valve was the first cardiac surgical intervention to benefit patients. Such interventions have now progressed, through open heart surgery to catheter techniques that can correct or replace defective valves (TAVI). The development of coronary bypass surgery, coronary artery angioplasty and stenting, have all provided real advances for patients.

Other surgical developments, like the development of pacemakers and defibrillators, heart transplantation, the development of extracorporeal pumps used during surgery, valve and arterial protheses, and mechanical hearts have all helped patients survive.

In Ugo Filippo Tesler's book: *A History of Cardiac Surgery: An Adventurous Voyage from Antiquity to the Artificial Heart*. (2020). Cambridge Scholars Publishing; ISBN: 978-1-527-54480-2, there is an excellent account of the developmental history of cardiac surgery.

The following highlights have marked the way. In 1912, the Nobel Prize for Physiology and Medicine, was awarded to Alexis Carrel, 'in recognition of his work on vascular suture and the transplantation of blood vessels and organs.' Carrel and Charles Lindbergh tried to develop a pump that would temporarily replace the heart and enable open cardiac repair.

The earliest experiences with closed cardiac surgery for treating mitral stenosis, were initially unsuccessful and often fatal, but evolved further. Early pioneers like Charles Bailey, Dwight Harken, and Russell Brock, developed better techniques for closed mitral commissurotomy.

John Gibbon pioneered the extracorporeal pump and oxygenator, to enable cardiopulmonary bypass. Twenty-five-years would pass, before the first successful closure of an atrial septal defect in 1953. Open heart, surgical developments followed. The first attempts at open heart surgery were those of John Kirklin at Mayo Clinic, Lillehei in Minnesota, and Cooley in Houston.

Lillehei used cross-circulation between a patient and a living related donor in 1954; this enabled intracardiac repairs in children. DeBakey designed the roller pump; Ake Senning, Viking Olov Björk, and later Earle Kay and Frederick Cross, invented the rotating disc oxygenator. Forest Dodrill, C. Walton Lillehei, and Richard DeWall developed the bubble oxygenator.

Complete heart block became a fatal complication of congenital heart defect repairs, so at Lillehei's request, Earl Bakken in 1957, developed an external pacemaker. Bakken later founded the company Medtronic, to manufacture pacemakers.

Collaboration between surgeon Albert Starr and engineer Lowell Edwards, led to the development of the first prosthetic heart valves (Starr-Edwards valves). Alain Carpentier, Magdi Yacoub, Tirone David, and Donald Ross, pioneered biological valves. They developed valve repair, aortic replacement, homograft valves and porcine bio-prosthesis implantation.

Charles Bailey, performed the first surgical procedures for coronary artery disease in 1956, using direct coronary endarterectomy. Two years later, Longmire repaired a right coronary artery using an internal mammary artery, thus performing the first coronary artery bypass graft (CABG). Vasilii Kolesov in Russia developed this further, and George Green in New York, made internal mammary artery bypass the preferred treatment for coronary disease.

In 1958, Mason Sones at the Cleveland Clinic, serendipitously injected contrast agent into a patients coronary artery, an act that lead to the development of diagnostic coronary angiography.

In 1967, René Favaloro developed the technique of CABG with saphenous vein grafts, attached to the aorta.

Percutaneous transluminal coronary angioplasty, including the first balloon angioplasty, was first performed by Andreas Gruentzig in 1977. This led to the development of coronary artery stents. The SYNTAX trial in 2009, compared the outcomes of CABG with those of percutaneous coronary intervention with stents. (Serruys, P.W., Morice, Marie-Claude, et al. 2009). Their conclusion was: *'CABG remains the standard of care for patients with three-vessel or left main coronary artery disease, since the use of CABG, as compared with PCI, resulted in lower rates of the combined endpoint of major adverse cardiac or cerebrovascular events at 1 year'.*

DeBakey and Cooley first reported using aortic homografts for abdominal aortic aneurysm repair in 1952. By 1955, they had performed 245 such repairs. DeBakey formulated a classification system for aortic dissections. He made the first Dacron grafts for arterial replacement on his wife's sewing machine. Cooley performed the first operations for repairing thoracic aortic aneurysms. Crawford devised a technique for repairing them, and a system of classification for thoracoabdominal aneurysms. It was Randall Griepp who developed hypothermic cerebral protection.

By 1967, Shumway and Lower had developed the steps needed for cardiac transplantation. In December 1967, Christiaan Barnard in South Africa, performed the first successful cardiac transplant. Although the first patient Louis Washkanskysurvived only eighteen days, the operation was a success and proved the concept. Early enthusiasm, however, turned to disappointment as subsequent transplant recipients died of infection or rejection. Shumway's group at Stanford, revived the technique by advancing the understanding of immunosuppression and transplant rejection. Cyclosporine for immunosuppression and endomyocardial biopsy helped improved patient survival.

As reported by The Heart of Cape Town Museum, Chris Barnard once said that getting old is one of life's greatest tragedies. Particularly hard, indeed, for a man 'who lived a life worth living'.

In Barnard's last interview before his death, he told Time magazine: "The heart transplant wasn't such a big thing surgically. The technique was a basic one. The point is I was prepared to take the risk. My philosophy is that the biggest risk in life is not to take a risk.

In 1964, the United States government funded a ten-year program in artificial heart research. DeBakey (at Baylor), Willem Kolff (at the Cleveland Clinic), and Kantrowitz (at Maimonides Hospital), became the lead developers. Vladimir Demikhov in Russia had already used a mechanical device in animals in 1937, but artificial hearts were first implanted in dogs in 1957 (Kolff and Akutsu). The first implantation of an artificial heart in a human by Cooley and Liotta, was in 1969. Further developments led Akutsu to make an artificial heart capable of keeping an animal alive, and to the Jarvik-7, a plastic and Velcro pump.

Axial-flow pumps for circulatory support and left ventricular assist devices (LVADs) were next to be developed, but a fully functioning totally artificial heart has yet to be made.

Cardiologists deal with many patients who have undergone CABG, those with implanted pacemakers and defibrillators, and those who have undergone angioplasty and coronary artery stenting. Only invasive cardiologists perform cardiac catheterisation, electrophysiology, angioplasty, pacemaker implantation and TAVI, many specialising in specific procedures. None of the latter are major surgical procedures, but they are all undertaken in an operating theatre, with X-ray fluoroscopy and angiography, physiological equipment to record ECGs, arterial and intracardiac pressures, and intracardiac electrograms. All require specific expertise. Dedicated textbooks describe each of these procedures, and the detail of every major surgical procedure.

In what follows I have focussed on only three common cardiac interventions: CABG, coronary angioplasty and stenting, and pacemaker implantation and function.

Coronary Vein By-Pass Grafting

Patients mostly offered CABG have widespread, three-vessel coronary disease and proximal left coronary stenoses. Surgeons use either saphenous veins from the legs, a left internal mammary artery (LIMA), or a radial artery graft. When the leg veins have been taken previously or they are unsatisfactory, a LIMA will need to be used. The surgeon will sew the proximal end of the graft into the aorta and the distal end into a coronary artery, beyond any coronary obstruction(s). Approximately 400,000 CABG procedures are undertaken each year (2023). (Bachar, J.B., Manna, B. (2023).*Coronary Artery Bypass Graft*. StatPearls. National Library of Medicine).

Patients who need CABG will usually have a strong atherosclerotic tendency, and their liability to develop more atheroma post-operatively will remain. This applies to both their native arteries and the bypass grafts. Regular follow-up is expedient to check both.

My practice was to perform an exercise test, eight to twelve weeks after any revascularising intervention, and thereafter every year. Some doctors may have to await the recurrence of ischaemic symptoms before they can review their patients.

Patients often complain more about the post-operative vein harvesting area of their legs, than about their chest wound. The healing of leg wounds is often slow and troublesome. Internal mammary artery harvesting is still used when the leg veins have been taken or are unsatisfactory.

Saphenous vein graft (SVG) failure is most common within thirty days of operation. Several factors are important: vein size, excessive graft length, slow flow, hypercoagulability and thrombosis.

A meta-analysis showed that after an average of one year, SVG and arterial graft failure, occurred in 33.7% of patients (16.6% of the grafts)(Guadina, M., Sander, S., et al. 2023). They found that graft failure was associated with an increased risk of myocardial infarction (8.0% in patients with graft failure, versus 1.7% in patients without graft failure). Arterial grafts, usually remain open longer, with patency rates exceeding 90% at ten years.

The other complications of CABG are stroke (1-2%), wound infection, renal failure, postoperative atrial fibrillation and death.

Sternal wound infections occur in about 1%. They relate to obesity, diabetes, chronic obstructive pulmonary disease (COPD), and the prolonged duration of surgery.

Postoperative renal dysfunction rates are 2% to 3%, with 1% requiring dialysis. Atrial fibrillation within the first five days of CABG is relatively common (20% to 50%), and associated with a higher risk of embolic stroke and increased mortality. Preoperative treatment with beta-blockade and / or amiodarone, reduces the postoperative occurrence of atrial fibrillation.

Percutaneous Coronary Intervention (PCI)

The atheroma in coronary arteries is either hard from calcification, or soft like wet sand. Once displaced by a balloon catheter (angioplasty) it will usually stay put for a while. To make sure it remains open, one can introduce a stent. These resemble chicken wire mesh, but made of stainless steel or cobalt chromium. Those impregnated with drugs to prevent clotting (drug-eluting stents) are the latest.

Schatz and co-workers developed the first stent and obtained FDA approval in 1987 (Palmaz-Schatz®; Johnson & Johnson, New Brunswick, NJ, USA). This was the first balloon-expandable, stainless-steel, tubular stent, widely used throughout the 1990s.

Cardiac catheterisation and coronary arteriography are now routine procedures, requiring arterial and venous access through the groin (femoral vessels) or wrist (radial artery). We guide catheters to the coronary arteries using fluoroscopy. The position and degree of any stenoses within the arteries, can be imaged from different angles, as radio-opaque dye is injected. One can replace this catheter with one that has a balloon and stent near its tip. All catheters come in many shape and sizes, and it takes some art and considerable experience, to choose one that will work best.

The outcome of angioplasty alone, can be poorer than angioplasty followed by stent insertion (Grines, C.L., Cox, D.A., et al. (1999). *Coronary Angioplasty with or without Stent Implantation for Acute Myocardial Infarction*. N Engl J Med 1999; 341:1949-1956). Since then, the stents used have developed and improved considerably.

Angioplasty alone had major drawbacks: artery closure from recoil, artery dissection, thrombosis, and restenosis from neointimal hyperplasia. Grines and Cox compared the procedures and found that six months after, fewer stented patients had angina (11.3%) compared to those who had angioplasty alone (16.9%).

They also showed, that at six-months follow-up, there were fewer cases of re-stenosis in the stented group (20.3%) than in the unstented group

(33.5%)(P< 001). Death, reinfarction, disabling stroke or the need for target-vessel revascularization, occurred in fewer of the stented group than in the angioplasty group (12.6 percent vs. 20.1 percent, P< 0.01). The six-month mortality appeared to be higher, however (4.2%) in the stented group as opposed to the unstented angioplasty group (2.7%), but the mean difference failed to reach statistical significance (P=0.27).

Interventional cardiologists have a variety of stents to choose from. Drug eluting stents are superior to bare metal stents, and are now considered the default option. These stents elute 'limus' drugs (calcineurin inhibitors). They form a complex with the cytoplasmic protein FKBP12. The complex inhibits two proteins: the p70 s6 kinase and 4E-BP1, the phosphorylation of which blocks calcineurin. The complex inhibits the nuclear factor activation of T-cells (NFAT).

Drug eluting stents made of cobalt/chromium or platinum/chromium are preferable, since they have thinner struts. Stents that elute everolimis, sirolimus, umirolimus or zatarolimus, all perform similarly. In patients with a high bleeding risk, Onyx or BioFreedom stents can be considered. No difference in early or late thrombosis, is apparent between stents. (Parker, W., Iqbal, J. 2020).

Not all studies have agreed with the use of angioplasty and stenting for patients with stable angina. One trial in particular, cast doubt on it use and concluded that medical treatment was favourable. The COURAGE study stated: 'There is no evidence of any reduction in mortality, or MI (heart attack) incidence, in those with stable angina' (*Optimal Medical Therapy with or without PCI for Stable Coronary Disease*. Boden W.E. et al. (2007). NEJM; 356 (15): 1503-1516).

As an example of trial analysis significance, and the clinical relevance of the conclusions drawn, consider the COURAGE trial in more detail. Its conclusions direct us to consider full medical treatment alone, and to avoid the stenting of coronary arteries in those with stable angina. According to this trial, coronary stents do nothing to improve the outcome. The problem for me was, few of their patient selection criteria matched my patients. In particular, most of my patients had ischaemia on exercise but were otherwise asymptomatic (many resulted from a diagnosis made on screening). Given a presumption that there are always devils in the detail, consider the patients they selected to include:

- *Only 24% of the their patients had a > 50% anterior descending stenosis. All of mine had a > 50% anterior descending stenosis.*
- *29% of their patients were smokers (none of mine were smokers).*
- *34% were diabetic (none of mine were diabetic);*
- *39% had prior cardiac infarctions (none of mine had infarcted before),*
- *67% were hypertensives (few of mine were hypertensive).*
- *All their patients had stable angina (only two of mine had angina).*

With these differences, I could hardly apply the results of the COURAGE trial to my patients. I do not doubt the conclusions for their particular patient group, but there were too many devils in the detail, for me to abandon PCI for my patients.

For decades before this study, my enthusiasm for stenting never waned. Because of its efficacy in relieving angina and ischaemia on exercise, I became an early advocate of balloon angioplasty and coronary artery stenting. I was qualified well before CABG became universally available, and coronary artery intervention (PCI) became routine. Before invasive therapeutic interventions, most doctors had personal experience of the natural history of coronary artery disease. Cardiac infarctions were common in the 1960s, and pharmacological treatment for angina was mostly ineffective.

I remained undeterred in my selection of patients for stenting, simply because it produced angina-free patients with an improved quality of life. I could not assess any effect on mortality. Relieving angina was a worthy advantage, but there was another, given how safe PCI became. It helped remove the worry of impending cardiac infarction (a sharp sword of Damocles for those with anterior descending lesions). Remarkably, none of my patients sustained an anterior infarction during thirty years of follow-up, and only a small number required further PCI interventions.

Pacemaker Implantation

Pacemaker implantation is mostly required to prevent syncope in complete heart block. It is a relatively simple invasive procedure, done under local anaesthetic. Leadless pacemakers now have the potential to make the procedure I will now describe, obsolete.

Having found the brachiocephalic vein (in the non-dominant arm), in the groove between the pectoralis and deltoid muscle, one can introduce a pacing electrode, guided under fluoroscopic control to the apex of the right ventricle. Sometimes subclavian vein puncture is used instead. An electrode, sited in the coronary sinus, can be used for atrio-ventricular pacing. The electrode position in the right ventricle is important; it must lie along the lower border of the RV, and be shown not to move with coughing or deep breathing.

Once the ventricular electrode is in position, a small subcutaneous pocket is made, anterior to the axilla, and 3-5cm below the clavicle. The pacemaker sits in this pocket, connected to the pacing electrode. The pacemaker itself, is best put under loose skin; it will then lie hidden from view.

Some doctors I tried to teach, had great difficulty in locating the brachiocephalic vein. Just occasionally, we located it using a die injection, but this was rarely necessary.

Types of Pacemaker

Pacemakers work in different modes, namely:

VVI (**V**entricular pacing; **V**entricular sensing; **I**nhibit Function).

In **VVIR**, the **R**ate is controlled by atrial sensing (so called physiological pacing).

AAI pacing is for sinus node dysfunction (**A**trial pacing; **A**trial sensing; **I**nhibition).

DDD (double chamber) pacing is needed for those with sino-atrial block and/or AV block combined. The acronym stands for: **D**ouble chamber pacing; **D**ouble chamber sensing; **D**ouble inhibition or triggering. It utilises four modes: **AsVs** (Atrial sensed, Ventricle sensed); **AsVp** (Atrial sensed, Ventricular paced); **ApVs** (Atrial paced, Ventricular sensed): **ApVp** (Atrial paced, Ventricular paced).

In **DDI** mode, inhibition is enabled.

In **DOO** mode, there is double chamber pacing with sensing and inhibition **O**ff.

AOO stands for atrial pacing with sensing with inhibition **O**FF.

VOO stands for **V**entricular pacing with inhibition and sensing **O**ff.

Pacemakers used for permanent complete heart block, were originally all VOO. Geoff Davies soon added VVI (with inhibition), working at St. George's Hospital, Hyde Park Corner, London.

Pacemakers have become smaller with time, and the choice of options extensive. There are now **leadless pacemakers** with all the parts inside one device. There is no separate battery and no electrodes since the whole device sits within the right ventricle.

R-Rate Response Pacing

Some patients have defective increases in heart rate with activity or metabolic demand that leads to exercise intolerance. SA node dysfunction is the usual cause. The pacemaker's sensing ability is defective, and no longer reflects the normal rate changes accompanying motion or ventilation. One can use R-Rate Response Pacing for this situation, called chronotropic incompetence.

PART 5: PREVENTATIVE CARDIOLOGY

Chapter Eighteen

The Early Detection of Heart Disease

In my practice every week, cardiac screening revealed at least one asymptomatic person (thereafter, a patient) with significant coronary artery disease (CHD). The anatomy and flow reducing character of their atheromatous plaques determined whether I prescribed prophylactic pharmaceutical treatment, stented them, or referred them for coronary artery bypass grafting (CABG).

Given the small number of patients involved, convincing statistical appraisal of my 20-year-old policy to stent asymptomatic patients with critical CHD, would not have the power to yield statistically significant results. Anecdotally, however, no patient of mine suffered an adverse long-term consequence of my stenting policy, although I found a few patients with atypical chest pain after they had been stented by cardiologists unknown to me.

I owe much of my policy's success to my colleagues at the Wellington Hospital, London: cardiologist Dr. David Lipkin and cardiac surgeon Stephen Edmondson. My patients experienced zero mortality and morbidity in their hands. This makes and important point. When considering the results of clinical trials, always give thought to those undertaking the procedures. I was lucky to work with world-class colleagues.

Patients should expect the most able doctors to treat them, although most cannot judge which doctors deserve such distinction. Performance figures published for and by doctors, can be misleading. Doctors can achieve exemplary figures by using ultra-selectivity; treating only those patients with the lowest risk.

My patients who were mostly affluent, were unwittingly pre-selected. The majority were wealthy, lean, fit, non-smokers. The preselection related both to their socio-economic class and their willingness to undergo cardiac screening. When considering the results of cardiac intervention in clinical trials, the influence of the health divide cannot be overlooked. Three to five times more cancers and cardiac infarctions occur among those from the lowest socio-economic groups. Since my patients were mostly wealthy, this will have had a significant influence on my results.

In the 1960s and early 1970s, before CABG became routine and aspirin was first used prophylactically, heart attacks and death from coronary atheroma were common monthly occurrences in most NHS general practices. Weight reduction, a quiet life, a low-fat diet and β-blockers, were all we could advise. Many post-infarction cases were anti-coagulated with vitamin K antagonists. CABGs, angioplasty, aspirin, 'statins', and ACE inhibitors all became available later, each with its own beneficial impact.

An important question remains. Can pre-symptomatic detection of atheroma and ischaemia improve cardiovascular morbidity and mortality? Although this sounds logical and seems obvious, it will remain controversial until large-scale studies provide the evidence. Until then, many high-risk patients will remain undetected, and the rich, strongly motivated as many are to survive, will keep an advantage.

The medical scientist who proposed the 'Polypill' (aspirin, a 'statin', a β-blocker, and an ACE inhibitor) suggested that we should all take it prophylactically. His claim, that a nation-wide reduction in morbidity and mortality from CVS disease would result, sounded reasonable. One of his initial mathematical arguments was faulty, however. He said, since each of the four drugs would reduce hearts attacks by 25%, taking them together would reduce strokes and heart attacks by 100%! When I spoke to him over lunch at a Royal College of Physicians advanced medical conference, he told me that trials were underway in India to prove his point. As a physician, treating individuals rather than entire nations, my preference was to treat only those with evidence of atherosclerosis.

Since *'1961 the UK's age-standardised death rate from heart and circulatory diseases (CVD) has declined by three quarters. Death rates have fallen more quickly than the actual number of deaths because people in this country are now living longer'.* (British Heart Foundation Factsheet, 2024). So there must be many more significant factors at work than early detection. Smoking reduction, dietary change, a gradual increase in vitamin intake and gymnasium attendance, could each have contributed to the reduction.

Expecting an advantage from cardiac screening (and any of the interventions that may follow), has its parallels. It is equivalent to assuming a benefit for cyclists wearing safety helmets, babies travelling in cars with safety seats, and adults wearing seatbelts in cars. These are all common sense, **BLOB** (**BL**indingly **OB**vious) suggestions. In academic terms, they all have a high antecedent probability of benefit, and a low index of suspicion for fooling us.

*Even if you are good at **BLOB** detection and are awash with common sense, a burden of proof always rests on the shoulders of anyone proclaiming benefit from any new investigation or clinical intervention.*

All screening strategies should aim to detect inherited and other risks. The results could help to formulate personal preventative strategies. At the moment, this only rarely includes genetic evaluation.

Screening asymptomatic patients allows the early detection of serious disease, with the potential to improve outcome. If no treatment is available for a detected condition, is there any point?

What benefit lies in detecting multiple sclerosis, or osteoarthritis before they become symptomatic; except for forewarning patients of the likely dangers, and to build a database of patients who might benefit in the future? It is easy to imagine the benefits that came to those with coeliac disease, pernicious anaemia, diabetes and pneumonia, once gluten-free diets, B_{12} injections, insulin and antibiotics became available.

In my lifetime, I witnessed a few impressive equivalents: H_2-receptor antagonists for peptic ulcer; $5HT_1$ agonists for migraine, and the part reversal of the inflammatory process in rheumatoid arthritis by anti-TNF (monoclonal antibody infliximab), pioneered by Ravinder N. ('Tiny') Maini at Charing Cross Hospital, London. In my field of work, the benefits of CABG, balloon angioplasty, and coronary artery stenting for angina were equally impressive.

Because there are now many effective cardiac interventions available, screening for cardiac disease is likely to benefit more patients than many other forms of investigation. Because cardiovascular diseases cause half of all middle-aged deaths in the western world, cardiac screening must have numerical, political, and economic advantages.

Because blood cholesterol does not accurately predict intimal cholesterol production, total blood cholesterol is virtually useless as a cardiovascular risk predictor in individuals (only 60% of those with atheroma have a raised total blood cholesterol. More have a low HDL cholesterol). Paradoxically, the mean total blood cholesterol in populations is a reliable predictor of large group, cardiac infarction occurrence. I discuss this 'statistical paradox' further in the chapter on clinical decision tools in my book, *'The Art and Science of Medical Practice'*.

For many years, I advocated that all my patients over the age of 35-years, should have their carotid arteries scanned for atheroma; especially if they had a family history of cardiovascular disease and/or hyperlipidaemia. If carotid atheroma was present, my recommendation was always to perform an exercise ECG. If the ECG remained normal and the patient showed no evidence of IHD (shortness of breath or chest discomfort), I would measure the rate of growth of any atheroma they had for two to five years. If the ECG was abnormal, but the patient remained asymptomatic, I would suggest a CT cardiac calcium score or CT angiogram. If the patient developed angina and/or had abnormal ECG changes on exercise, I would proceed directly to coronary angiography.

All those with a normal exercise test and a very high calcium score, I offered a CT or coronary angiography. Although expensive, this pathophysiological based strategy, is more predictive of coronary artery disease than any blood lipid measurement (second guessing the existence of atheroma-promoting genes, except for a low HDL-cholesterol), fibrinogen level or inflammatory marker.

In higher doses than that which lowers blood cholesterol, I observed 'statins' reducing or halting, the progress of atheroma. I observed the rate of atheroma progress in hundreds of patients, over a twenty-year period. I suggest we should use the rate of progress of carotid atheroma as a measure of pharmacological efficacy, and not just how effectively blood lipids become lowered. Now one can get carotid images using an iPhone ultrasound transducer, this has become more practicable. This direct test for atheroma is likely to be a more reliable predictor of cardiovascular morbidity and mortality in individuals, than any indirect test like blood lipids.

Traditionally, it is the symptoms and signs of disease that have led doctors to make diagnoses. At the beginning of my career, many thought it unprofessional to enquire beyond the remit of helping patients to overcome symptoms. Some members of the medical profession saw this as unethical personal intervention. The routine taking of blood pressure and cervical smears for women then became advised. Fortunately, attitudes have progressed towards prevention, early disease detection, and early intervention in asymptomatic patients. While the NHS has accepted the prophylactic benefits of cervical smears, mammography, and PSA measurements, only blood lipid measures and QRisk3 estimates are offered for heart disease risk detection among the asymptomatic. For this reason I started my private cardiac screening centre in 2000, although I had previously started a general screening and diagnostic centre much earlier, in 1973.

The inherent human need to persevere and reach safety, might explain why in evolutionary terms, some patients choose to deny symptoms. In ischaemic heart disease, without a collateral circulation, coronary artery stenoses > 80-85% will usually restrict blood flow enough to cause angina (Poiseuille's formula and Bernoulli's equation). Patients can learn to avoid it by slowing their activities; some will then deny its existence. Avoidance and denial both allow patients to carry on their usual activities. Unfortunately, this strategy reduces the chance of an early diagnosis and prophylactic intervention.

I screened many patients repeatedly for nearly five decades, at intervals of several years, each time entering their clinical information onto a standard data collection form. Retrospective analysis of these standardised records revealed something of interest. Among early symptoms, lie clues to later diagnoses. Sometimes, it takes years or decades for a condition to become symptomatic or for a diagnosis to be confirmed. Significant early symptoms can be inconsistent and undramatic, but nearly always appear before physical signs. False positive symptoms abound. One aim of repeated screening examinations is to recognise symptom consistency. Symptoms can be transient and lack explanation, but when they persist they are usually of clinical significance. Managing patients on a long-term basis allows one to assess a patient's symptom reliability.

Up to a point, many types of pathology originate and progress without symptoms; in fact, symptoms may arise only at the end of the expected natural history. Neither atheroma nor hypertension generate early symptoms, despite their lethal potential. By discovering them decades before symptoms arise, 'pre-patients' will probably benefit from routine screening: repeated blood pressures, artery scanning (ultrasound, and CT), ECG exercise testing, and echocardiography. All need to be repeated every 2–5 years if one is to detect progressive pathology. Both initial and repeated investigations of asymptomatic patients can reveal those, close enough to a cardiovascular event, to make urgent intervention advisable. From a practical point of view, a large majority will benefit only from re-assurance; a not insubstantial matter to many patients.

In my early days of screening (1970s), most doctors regarded it as a waste of time. This was mainly because its 'useful' diagnostic yield was low. Most of my patients expressed a different view. They regarded screening as sensible for two reasons. Many thought it prudent to rule out significant early disease and also wanted the opportunity to benefit from any available treatment. They regarded screening as insurance, even if the benefits were speculative.

There are key academic questions about screening to be answered.

- How often are prognostically relevant early symptoms and signs revealed by screening?

- Does early detection of disease lead to improved outcomes— reductions in patient morbidity and mortality?

Significant early symptoms of heart disease do occur, but they are uncommon among the many 'healthy' subjects who present for screening. Some early symptoms worth detecting are:

Slowly progressive shortness of breath (with or without angina), and
Progressive tiredness, occurring early in ischaemic heart disease and heart failure.

Routine questioning will detect the occurrence of these symptoms at a time well before any patient would openly complain of them or be aware of their significance. Using a standardised pro-forma checklist of symptoms each time, allows the verification of symptoms that have persisted (often contrary to what patients remembered), as well as symptoms that are new.

Every general physician knows not to ignore constant dysphagia, a persistent cough, or a skin mole that has changed colour and shape. Unfortunately, these symptoms may not get reported early enough to make a difference to the outcome. With the best of intentions, pursuing early symptoms may not always alter the outcome. I found this to be the case for those with early oesophageal and lung cancer, melanoma, some brain tumours, and a few other malignant conditions. Cardiac screening is different: early asymptomatic detection can offer real benefit.

There are practical problems with prevention. Too few people (potential patients) ask: *'is there something wrong with me that could be managed better if detected early?'* In my practice, many came for cardiac screening after a friend or relative had died. Our screening unit mostly saw anxious, low-risk patients, rather than those whose life was at risk. Couples often presented themselves for screening: the patient and their anxious protagonist. It was not uncommon for the asymptomatic protagonist to become the beneficiary.

Does screening reduce mortality or morbidity? Despite the lack of definitive answers, my patient's desire for medical and cardiac screening never dampened. Given their responsibilities to others (their family and employees), many saw screening as dutiful.

The wealthy often see the cost of screening as a small price to pay for the chance to preserve their life and lifestyle. As business people, many will have made similar 'common-sense', strategic judgements about their business, based on much less evidence. The chance to acquire a better future for themselves, their families, and their employees, motivated these decision-making patients to do whatever they could. Justifying action in the 'public interest' or for the general good, was not their concern. They were prepared to gamble and pay for the promise that early detection might lead to early intervention and an improved outcome. They saw this as a worthy gamble (in the absence of solid proof, and given they could afford it). Many of my patients regarded our academic controversies about screening as time-wasting, and economics related. The personal stand they took on issues of future significance (without sufficient evidence), had led many of them to become successful in business.

Before driving their car, there are those who never give thought to their tyre pressures, engine oil level, or radiator fluid level, and will set off on a long road trip without checking them. Others give no thought to saving money for their retirement, believing that winning the National Lottery, or receiving a windfall from a distant relative will provide for them in later life. A little intelligent forethought will more often prove better; more successful than hoping for the best, ill-informed gambling, disinterest and absent-mindedness.

During the tragic bush fires in Australia in late 2019, Steve Harrison, a resident of Balmoral, New South Wales, constructed a ceramic coffin-sized kiln, just in case a bush fire overwhelmed his property. Because he put a fire extinguisher and several bottles of water in the kiln, many thought him over-anxious. His forethought saved his life. He hid inside the kiln for 30-minutes while a raging firestorm swept by.

Apart from the relief of suffering and curing illness, medical practice now has secondary aims— to reduce patient morbidity and mortality. Some wealthy nations have gone further than the relief of symptoms; they have invested in prevention and the early detection of disease. The statement *'prevention is better than cure'* is a well-worn cliché, but one many patients hold dear.

A Full Cardiac Screen must include an ECG with exercise as a 'road-test'. For those patients with high blood pressure, blackouts, heart failure, or heart valve problems, a 24-hour ECG and echocardiography are required. Taken together, these are the basic screening requirements for astronauts, and others where data compromise is untenable.

The clinical data obtained provides a baseline database for future reference. From this, future changes can be flagged and diagnoses made more easily.

Patients should know to ask:

- Do I have relevant symptoms?

- Is heart disease in my family?

- When did I last check my heart and arteries?

- Am I overweight?

- Am I diabetic?

- Am I unfit?

Investigations

Routine blood fats (especially LDL and HDL), homocysteine, lipoprotein 'a', clotting (fibrinogen) and inflammatory markers (ESR and CRP), cardiac enzymes and troponin, can be part of the cardiac screening evaluation, although for most individuals, the results will prove more of academic significance than useful.

Triglyceride levels, often raised in the overweight and unfit, is often abnormal in diabetics and pre-diabetics. The combination, referred to as the 'Metabolic Syndrome', relates to increasing insulin resistance. Its presence adds to the calculated risk of coronary artery disease. Tests for diabetes include random blood sugars, HbA1c, and a glucose tolerance test. Not all diabetics have atherosclerosis, and only a few with atherosclerosis are diabetics. Insulin production and resistance in relation to glucose and fat metabolism, provides a link to atherosclerosis.

The clotting risk factors, fibrinogen, Leiden factor V, etc., are of possible significance in some individuals. Raised levels can mean a liability to CVS events. The significance of raised, non-specific inflammatory markers, like ESR and CRP is of questionable significance in individuals, although atherosclerosis is accepted to be a partly inflammatory process. Although these tests may be significant in large groups, they are of little or no diagnostic use in individuals. I prefer to detect atherosclerosis directly, and not indirectly with tests like blood fats and inflammatory markers.

ECGs, 24-48 hour ECGs and exercise ECGs, can all be helpful, albeit with problems caused by false negativity and positivity. Occasional, unexpected heart rhythms may be revealed. Those with rarer, specific electrical abnormalities, may need electrophysiological studies.

An exercise ECG test can reveal more than just electrical changes. Unfitness and unusual shortness of breath are soon apparent. Sometimes chest pain and faintness will put an early end to the test. ECGs recorded at the same time as exercise are essential. Because of physical disability, some patients will not be able to perform the test. One can then use intravenous dobutamine instead. By speeding the heart, ECG changes equivalent to those on exercise may become apparent. Simultaneous echocardiography can be valuable, showing not only normal myocardial function, but ischaemic and damaged myocardium.

Chest X-rays (CXRs) can be of value, although I now regard them as optional. A chest X-ray can be informative in certain types of heart trouble. A CXR is essential for all those with shortness of breath and heart failure.

All patients found to have carotid atheroma, and those with a positive cardiac family history despite no sign of past or present disease, need further testing. It was my policy to advise a cardiac CT scan, looking for calcium inclusions in the coronary artery walls (a coronary calcium score). Those with more than a slight suspicion of IHD or angina, should have a CT angiogram; those with some evidence of IHD should bypass such tests and have a coronary angiogram. This enables the anatomy and severity of any coronary artery atheroma to be assessed, and the need for artery stenting or coronary bypass surgery discovered.

Echocardiography is essential for all those with hypertension (to detect heart muscle hypertrophy). It is also essential for all those with heart murmurs and palpitations. A stress echocardiogram (with exercise or dobutamine) can help to make a functional diagnosis of IHD.

Over the course of fifty years, I found many asymptomatic people with severe coronary heart and valvular heart disease. Many of these 'pre-patients' subsequently needed coronary artery stenting, coronary bypass surgery or pacing. My experience of patients, decades before screening became acceptable, left me with little doubt that cardiac screening will advantage many patients. Without early detection, a few of my cases with severe coronary atheroma would probably have died without warning. Many others, survived long enough for their condition to progress and produce symptoms.

As a private cardiologist in the UK, I had a big advantage —long-term, patient continuity. I often saw my patients annually or biannually; some for forty years. My personal knowledge of them made early clinical changes easier to identify. Early diagnosis and management plans could be based on my personal knowledge of them and their circumstances.

I always used my research findings and experience, to support my management advice. The mutual trust I developed with my patients, sanctioned me to use both the art of medicine, alongside medical science. What my patients wanted was advice, tempered by professional experience and some personal knowledge of them and their circumstances. With a physical and mechanical subject like cardiology, one can achieve much with one-off, impersonal consultations. Gaining patient satisfaction and trust is another matter.

To achieve the best results, doctors and nurses must employ some of the art of medicine. Only by establishing trusting relationships and knowing patients well, can we expect our medical advice to be fit for purpose. With a patient's life at stake, applying some of this art can be crucial to clinical success and patient satisfaction. In my book, *'The Doctor's Apprentice. The Art and Science of Medicine'*, I examine many other aspects of patient doctor relationships.

Chapter Nineteen

The Inheritance of Heart Disease

Family History

For every machine, road, or building to function effectively, needs a purposeful design before construction starts. A detailed plan is essential for construction workers. In some cases there must be freedom to remove redundant features and to expand the worthwhile features. Each of us inherits a set of genetic codes, equivalent to a construction plan. Like most plans, deletions, additions, alterations and transformations (mutations) can occur as time passes.

Taking a family history is simple enough, and one way to access the possibilities of a patient's inherited predispositions. The accuracy of the information and its completeness of this third-party information, can be a major source of error.

Finding a positive family history of patients with coronary heart disease and high blood pressure, is often relevant. For those with a negative family history, the notable absence of high blood pressure or IHD, can also prove reliable, even when the symptoms suggest otherwise. Although it is difficult to assess the reliability of a family history alone when predicting coronary artery disease and hypertension, in my experience it helps in 80% of cases (tossing a coin helps in only 50% of cases).

If a patient has one parent with coronary heart disease or high blood pressure, one in four of their children are likely to develop the same. If both parents have the condition, at least eight of ten children could inherit it. These, at least, are my anecdotal impressions gained after taking thousands of family histories over many decades.

My anecdotal experience suggests the strong inheritance of atherosclerosis and the arteriosclerosis associated with hypertension. Many mistakenly believe that 'unhealthy' lifestyles and diet alone, cause heart disease; media presentations have influenced this belief. At least, it might be partially true that lifestyle and diet can influence CVS disease progress.

For those who have already had a heart attack:

- Eating a Mediterranean diet can substantially reduce the chance of a second one (by 70%, when compared to control subjects who made no change to their diet). (See: de Logeril, M, et al. (1999). *The Lyon Diet.* Circulation; 99(6): 779-85).

- A similar level of benefit applies to exercise (20-29% reduction). See Darden, D., Richardson, C., Jackson, E.A. (2013) *Physical Activity and Exercise for Secondary Prevention among Patients with Cardiovascular Disease.* Curr. Cardiovasc. Risk Rep 7(6): 10.1007.

- Smoking cessation reduces secondary end-points by one third. See Wu, A.D., Lindson, N., et al. (2022). *Smoking Cessation for Secondary Prevention of Cardiovascular Disease.* Cochrane Database Syst. Rev. 2022(8) CD014936).

These positive notions will bring comfort to those who want to feel in control of their lives.

Without inheriting the adverse genes that lead to heart disease, I doubt few develop it, whatever they eat or do. Patients need individual evaluation, not statistical likelihood, even though that may be the best we can do at present.

In order to define which dietary nutrients might affect atherosclerosis, I researched the relationship between diet and arterial pathology defined by animal experiments. I published my observations and comments for non-academic readers in two books: *Eat to Your Heart's Content. The diet and lifestyle for a healthy heart.*(2003). HeartShield. ISBN: 0-9551072-0-2, and *HeartSense. How to look after your heart.*(2006) . HeartShield. ISBN: 0-9551072-1-0.

Animal experiments suggest a preventative role for the amino-acids taurine and arginine; the minerals Se, Zn, Mg, Mn; the vitamins B_{12}, B_6, and E, and the fats Omega 3, 6 and 9. How effective they are, remains to be established. Many animal experiments have confirmed the deleterious effects of saturated fat. Where are these nutrients to be found in food, and what arterial benefit can they have when combined with atherogenic nutrients? If de Logeril's analysis is correct (see above), the detailed nutrient analysis of food should be important for prevention strategies.

How Does Inheritance Work?

Strings of genetic instructions (genes) make up the chromosomes in the nucleus of each living cell. Humans have 23 pairs of chromosomes, formed from only four amino acids. Genes act as strings of instructions directing all the biochemical processes within our cells. The four amino acids involved, labelled A-T-C-G, are: A=adenine. T=thymine, C=cytosine and G=guanine occur in various combinations. Genetic genome research aims to identify these combinations as codes that translate into biochemical processes. Many codes leading to disease have been identified, but a long road lies ahead to identify them all.

Much of the code present on chromosomes is redundant; the equivalent of gobbledegook, filling the space between meaningful code. Mutations occur when the strings of functioning code are added to, or deleted. Some will critically change the instructions given to cells, causing their malfunction and disease. Because viruses are formed from RNA or DNA, they have the potential to interfere with critical encoded instructions. Fortunately, most viruses are transient and inconsequential. Occasionally, viruses arise that cause death and disease, like poliomyelitis, smallpox, myocarditis, encephalitis and COVID-19.

Each of our forty-six chromosomes (two strings of twenty-three), consist of DNA. At the atomic level, the two strings run in parallel, forming a spiral or double helix of three billion base pairs. The only pairs connecting each DNA strand are: A-to-T and G-to-C, making each strand complimentary to the other in a predictable way. James Watson and Frances Crick, published the details in 1953. (see: Watson, James D., and Francis H.C. C rick. *Molecular Structure of Nucleic Acids.* Nature 171 (1953): 737–; Watson, James D., and Francis H.C. Crick. *Genetical Implications of the Structure of Deoxyribonucleic Acid. Nature* 171 (1953): 964–7).

Distinct from the nuclear chromosomes are mitochondrial chromosomes. Although only inherited from mothers, they direct the many energy-related biochemical processes within each cell.

Mitochondrial DNA (mtDNA) has approximately 16,500 base pairs; only a small fraction of the total DNA in each cell. Mitochondrial DNA has thirty-seven genes, responsible for normal mitochondrial function, thirteen of which control oxidative phosphorylation and the production of ATP from simple sugars. Together with messenger RNA (mRNA), the remaining genes provide instructions for assembling proteins from amino acids. For this, transfer RNA (tRNA) and ribosomal RNA (rRNA) are required.

Pathogenic Genetic Variants

Many genetic variants underlie congenital heart disease. Discovering them, can concern both patients and their relatives, so personal counselling has an important role to play.

Congenital heart disease can derive from:

- Major chromosomal abnormalities, such as trisomy and chromosome deletions.

- Single gene variations like deletion and duplication.

- Autosomal genes (dominant or recessive), or

- X-linked abnormalities.

Benign genetic variants occur in more than one in twenty people. Their relevance has yet to be understood.

Several types of genetic testing are now available commercially. Single gene testing is cost effective; gene panel testing can detect the genes implicated in known phenotypes. Exome testing sequences the coding regions (the exome) of genes.

Consider some commonly occurring clinical conditions.

Several **storage diseases** have a genetic basis and can affect cardiac function. **Fabry disease** is a lysosomal storage disorder. Without alpha-galactosidase (X-linked, alpha-galactosidase deficiency), sphingolipids build up in the heart, brain and skin. LVH and arrhythmias can result.

In **inherited haemochromatosis** (autosomal), the bowel absorbs too much iron which can then locate in the heart, brain and skin. It causes cardiomyopathy affecting diastolic function. Later on, it can cause dilated cardiomyopathy and arrhythmias.

Trisomy 21 (Down's syndrome) can be associated with an ASD, VSD, or PDA. It is also associated with Fallot's tetralogy (an overriding aorta, PS, RVH and VSD).

Turner syndrome occurs only in females, where one X chromosome (or part of it) is missing. Its cardiac associations are coarctation, aortic aneurysm and bicuspid aortic valve.

In **Fragile X syndrome** (FMR 1 gene on the X-chromosome), mitral valve disease can occur.

Autosomal dominant abnormalities. Both **Holt-Oram syndrome** (ASD, VSD, conduction problems, missing thumb and bone developmental abnormalities), and **Noonan Syndrome** (pectus excavatum, PS and HCM) have a dominant autosomal basis. The same genetic basis applies to **Marfan Syndrome** (FBN1)(AI, aortic dilatation, tall stature, hyper-mobile joints, lens dislocation, high arched palate and pectus excavatum), and **Ehlers-Danlos Syndrome** (COL3A1). The latter can cause aortic dissection, joint hypermobility and thin skin.

Many genetic variants are associated with **HCM.** They are mostly autosomal dominant. They relate to myosin binding (protein 3)(MYBPC3 gene); myosin heavy chain 7 (MYH7); troponin T and I (TNNT2 and TNNI3 respectively); tropomyosin 1 (TPM1); actin alpha (ACTC1), and myosin light chains 2 and 3 (MYL2 and MYL3).

Dilated cardiomyopathy has been associated with genes SCN5A, TTN and lamin A/C (X-linked), and dystrophin (DMD). It also occurs in Alström syndrome (obesity, diabetes, deafness).

In **Arrhythmogenic RV dysplasia** (fibro-fatty infiltration), **Brugada syndrome** (VT, VF sudden death) and **long QT syndrome** (syncope and sudden death), the genes KCNQ1, KCNH2 and SCN5A occur in 97% of cases. All have autosomal dominant inheritance.

Genetic Analysis and Cardiac Risk

Although many accept the inherited nature of coronary artery disease and hypertension, clinically useful, specific genetic factors have yet to be identified. Although many clinical and genetic factors have been used to calculate the likelihood of coronary artery disease (CAD) and hypertension, none are yet accurate enough for individual prediction. This is a work in progress, with much already achieved. Some important milestones are worth noting.

One analysis quantified CAD heritability using updated genome-wide approaches; the estimate was between 40 and 50%. (Won H.H., Natarajan P., et al. (2015) *Disproportionate Contributions of Select Genomic Compartments and Cell Types to Genetic Risk for Coronary Artery Disease*. PLoS Genet. 2015;11:e1005622.)

In the Framingham Heart Study, a family history of cardiovascular disease in a parent or sibling, strongly predicted CAD (a 55% increase in risk). (Murabito J.M., Pencina et al. (2005) *Sibling cardiovascular disease as a risk factor for cardiovascular disease in middle-aged adults.* JAMA. 294:3117–3123. doi: 10.1001/jama.294.24.3117).

The familial risk of CAD was first described in studies involving twins and prospective cohorts (Marenberg M.E., Risch N., Berkman L.F., et al. (1994). *Genetic susceptibility to death from coronary heart disease in a study of twins.* N. Engl. J. Med.;330:1041–1046. doi:10.1056/NEJM199404143301503).

Common variant association studies, genome-wide association studies, meta-analyses and genetic risk scores, have allowed a better understanding of the genetic risk factors driving CAD. The genes of interest affect endothelial function. They relate to lipid and carbohydrate metabolism, those that regulate the function of the endothelium and vascular smooth muscles. Others influence the coagulation system or affect the immune system.

Both mouse and human genetics have been studied. Apart from genes predicting dyslipidaemias, those associated with atherosclerosis include the expression of PEAR1 and SVEP1 in platelets and endothelium. A variety of disease-relevant cell-types express PEAR1 within coronary arteries, including fibroblasts and smooth muscle cells.

Many immune cells such as leukocytes, macrophages, and lymphocytes, are involved in atherosclerosis; they produce mediators capable of increasing and maintaining inflammation. Many cytokines, chemokines, metalloproteinases and growth factors, are involved. The results of some studies suggest that 'statins' reduce inflammation in the vasculature and are not only because of their lipid-lowering effect.

Ongoing inflammation within the endothelium can cause the loss of normal functioning, with excessive platelet aggregation and local thrombotic lesions occurring. In addition, a decrease in vasodilatory mediators like nitric oxide, and an increase in the synthesis of vasoconstrictive mediators like endothelin, can occur. The subendothelial accumulation of lipoproteins can aggregate into atherosclerotic plaques.

Macrophages located in the endothelium, absorb oxidised lipoproteins. They then secrete several cytokines and adhesion factors, such as intercellular adhesion molecule 1 (ICAM-1), E-selectin, and vascular cell adhesion molecule 1 (VCAM-1). All can increase inflammation in the vessel wall and promote the development of atherosclerotic lesions. Because of adhesion molecules, circulating monocytes will adhere to the endothelium and differentiate into macrophages. They secrete several pro-inflammatory cytokines involved in the atherosclerotic process (IL-1α, IL-1β, IL-6, IL-15, IL-18, TNF-α), as well as anti-inflammatory cytokines (like interleukin 4,

10, and 13, and transforming growth factor β (TGF-β)), all of which can inhibit inflammation and atherosclerosis.

The development of inflammation and atherosclerosis results from an imbalance in the production of pro- and anti-inflammatory mediators. There are several key factors:

Macrophages play an important role.

Macrophages absorb oxidized low-density lipoprotein C (LDL-C) to form cholesterol-containing foam cells. With inflammation, they can undergo apoptosis, and stimulate the migration of vascular smooth muscle cells into the intima.

Ongoing inflammation can make unstable atherosclerotic plaques prone to rupture and thrombosis formation.

Hypoxia can cause neovascularisation, further increasing the instability of atherosclerotic plaques.

The inflammatory process is associated with the secretion of metalloproteinases. By causing collagen degradation, they can also make atherosclerotic plaques unstable.

The chromosome 9p21 locus is the most widely replicated genetic CAD risk locus. It increases risk by 15% to 35% in carriers of the variant allele (Palomaki G.E., Melillo S., Bradley L.A. (2010). *Association between 9p21 genomic markers and heart disease: A meta-analysis.* JAMA. 2010;303:648–656. doi: 10.1001/jama.2010.118).

Genome-wide association studies (GWAS), have been used to identify **single-nucleotide polymorphism (SNP)**. A SNP (or 'SNIP') can mark disease susceptibility. GWAS have been used to create genetic risk scores, said to improve CAD risk prediction. These studies use high-throughput genotyping technologies, mapping thousands of SNPs in the human genome and correlating them with clinical conditions and measurable traits. GWAS are very useful in discovering genetic variants related to different diseases, but they have limitations. They provide no new information on any biological pathways linked to disease.

Many CAD-associated, single-nucleotide polymorphisms are now known, but with unidentified functions. Combining candidate risk SNPs with other well-known clinical risk factors, can create a genetic risk score (GRS). Using GWAS, over 230 SNPs associated with CAD have been identified. The simultaneous use of the most common and strongest risk marker SNPs, and clinical risk factors like high LDL cholesterol, low HDL cholesterol, high blood pressure, family history, diabetes, smoking, etc., have yet to distinguish diseased from healthy subjects.

Mean elevated LDL levels, are causally associated with increased group CAD risk. In familial hypercholesterolemia, mutations in the LDLR gene can be involved. Genetic variants in the PCSK9, NPC1L1, and HMGCR genes, are associated with elevated serum LDL levels and are useful predictors of CAD risk in population studies.

An important mediator of inflammation, is transforming growth factor TGFβ1. TGFβ1 plays an important role in the processes leading to CAD, by stimulating the chemotaxis of macrophages and fibroblasts, as well as increasing extracellular matrix synthesis.

Transforming growth factor (TGF) β1, has a regulatory function in endothelial cells, vascular smooth muscle cells, and the extracellular matrix. The exact mechanisms controlling the TGFβ1 signalling within the arterial vasculature, remain to be discovered. A low level of serum TGFβ1 protein, is a biomarker for the diagnosis and risk stratification of CAD.

Both elevated levels of TGFβ1 in peripheral blood and the **rs1800470 genotype**, have been associated with a higher risk of developing cardiovascular disease (cerebral infarction, myocardial ischemia in patients with diabetes, and complications of CAD). In a Mexican population, the rs1800470 polymorphism, has been associated with the risk of developing re-stenosis after coronary stent implantation.

Other polymorphisms of the TGFβ1 gene that affect its expression, include the rs1982073 polymorphism. Meta-analysis of the risk of CAD in Caucasian populations, showed an association between rs1800469 and rs1982073 in the TGFβ1 gene, confirming that TGFβ1 signalling is probably involved in the pathogenesis of CAD.

TGFβR2 could be a tumour suppressor gene. Mutations in TGFβR2, are associated with many cardiovascular diseases, such as autosomal dominant Loeys–Dietz syndrome (bifid uvula, club foot, familial thoracic aortic aneurysms), and sudden cardiac arrest in CAD patients. TGFβR2 in the gastrointestinal tract, acts as a tumour suppressor gene; approximately 30% of colorectal cancers carry TGFβR2 mutations.

Mutations in TGFβR2 disrupt TGFβ signalling, and can create genetic conditions that predispose to Marfan syndrome, thoracic aortic aneurysm and dissection. Signalling through the TGFβR2 receptor in endothelial cells, plays an important role in cardiac development, also promoting fibrosis and myocardial remodelling in CAD.

TGFβR2, rs6550004, has been associated with Kawasaki disease, and rs1495592 has been associated with coronary artery lesions in children (Choi Y.M., Shim K.S., et al. (2012). *Transforming growth factor beta receptor II polymorphisms are associated with Kawasaki disease.* Korean J. Pediatr. 2012;55:18–23. doi: 10.3345/kjp.2012.55.1.18).

Defining the role of specific pro-inflammatory mediators in the development of CAD, may enable the development of new therapeutic strategies for CAD; in particularly, the development of new antiplatelet drugs and strategies for preventing stent occlusion.

Relaxation and psychological stress might modify atherosclerosis. The main parasympathetic mediator—acetylcholine—inhibits the activity of macrophages, blocking the secretion of pro-inflammatory cytokines. Relaxation, and increased parasympathetic activity, rather than stress, could inhibit plaque inflammation. Sympathetic nervous system stimulation in stressful situations, accompanied by decreased parasympathetic influence, could increase plaque inflammation, but to what extent, I wonder. Sympathetic nervous system stimulation can also result in the excessive secretion of vasoconstrictive mediators such as endothelin and decrease vasodilatory mediators like nitric oxide.

Hypertension and Genetic Profiles

A family history is common in those with hypertension. In practice, the association is strong enough to notice hypertension occurring mainly in those with a family history of hypertension and haemorrhagic stroke. Defining hypertensive genetic traits is difficult because renal and endocrine functioning are both involved. A predisposition to arteriosclerosis is of importance. It could well be that the inheritance of arteriosclerosis and LVH, is key. Hypertensives with LVH, are liable to death from haemorrhagic stroke.

To quote Cowley, 'Defining the genetic basis of susceptibility to hypertension is challenging, if only because of the complex polygenic nature of arterial blood pressure which is a quantitative trait that is influenced by multiple variants, gene-gene interactions and environmental factors.' Cowley, A.W. (2006). *The genetic dissection of essential hypertension.* Nature Reviews Genetics: 7; 829-840.

Chapter Twenty

The Prevention of Heart Disease

Many recommendations are made for primary and secondary prevention (sometimes tertiary). All primary prevention measures, apply to secondary prevention.

PRIMARY PREVENTION

For those who do not yet have cardiovascular disease, primary prevention aims to stop it occurring. Apart from suggesting that we should all medicate ourselves with aspirin, minerals and vitamins or a polypill every day, primary prevention rests on lifestyle advice. So far, no genetic engineering interventions exist.

Doctors agree on many preventative measures, but despite this, 35 million people world-wide experience cardiovascular incidents each year (World Heart Federation, 2024). The advice we offer to western populations is now well-worn, and little more that yawn-inducing cliché. It comprises:

- Don't smoke.

- Don't drink too much alcohol.

- Eat less. Exercise more.

- Lose weight.

- Get fit. Keep fit (preferably to an athletic standard).

- Eat a balanced diet.

I would add, avoid stress. Change from having a Type A personality, to being a Type B (Rosenman, R.H., Friedman, M. (1974) *Type A Behaviour and Your Heart*. Knopf Doubleday Publishing Group). They described being a Type A (time urgency; rushing around in ever decreasing circles, trying to achieve more in less time), as opposed to being the more relaxed opposite Type B. The change from Type A to Type B, could possibly halve CVS risk.

A change in socio-economic status might help more. The poor risk five times more cancer and cardiovascular disease than the rich. There are many possible reasons that relate to food, smoking, exercise and stress. Successful sports people and those with profitable businesses often change status, as do those who start at the bottom in a corporation and rise to senior management. As people rise in status though promotion in an organisation, the CVS risk lessens (Marmot, M.G., Smith, G.D., et al. (1991) *Health inequalities among British civil servants: the Whitehall II study*. Lancet;337(8754):1387-93).

Many factors are interlinked and their individual potency for CVS prevention becomes difficult to sort. Middle-class, high earning people, are often better educated, exercise more, eat more protein, carry less weight, smoke less and take more holidays. Most will have escaped wondering where the money will come from to feed their children and pay energy bills. Life circumstances and the associated lifestyle options, have consequences. Tackling the health divide is a major political and anthropological problem. Only a few rich States in the world, like Monaco and Brunei, have addressed it successfully. The failure of politicians to reduce the health divide has consequences that have fallen to the medical profession to solve. Doctors must necessarily limit themselves to lame advice—advice that few poor people want to follow, or can follow.

Although medical advice will mostly have a proven epidemiological basis, many will see such advice as unrealistic. Why might we advocate smoking cessation for individuals over 70-years of age (assuming they smoke in isolation)? Why suggest a Mediterranean diet to include smoked herring or salmon, with five pieces of fruit or vegetables every day, if unaffordable or unacceptable (see: Prof. Helen Stokes-Lampard, Chair, RCGP Council)?

Here are a few key facts about individual primary preventative measures:

Weight Loss

The poor more often eat cheap, high carb fat-rich food. The rich are more likely to eat high protein, nutrient-rich food. As a group, the poor are fatter than the rich. This makes exercising more difficult.

The lowest all-cause risk of death, occurs among those with a BMI between 20 and 24.(Aune, D., Sen, A. (2016). *BMI and all-cause mortality: systematic review and non-linear dose-response meta-analysis of 230 cohort studies with 3.74 million deaths among 30.3 million participants. BMJ;353:i2156*).

Important body weight related facts are:

- For Class I obesity – 'overweight' (BMI 25-30), the risk rises by 40%.

- For Class II, 'obese' (BMI 30-40), the risk multiplies 2.5 times.

- For the extremely obese, Class III (BMI > 40, the risk multiplies three-fold.

- For those with a BMI > 45, the risk of all-cause death multiplies five-fold.

In the US, cardiovascular deaths have been dropping for decades, while body weights have been increasing. This suggests that there are more important factors than obesity affecting CVS outcomes.

For more details, and practical advice for those wanting to lose weight, see my book entitled, *Who Loses Wins. Winning Weight Loss Battles. A 'Fat Mentality versus a 'Fit Mentality'.(2024).*
ISBN: 978-1-7385207-1-8. E-book from https://stan.store/drdhd001 001.

Smoking, Lung Cancer and Heart Disease

Many more non-smokers than smokers reach the age of 65 years. First published in 1950, Doll and Hill used the results of a questionnaire to show that smoking and not traffic exhaust inhalation or tarmac, caused lung cancer. (Doll, R., Bradford-Hill, A. (1950). *Smoking and Carcinoma of the Lung.* BMJ; 2:739). This was partly a chance finding. They gathered data from those living on main roads and those living in rural communities. The thesis was that vehicle exhaust inhalation or road surface tar-based chemicals caused it. Both proved incorrect. Those living around the North Circular Road had no greater chance of lung cancer, unless they smoked.

Does smoking cause CHD? In mouse models, smoking accelerates atherosclerosis and adversely affects many other processes. The full causal picture awaits further definition, although the epidemiology is clear. Smoking determines 30% of all CVS deaths and doubles the 10-year risk of fatal CVS death. Secondary smoking increases the risk of stroke by 20-30%. Females who smoke have a 25% greater chance than males of developing CHD, a phenomenon possibly related to thrombin signalling. The cessation of smoking before the age of 40-years, reduces the risk of death by 90%. (Gallucci, G.,Tartarone, A., et al.(2020). *Cardiovascular risk of smoking and benefits of smoking cessation.* J. Thoracic Dis.; 12(7): 3866-3876).

Exercise and Heart Disease

Because exercise improves many of the traditional CVS risk factors, such as blood lipids and blood pressure, it will contribute to group primary prevention. Extreme aerobic exercise, however, can be associated with calcific atheroma and AF. (Mehta, A., Kondamudi, N., et al. (2020). *Running away from cardiovascular disease at the right speed: The impact of aerobic physical activity and cardiorespiratory fitness on cardiovascular disease risk and associated subclinical phenotypes.* Prog. Cardiovasc. Dis. 63(6):762-774).

I have always wondered whether aerobic training is better than weight training, for primary and secondary prevention. The physical effects are different, so perhaps the benefits differ. I have found no satisfactory answer to this question.

Food and Heart Disease

A bias that is mostly incorrect, is now commonplace. The 'healthy food bias', influences many to believe that health and disease prevention, depend entirely on a healthy diet.

During the North African desert campaign of World War 2, Winston Churchill visited Field Marshall Montgomery ('Monty') in his caravan. They ate a meagre lunch. Monty announced: 'I neither smoke nor drink alcohol, and I am 100% fit'. Churchill, the epicurean, replied: 'I drink and smoke, and I am 200% fit!'

Churchill lived for 90-years, 'Monty' for 88 years. Although Churchill suffered a few cardiac and cerebral infarcts, I very much doubt he regretted smoking his cigars or eating a gourmet diet.

The idea of 'we are what we eat', dates back at least to 1942, and Victor Henry Lindlahr's book, *You are What You Eat*, (National Nutrition Society, Inc.). He linked better health to a healthy diet. But what of disease? Is that related?

Diet, lifestyle, and mental health can all influence health, but can they influence disease predisposition? Genetic profile and biochemical factors at a cellular level are more likely to predict liability to disease than food. It is sometimes obvious clinically, however, that diet, lifestyle, and mental health can affect the course of disease.

The power of genetic influence is contrary to many views expressed in modern media broadcasts. There is a good psychological reason for this. Humans need hope, many feeling a need to control their destiny. Freedom to manipulate diet and lifestyle allows this. That our fate is pre-determined by genetic profile is likely to remain unacceptable to many.

Health is something all humans appreciate, although it is more difficult to define than the absence of disease. It includes having sufficient physical and mental energy for what we wish to accomplish; normal bodily functioning despite any disability (compared to others); adequate sleep and a disposition unhindered by depression or failures in life; the absence of any symptoms impairing our quality of life or interfering with our desired life narrative. Is it acquired, inborn or both? From personal experience, and from my association with thousands of patients, I would say both apply. There is no doubt about what can foster healthy feelings. They arise from exercise and attaining fitness, and from the contentment arising from the accomplishments of a fulfilled life. Unhealthy feelings can come from sloth, and the overconsumption of food and alcohol; from depression and a dissatisfied life. All result in what we perceive to be deficient personal energy. Could this relate to mitochondrial dysfunction, and if so, how?

The acquisition of health has ancient origins. In Greek mythology, *Hygieia* daughter of *Asclepius*, played an important part in her father's cult. Her father was associated with healing, while she focussed on the prevention of sickness and the continuation of good health. The Romans imported her as *Valetudo*, the goddess of personal health. In time she slowly became identified as the ancient Italian goddess of social welfare, *Salus*.

In primary prevention, providing medical prophylactic measures are the *raison d'être* of many companies promoting food supplements. Even the double Nobel prize winner, Linus Pauling advocated vitamin C (as an anti-oxidant) as beneficial for cancer and heart disease prevention (with no substantial proof yet forthcoming).

For prevention and diet, the British Heart Foundation dietician Victoria Taylor suggests:

> *'It is a good idea to avoid any multivitamins that include a mix of antioxidant vitamins such as vitamin A, vitamin E and beta-carotene. This is because research shows these supplements do not improve cardiovascular disease and are linked to a raised risk of death.*

Some supplements can also affect the medicines you are taking, causing side effects. For example, vitamin K, omega-3 and St. John's wort all interact with the blood-thinning drug warfarin.

. . . it is recommended we all take 10 micrograms a day of vitamin D supplement in the autumn and winter, because vitamin D is mainly made in our bodies with the help of sunlight.'

The ARIC study [2021; 1985 onwards]) is an ongoing population study in the US. The results suggest a reduction in myocardial infarction and CHD from using a vegetarian diet, alcohol, polyunsaturated fat, vitamins, minerals, garlic, Q10, phytosterols and polyphenols.

Not all suggested preventative measures have a research backing with worthwhile evidence of atheroma prevention and any significant reduction in CVS morbidity and mortality. There is a lot more evidence for speculation, commercial promotion, corporate profit and hearsay.

Many claims made are based on the anti-oxidant effect of food nutrients. Since oxidised LDL (ox-LDL, the product of free oxygen radicals) stimulates atheroma generation, it logical, but inductive reasoning, to think that nutrient anti-oxidants must be beneficial. Naturally occurring, arterial nitric oxide, has important roles. It is a vasodilator, and as an anaerobic agent, it will oppose the oxidation of LDL and atheroma formation, by displacing oxygen radicals. There is much detailed scientific justification for the idea, but little hard evidence to justify the claims and speculation of those promoting food supplementation.

Individual nutrients have not always been studied. The colour of fruit and vegetables (white, green, orange and red, related to flavonoid, carotenoid and anthrocyanidin content) is of significance to stroke risk. Linda Griep et al., (2011), found that only white fruit intake related to reduced stroke incidence over a ten year period (RR: 0.48). Also, 'Each 25-g/d increase in white fruit and vegetable consumption was associated with a 9% lower risk of stroke (HR, 0.91)'. In the same publication, Wersching commented:'The direct interplay between nutrients in whole fruits and vegetables may be more important than the unique nutrients on their own in the reduction of risk in cardiovascular disease'. The primary preventative directive to have five pieces of fruit per day remains. (Wersching, H. (2011) *An apple a day keeps stroke away? Consumption of white fruits and vegetables is associated with lower risk of stroke.* Stroke. 2011;42:3001-3002. doi: 10.1161/STROKEAHA.111.626754).

The Vegetarian Diet

Thorogood, M., Mann, J., McPherson, K. (1994). *Risk of death from cancer and ischaemic heart disease in meat and non-meat eaters.* BMJ; 308(6945):1667-70, showed that UK vegetarians had 40% fewer cancer deaths and 30% fewer CVS deaths than meat eaters, over a twelve-year period. The ARIC study is consistent and the World Health Organisation (WHO) recommends one glass of wine per day and nuts as cardioprotective.

Alcohol

Wine is said to loosen the tongue and warm the heart! Its effects are complicated. From meta-analyses, daily doses of alcohol between 2.5g and 60g daily can reduce both CVS mortality and mortality. (Piano, M.R. (2017)*Alcohol's Effects on the Cardiovascular System.* Alcohol Res; 38(2):210-241). Hypertension is common among regular drinkers. To find if this applies, get your patients to stop drinking for ten days. An intake of over 60-grams of alcohol daily, increases stroke risk (Ronskley P.E. et al. 2011).

Wine is an important component of the Mediterranean diet, and can raise HDL, lower LDL and reduce platelet adhesiveness (Gaetano, G de., Cerletti, C. et al. (2001) European Project. FAIR CT 97 3261 Project participants, in Nutrition, Metabolism and Cardiovascular Diseases; 11:47-50). In rabbits, wine can reduce atheroma generation (da Luz, P.L., Serrano Jr, C.V., et al.(1999). *The effect of red wine on experimental atherosclerosis: lipid-independent protection.* Experimental and Molecular Pathology, 65(3), pp.150-159).

Red wine contains resveratrol, arginine and polyphenols (all anti-oxidants); white wine only contains arginine. All are beneficial. Red wine extract provides a non-alcohol alternative. Because the amount of alcohol needed for any beneficial effect is likely to be 20-30grams per day (1/4 to 1/3 of a bottle of 14% alcohol wine), one must consider the risk of liver disease. The question is, how much do polyphenols (flavonoids) contribute to the protective effect of wine.

Polyunsaturated Fats

These dietary fats have unsaturated chemical bonds and are liquid at room temperature. We know that they reduce heart attack rates (GISSI [1999] and DART [1998] studies). Albert, C.M., Hennekens, C.H., et al. (1998). *Fish Consumption and Risk of Sudden Death*. JAMA; 279:23-28) also showed that they reduced sudden death, but not heart attack rates.

Vitamins

Vitamin E, an anti-oxidant and anti-coagulant, has been shown in animals to reduce early atheroma progression (Cyrus, T., Yao, Y., et al. *Vitamin E reduces progression of atherosclerosis in low-density lipoprotein receptor-deficient mice with established vascular lesions.* Circulation.4;10 7(4):521-3. In humans, its associated with brain haemorrhage has caused concern.

Folic acid and vitamins B_{12} and B_6, can reduce blood homocysteine, but only B_6 is known to reduce CVS risk (Huang, J., Khatan, P., et al. (2023) *Intakes of folate, vitamin B_6, and vitamin B_{12} and cardiovascular disease risk: a national population-based cross-sectional study.* Cardiovasc. Med. 14;10:1237103).

Because CAD is more common in the more northern and southern latitudes on Earth, there could be a connection between deficient sunlight and vitamin D status. Vitamin D supplementation in older adults, could reduce the number of major CVS events, but more research is required. (Thompson, B., Waterhouse, M., et al. (2023). *Vitamin D supplemen-*

tation and major cardiovascular events: D-Health randomised controlled trial. BMJ. 2023;381:e075230.)

A preliminary report from Denmark (1978), showed vitamin D blood levels to be considerably lower among myocardial infarction and angina patients, when compared to controls (Badskjaer, L.B., et al. 1978).

Vitamin D can inhibit LDL foam cell formation and the generation of atheroma, but the advantages of supplementation are not definite. For a review see: Haider. F., Ghafoor, H., et al. Vitamin D and Cardiovascular Diseases: An Update. Cureus. 2023 Nov 30;15(11). Modulation of endothelial function can suppress vascular calcifications and prevent plaque formation. Vitamin D also helps to regulate visceral and ectopic fat deposition, and deficiency can contribute to cardiometabolic dysfunction.

A recent study of vitamin D supplementation (daily dose of 2,000 IU) in 25,871 patients older than 50 years, found no subsequent advantage to cardiovascular outcomes. (Manson JE, et al. 2019).

There is some evidence that vitamin D reduces hypertension. Its negative RAAS regulation could signify an advantageous role for vitamin D in hypertension treatment. It helps to control endothelial and vascular smooth muscle cell proliferation.

Research into the CVS benefits of vitamin D continue.

Minerals

The minerals Mn, Mg, Se are contenders for primary prevention. Manganese can reduce atheroma in rabbits. Coronary atheroma in pigs is associated with low blood magnesium levels (Ito, M., Toda, T., et al. (1986). *Effect of magnesium deficiency on ultrastructural changes in coronary arteries of swine.* Acta Pathologica Japonica; 36: 225-234). The progression of atheroma is associated with low blood selenium levels (Salonen, J.T., Ylä-Herrtuala, S., et al. (1992). *Auto-antibody against oxidised and progression of carotid atherosclerosis.* Lancet; 339: 883-6).

Garlic

Steiner, M., Kahn, A.H., et al. (1996) *A double-blind crossover study in moderately hypercholesterolemic men that compared the effect of aged garlic extract and placebo administration on blood lipids.* Am. J. Clin. Nutrition;64(6):866-70), showed that 7.2 grams/day of aged garlic could reduce LDL by 4% and BP by 5%. Whether it affects CVS morbidity and mortality remains unproven.

Coenzyme Q_{10}

Commended by many, it is an essential compound found in the human body. Synthesised in the mitochondrial inner membrane, it is involved in ATP production. COQ10 exists in two forms: oxidized (ubiquinone) and reduced (ubiquinol). The heart contains a high concentration. It is a powerful intracellular antioxidant, helping to sustain endothelial NO, and allowing vasodilatation in hypertensive patients with some improvement in BP. (Ho M.J., Bellusci A., Wright J.M. 2009). Any benefit to CHD remains theoretical, and relates to its antioxidant, anticoagulant and anti-inflammatory effects. No trial has yet shown it to prevent heart attacks or strokes.

Phytosterols

The phytosterols: sitosterol, campesterol and sigmasterol and stanols, are plant-based chemicals, similar in structure to cholesterol. Like cholesterol, absorption from the bowel is poor. They have a claim for a role in primary CVS prevention, because they reduce the absorption of cholesterol from the bowel by up to 50%. Two and a half grams per day (found in Benecol® and Flora ProActive®) can lower blood cholesterol by 0.5mmols/l.

Polyphenols

These substances are manufactured partly as a plant defence mechanism. There are several subclasses: flavonols, flavones, isoflavones, flavanones, anthocyanidins. We find them in olive oils and various berries. They are anti-oxidant, anti-inflammatory, and can lower LDL and reduce platelet aggregation (as a COX 1 inhibitor). They can have a pro-oxidant effect in the presence of iron and copper.

In the endothelium, resveratrol can stimulate eNOS activity and increase the amount of NO. It can also prevent hypertrophy and re-modelling in hypertension. As such, polyphenols might restrain atherosclerosis development, but I have seen little direct proof of it, except for curcumin reducing the fatty streaks appearing in rabbit arteries (Majeed, M.L., Ghafil, F.A., et al. (2019) *Anti-Atherosclerotic and Anti-Inflammatory Effects of Curcumin on Hypercholesterolemic Male Rabbits.* Indian J Clin Biochem.; 36(1): 74-80).

Resveratrol, epigallocatechin gallate (EGCG) (from green tea), and curcumin (with turmeric as Indian spices) have beneficial effects on cardiovascular health, some acting against aging. (Khurana, S., Venkataraman, K., et al. (2013) *Polyphenols: Benefits to the Cardiovascular System in Health and in Aging.* Nutrients.Sep 26;5(10):3779–3827). Heavy green tea drinkers appear to have fewer cardiac infarctions (Geleijnse J.M. , Launer L.J., et al.(2002). *Inverse association of tea and flavonoid intakes with incident myocardial infarction: The Rotterdam Study.* Am. J. Clin. Nutr. 2002;75:880–886).

One can find resveratrol in red grapes, blueberries, peanuts, itadori tea, hops, pistachios and in grape and cranberry juices. Pinot noir is the grape yielding most.

Quercetin is another polyphenol; found in capers, cocoa powder, broccoli and green and black tea and plums. In high dosage it can slightly reduce blood pressure in hypertension, perhaps acting as an ACE inhibitor

(Balasuriya N., Rupasinghe H.P. (2012). *Antihypertensive properties of flavonoid-rich apple peel extract.* Food Chem. 2012;135:2320–2325).

Polyphenols are in berries (blueberry, bilberry, ligonberry, black currant or blackberry, raspberry, cranberry and strawberry) The anthocyanins are responsible for the dark colour of berries (blackberries and grapes) and vegetables like red cabbage and pomegranates. Most is in the skin. Grapes contain high amounts of polyphenols in their skin, flesh and juice. In the Zutphen study of older male adult subjects, an inverse correlation was found between the consumption of flavonoid rich fruits and vegetables, and mortality from coronary heart disease. (Hertog M.G., Feskens E.J., et al. *Dietary antioxidant flavonoids and risk of coronary heart disease: The Zutphen Elderly Study.* Lancet. 1993;342:1007–1011).

Polyphenols are in olive oil; more in the extra virgin variety, and those made from less ripe olives. The presence of protective monosaturated fatty acids, is perhaps more significant.

Aspirin

Patients have long thought that taking 50mgs of aspirin daily, will prevent strokes and heart attacks, but can this be substantiated?

When trialled on American physicians, with 22,017 participants over a five year period, the trial was stopped early. A 44% primary reduction of myocardial infarctions, occurred among the physicians who took 325mgs aspirin on alternate days. There was a slight increase stroke occurrence, but the difference failed to reach statistical significance. Aspirin caused no reduction in CVS deaths, but did increase stomach ulcers by 22%. The reduction in myocardial infarction occurred most among those over 50-years-of-age, and those with a low blood cholesterol (Steering Committee of the Physicians' Health Study Research Group (no authors listed). 1989 . *Final report on the aspirin component of the ongoing Physicians' Health Study.* NEJM.; 321(3):129-35).

A meta-analysis study 26 years later, found a 38% increase in stroke rate, a 6% reduction in all-cause mortality and a 10% reduction in major cardiovascular events, from prophylactic aspirin (dose range not stated).

Also found was a reduced incidence of cancer (Sutcliffe, P., Connock, M., et al. (2013).*Aspirin for prophylactic use in the primary prevention of cardiovascular disease and cancer: a systematic review and overview of reviews.* Health Technol. Assess.;17(43):1-253). On this basis, any benefit from a reduction in myocardial reduction risk, is more than offset by the increased stroke risk; aspirin as a primary preventative strategy is thus unacceptable. I have always discouraged hypertensive patients taking it without good reason.

A Japanese study of patients aged 60 to 85-years of age, found that aspirin (100mgs daily) 'did not significantly reduce the risk of the composite outcome of cardiovascular death, nonfatal stroke, and nonfatal myocardial infarction among Japanese patients 60 years or older with atherosclerotic risk factors.' (Ikeda, Y., Shimada, K., et al., (2014). *Low-dose aspirin for primary prevention of cardiovascular events in Japanese patients 60 years or older with atherosclerotic risk factors: a randomized clinical trial.* JAMA, 17;312(23):2510-20).

By comparison, the prophylactic use of aspirin in secondary prevention has undoubted benefits.

The Polypill

The idea of giving a fixed dose combination of drugs for primary and secondary CVS prevention is now twenty-years old. Wald and Law introduced the idea in 2003. (Wald, N.J., Law, M.R. (2003) *A strategy to reduce cardiovascular disease by more than 80%.* BMJ (Clinical research ed.). (2003) 326:1419). Many trials have assessed its efficacy, dangers and cost-effectiveness. They advocated giving a fixed dose of aspirin, a beta-blocker, an ACE inhibitor and a statin to high-risk patients. Their original claim was for an 80% reduction in adverse CVS outcomes; a claim that proved to be doubly exaggerated.

Although understandable as a public health measure, it is less appropriate in medical practices where treatment can be customised, rather than generalised. It may be convenient to take one, rather than several pills at once, but the doses of each component, cannot be altered when required.

Although a sound idea for high-risk patients, physicians should be sceptical of any one-size-fits-all, treatment policy.

One meta-analysis that included 20653 people, revealed a 29% primary risk reduction of fatal and nonfatal major adverse cardiovascular events while taking the Polypill (a statin and two anti-hypertensive drugs; some with aspirin). For high-risk people, the reduction was 37%. (Kandil, O., Motawea, K.R., et al., (2022). *Polypills for Primary Prevention of Cardiovascular Disease: A Systematic Review and Meta-Analysis.* Frontiers Cardiovasc. Med.; 14: 9:880054).

There is also high-quality evidence that fixed-dose, combination therapy, reduces all-cause mortality by 10%. (Rao, S., Siddiqi, T.J., et al. (2022). *Association of polypill therapy with cardiovascular outcomes, mortality, and adherence: A systematic review and meta-analysis of randomized controlled trials.* Prog. Cardiovasc. Dis.; 73:48-55). In addition, the Polypill is cost effective.(Jahangiri R, Rezapour A, et al. (2022). *Cost-effectiveness of fixed-dose combination pill (Polypill) in primary and secondary prevention of cardiovascular disease: A systematic literature review.* Plos One. 2022;17(7):e0271908).

'Statins'

Most physicians accept the benefit of prescribing 'statins', but the evidence is not consistent. The aim is to reduce blood cholesterol as a risk factor, to below 5mmols/l (especially LDL-cholesterol to below 1.8m mols), by inhibiting hepatic production. As it happens, 'statins' also reduce atheroma formation, a more pertinent measure of cardiovascular risk (CLAS Study. Azen, S.P. et al. 1996).

'Statins' are hydroxymethylglutaryl-coenzyme A (HMG-CoA) reductase inhibitors. They affect cholesterol biosynthesis in a reversible, rate-limiting way by inhibiting the conversion of HMG-CoA, to l-mevalonate + CoA. They competitively bind to the catalytic domain of HMG-CoA reductase.

Several variants are available, among which are: atorvastatin, rosuvastatin, pravastatin, and simvastatin, to treat primary hypercholesterolemia. Red yeast rice contains insignificant amounts of simvastatin.

Beware of interactions with common drugs (erythromycin, clarithromycin, diltiazem, verapamil, gemfibrozil and sildenafil).

After reviewing pooled analyses, the US Preventive Services Task Force, reported on 'statin' therapy in 2022. They reported:

- A decreased risk of all-cause mortality in 18 trials (n=85,816; relative risk reduction 8%; absolute risk reduction −0.35%).

- A decreased risk of fatal or nonfatal stroke in 15 trials (n=76,610), relative risk reduction 22%; absolute risk reduction, -0.39%).

- A decreased risk of fatal or nonfatal myocardial infarction in 12 trials (n=76498; relative risk reduction 33%; absolute risk reduction. -0.89%).

- A decreased risk of composite cardiovascular outcomes in 15 trials (n=74,390; relative risk reduction 28%); absolute risk reduction, −1.28%).

Twelve trials (n=75,138) reported on cardiovascular mortality, but only one trial, WOSCOPS (West of Scotland Coronary Prevention Study, n=6595), reported a statistically significant difference between statin and placebo in the risk of cardiovascular mortality (Relative Risk Reduction, 32% at 6 years; absolute risk reduction, −0.70%.In pooled analyses of all 12 trials, 'statin' therapy was associated with only a slight reduction in cardiovascular mortality risk (at 2 to 6-years) that was not statistically significant (relative risk reduction, 9%; absolute risk reduction, −0.13%).

'Statins' are effective in those with one or more CVS risk factors (dyslipidaemia, diabetes, hypertension, or smoking), but only between the ages of 40 and 75-years of age (US Preventive Services Task Force (2022). *Statin Use for the Primary Prevention of Cardiovascular Disease in Adults. US Preventive Services Task Force Recommendation Statement.JAMA.* 2022;328(8):746-753. doi:10.1001/jama.2022.13044).

Many treated patients report side-effects. Ten to 15% of those taking them will experience muscle pains; 30% stop taking them within the first year. Changing the drug often helps; in my experience, atorvastatin and rosuvastatin are best tolerated in my experience. Do patients now expect to get muscle pains? The level of intolerance I have seen in stoic patients, does not suggest a nocebo effect. Cataracts and cognitive impairment have been reported, but remain unconfirmed as consequences of 'statin' therapy.

Rhabdomyolysis is rare (0.15 per million prescriptions). It is associated with muscle pains, red-brown urine, renal impairment and raised enzyme blood levels (CK, CK-MB, AST, ALT and troponin).

SECONDARY PREVENTION

All primary preventative measures apply to secondary prevention. The impact of secondary prevention can, however, be more impressive.

Secondary prevention targets those with established cardiovascular disease. It includes any strategy that reduces the probability of another cardiovascular event. It includes continuing with all the primary intervention methods, and in addition reviewing patients for signs of deterioration, the efficacy of their drug therapy, and any need for surgical intervention and counselling. The aims are to reduce morbidity and mortality.

What level of secondary reduction should one accept as valuable? In population terms, even a small reduction (5%), would account for substantial numbers of people benefitting. A 3 : 1 chance of likely benefit, would not be unreasonable for any individual considering a prophylactic tablet or lifestyle measure change. Any prophylactic measure advised, should therefore reduce CVS outcomes by 30%. Some interventions I have reviewed so far, offer a 50% reduction (a 2 : 1 chance of benefit), but a 20-30% benefit is more usual (odds of 5 : 1 and 3 : 1 for achieving benefit).

Screening and Secondary Prevention

Because cardiovascular symptoms can occur late in the natural history, screening asymptomatic patients allows early disease detection and offers a chance to improve outcomes. If no treatment is available for the condition being detected, is there any point to it? We can, at least, build a database of patients who might get some benefit in the future.

The quest for effective secondary prevention continues, even for those where screening has detected any early disease (atheroma, hypertension, adult congenital heart disease, etc.). The results can help to re-direct patient strategy at an early stage in the natural history of their condition. At the moment, this rarely includes genetic evaluation (or modification), but it will in the future.

Cancer Research UK, published an article that demonstrated the reduced cost of treating cancer early (see Birtwhistle, Mike (September

2014): *Saving lives and averting costs? The case for earlier diagnosis just got stronger.* Cancer News). In all corporations, including the NHS, financial considerations greatly influence patient management. For instance, what is the value and cost of adding a routine ECG to a screening evaluation? Routine ECGs for athletes, done to detect a liability to sudden death, would cost $62,000 to $130,000 per life-year saved (Wheeler, M .T. et al. 2010. *Cost effectiveness of pre-participation screening for prevention of sudden cardiac death in young athletes.* Ann. Int. Med: 152(5): 276-286)doi: 10.1059/0003-4819-152-5-201003020-00005. Those who can afford it, can choose for themselves, thus maintaining the health divide!

When spending public money, medical bureaucrats rightly insist on some measure of value (ignoring the large sums spent on their salaries and administration). Accountants working with them will be directed to calculate how much it costs to save or extend a life. As a physician, I would sooner waste money than waste lives; an outlook, resource restricted public service bureaucrats, cannot be seen to encourage.

The value of currency now exists only by consensus; the value of life and death is a tangible reality. Compare the money we spend on projects detached from reality or practicality. The Europa space project, for instance, cost $5 billion! It hopes to discover the ingredients of life on Europa, one of the remote icy moons of Jupiter. Who will dare openly discuss its relevance to humanity, and its value for money, given that the promotion of science is so important? Whatever is found, someone is bound to say, 'How interesting!', then struggle to its justify human relevance.

Patient Involvement in Secondary Prevention

Educating patients is an important secondary preventive strategy. Patients who know nothing about medicine, and have no wish to know about their own medical condition, are playing a game of chance with their risks. Some see no need to comply with advice, keeping appointments, or taking medication. Many think they know best (despite their lack of medical knowledge and experience). Some of them will recognise a person whose opinion they value more than any medical professional. Consider educating them, and if this fails, banishing them from your practice. Safe patient doctor relationships work best with mutual respect, patient education and cooperation.

Perhaps one should consider a double-blind trial to measure the prophylactic effectiveness of patient knowledge and education on their morbidity and mortality.

In contributing to their medical management, many patients are capable of effective co-operation, understanding, and compliance with agreed management suggestions. It is usually obvious which patients are adaptable, but surprises await. Some patients need a stimulus, like a close friend having a stroke, heart attack or dying, before they will request a medical evaluation and come to regard early diagnosis and preventative medicine as expedient. Some doctors, whose only wish is to follow the established medical model (each patient complaint followed by a specific solution), may not want to embrace early detection. Others will favour secondary prevention, but only after experiencing preventable medical events themselves, or among their family and friends.

Mr. C.F. had heart failure. He had SOB, and pitting oedema up to his thighs. His JVP was raised, but only after brief exercise. He needed diuretics, but which one, and in what dose? I weighed him, started him on a daily regime of frusemide 20mgs, and saw him again after five days. He had lost 3 kilos, had less oedema, and had passed urine frequently. Because his JVP was still raised, my guess was that he needed to lose another two kilos of fluid.

Mr. C.F. was a highly capable person. He continued to run his business while unwell. We decided he should weigh himself every morning (after toilet procedures). We agreed, he was to keep his weight steady at 100 Kgs. If it increased overnight, or he felt breathless lying flat, he would take a further dose of frusemide 20mgs a.s.a.p. He later remained on a daily maintenance dose of bendrofluazide 5mgs. His intelligent co-operation ensured optimum hydration and cardiac output.

Intelligent patients can take their own BP and pulse, and co-operate beneficially. Patient empowering, works well for some. The insight some patients gain from learning how therapy works, can significantly improve their engagement. The successful management of their condition often results.

Some patients, more inclined to arrogance and ignorance, will demand tests and treatments they have heard about at their local pub, on the Internet or in a newspaper. It is important to recognise that the combination of ignorance (no discernible education, or insight), and arrogance, can be fatal. Attempts to enlighten them could prove futile or be very hard work. Consider referring them elsewhere. Never encourage them unless they have a legitimate claim to arrogance (sometimes when medical professionals become patients) or hold verifiable beliefs (beware of the Halo bias).

Those seeking the early detection of heart disease must at least, rule out high blood pressure and atherosclerosis. Most doctors insist on blood lipid measurements, etc., without acknowledging that they lack specificity enough to rule out coronary disease in individuals. A painless ultrasound scan of carotid and other arteries will, however, detect atherosclerosis as a trait. My unpublished studies show that over 95% of those with proven coronary artery disease have carotid atheroma, while only 60% have hypercholesterolaemia.

Those with a family history of high blood pressure, should appreciate that it can change from normal to constantly raised, within one year. It would seem that the controlling genes possess a clock mechanism directing the age of onset. I have seen the onset of hypertension arising within one year, many times. It occurs especially in women as their menopause progresses. It is much less predictable in men.

Medical Prophylaxis

Aspirin

It was the ISIS-2 trial in 1988 that established secondary benefits of aspirin (160mgs daily). (*Randomised trial of intravenous streptokinase, oral aspirin, both, or neither among 17,187 cases of suspected acute myocardial infarction:* ISIS-2. ISIS-2 (Second International Study of Infarct Survival) Collaborative Group. Lancet; 2(8607):349-60).

A meta-analysis of the long-term effects of anti-platelet therapy (including aspirin 75-150mgs for two years) showed a 25% odds reduction in second heart attacks, and an 11% reduction in second strokes. Antithrombotic Trialists' Secretariat, Clinical Trial Service Unit, Radcliffe Infirmary, Oxford. *Collaborative meta-analysis of randomised trials of antiplatelet therapy for prevention of death, myocardial infarction, and stroke in high risk patients.* BMJ 2002; 324:71

For those with stable angina, 325mgs of aspirin on alternate days, reduced myocardial infarction rates by 87% (Ridker, P.M., Manson, J.E., et al. (1991).*Low-dose aspirin therapy for chronic stable angina. A randomized, placebo-controlled clinical trial.* Ann. Int. Med. 15;114(10):835-9).

No difference was found in side-effects or major CVS incidents, between those taking 81mgs or 325mgs of aspirin daily. (Jones, W.S., Mulder, H. et al. (2021). *Comparative Effectiveness of Aspirin Dosing in Cardiovascular Disease.* NEJM; 384:1981-1990.

The Polypill

Polypill use for secondary prevention, was examined by the SECURE trial. They recruited 2499 participants over the age of 65-years, with a previous myocardial infarction or a previous stroke within six months.

The polypill used contained aspirin (100 mg), ramipril (2.5, 5, or 10 mg), and atorvastatin (20 or 40 mg). A 30% relative reduction in CVS events (cardiovascular death, nonfatal myocardial infarction, or nonfatal ischemic stroke), was seen within 36-months of follow-up. (Castellano, J.M., Popcock, S.J., et al. (2022). *Polypill Strategy in Secondary Cardiovascular Prevention.* N Engl J Med 2022;387:967-977).

'Statins'

In secondary prevention, once a major CVS event has occurred, or atheroma and hypertension have been identified, statins are of proven benefit.

NICE (Quality statement 6: Statins for secondary prevention) states:

'High-intensity statins are the most clinically effective option for the secondary prevention of CVD – that is, reducing the risk of future CVD events in people who have already had a CVD event, such as a heart attack or stroke. Evidence shows that atorvastatin 80 mg is the most cost-effective high-intensity statin for the secondary prevention of CVD, which can improve clinical outcomes.'

A meta-analysis of secondary prevention in older patient over 65-years of age, and a previous MI or CHD, haven taken simvastatin (20-40mg), pravastatin (40mg) or fluvastatin (80mg), showed:

- A reduction in all-cause mortality of 22% (RR 0.78);
- A reduced CHD mortality of 30% (RR 0.7);
- Non-fatal MI was reduced by 26% (RR 0.74);
- The need for revascularisation reduced by 30% (RR 0.70);
- Stroke incidence reduced by 25% (RR 0.75).

(Afilalo J, Duque G, et al.. *Statins for secondary prevention in elderly patients: a hierarchical Bayesian meta-analysis.* J Am Coll Cardiol. 200 8;51:37–45.).

These results are representative of many other trial results available, although the intensity of treatment varies between them.

It has been concluded that: '. . . there is a graded association between intensity of statin therapy and mortality in a national sample of patients with ASCVD(atherosclerotic cardiovascular disease). High-intensity statins were associated with a small but significant survival advantage compared with moderate-intensity statins, even among older adults. Maximal doses of high-intensity statins were associated with a further survival benefit.' (Rodriguez F., Maron D.J., et al., (2017) *Association between intensity of statin therapy and mortality in patients with atherosclerotic cardiovascular disease.* JAMA Cardiol 2017; **2**: 47).

Their patients were aged 18 to 85-years old. In this study, they defined the treatment received as follows:

Low-intensity statin therapy: fluvastatin 20 to 40 mg; lovastatin 20 mg; simvastatin 10 mg; pitavastatin 1 mg; and pravastatin 10 to 20 mg.
Moderate-intensity therapy was atorvastatin 10 to 20 mg; fluvastatin 40 mg, twice daily or 80 mg once daily; lovastatin 40 mg; pitavastatin 2 to 4 mg; pravastatin 40 to 80 mg; rosuvastatin 5 to 10 mg; and simvastatin 20 to 40 mg.
High-intensity statin therapy was atorvastatin 40 to 80 mg, or rosuvastatin 20 to 40 mg.

To get the best results, patients must take their medication consistently.

I would like to see a forthcoming trial that separates those with hyperlipidaemia alone, from those with atheroma and hyperlipidaemia, and those with atheroma and no hyperlipidaemia. I strongly suspect that 'statins' will bring most benefit to those whose with atheroma, justifying the need for artery scanning as an early diagnostic measure.

Beta-Blockade

A meta-analysis of 202,752 patients treated with beta-blockade after myocardial infarction found:
Benefit among the elderly, those with pulmonary disease and those with a reduced ejection fraction.
A 40% reduction in mortality.
The same applied to patients with non-Q-wave infarction and those with chronic obstructive pulmonary disease.

(Gottlieb, S., McCarter, R.J., Vogel, R.A. (1998). *Effect of beta-blockade on mortality among high-risk and low-risk patients after myocardial infarction.* NEJM339(8):489-97).

The latest advice, expressed at the American College of Cardiology's Annual Scientific Session, April 2024 (Nicole Napoli, nnapoli@acc.org, 202-669-1465), was:

'Taking beta blockers after a heart attack did not significantly reduce the risk of death or a second heart attack among people with normal heart pumping ability, as indicated by an ejection fraction of 50% or higher'.
This findings calls into question the routine use of beta blockers for all patients following a heart attack, which has stood as a mainstay of care for decades. Approximately 50% of heart attack survivors do not experience heart failure. Among such patients, the study found no difference in the composite primary endpoint of death from any cause or new non-fatal heart attack between those who were prescribed beta blockers and those who were not.'
How long should beta-blockade be continued after cardiac infarction? In those with an ejection fraction> 40%, in one recent trial, no difference in major events occurred between those taking beta-blockade for 1.2 or 6.4 years. (Silvain, J., Cayla, G., et al. (2024). *Beta-Blocker Interruption or Continuation after Myocardial Infarction.* N Engl J Med 391:1277-1286.)

ACE Inhibitors and ARBs

Both drug groups benefit those with hypertension and heart failure. But how do they effect the occurrence of myocardial infarction, all-cause mortality, and major CVS events?

ARBs significantly reduce systemic blood pressure, stroke occurrence and the subsequent development of heart failure and diabetes mellitus. Considerable controversy once existed about the secondary preventive effect of ARBs on myocardial infarction.

No such controversy exists for ACE inhibitors. The evidence has consistently supported them reducing the incidence of myocardial infarction, all-cause mortality, CVS deaths and stroke.

From a MEDLINE and Cochrane database searches, Saha, S.A., et al. 2007), extracted data from four trials (n=555). Those using ramipril, perindopril, quinapril, or trandolapril had reduced:

- All-cause mortality by 13% (RR: 0.87);

- Cardiovascular mortality by 17% (RR: 0.83);

- Non-fatal acute myocardial infarction by 16% (RR: 0.84);

- Fatal and non-fatal acute myocardial infarction by 17% (RR: 0.83);

- Stroke by 23% (RR: 0.77);

- Any need for percutaneous coronary intervention/coronary artery bypass graft by 5% (RR: 0.95);

- Hospitalisation for congestive heart failure by 22% (RR: 0.78), and

- New-onset diabetes by 21% (RR: 79).

There was no statistically significant difference between placebo and angiotensin-converting enzyme inhibitors for the numbers admitted to hospital with angina. (Saha, S.A., Molmar, J. Arora, R.R. (2007). *Tissue ACE inhibitors for secondary prevention of cardiovascular disease in patients with preserved left ventricular function: a pooled meta-analysis of randomized placebo-controlled trials.* Database of Abstracts of Reviews of Effects (DARE): Quality-assessed Reviews. Bookshelf ID: NBK73586).

Another review of the literature, found similar results for ACE inhibitors, namely:

- A 14% decrease in all-cause mortality;

- A 19% reduction in cardiovascular mortality;

- An 18% reduction in myocardial infarction, and

- A reduction of stroke by 23%.

(Danchin, N., Cucherat, M., Thuillez, C. (2006). Angiotensin-Converting Enzyme Inhibitors in Patients With Coronary Artery Disease and Absence of Heart Failure or Left Ventricular Systolic Dysfunction. An Overview of Long-term Randomized Controlled Trials.Arch Intern Med. 2006;166(7):787-796).

Results of the HOPE trial (Heart Outcomes Prevention Evaluation) and the EUROPA trial (*European Trial on Reduction of Cardiac Events With Perindopril in Patients With Stable Coronary Artery Disease*), showed a significant benefit from ACE use, although the IMAGINE trial (

Ischemia Management With Accupril Post-Bypass Graft via Inhibition of the Converting Enzyme), failed to show a benefit from ACE use.

The QUIET Trial hypothesis (Pitt, B., O'Neill, B. et al. 2001), was that 'quinapril 20 mg/day would reduce ischemic events (the occurrence of cardiac death, resuscitated cardiac arrest, nonfatal myocardial infarction, coronary artery bypass grafting, coronary angioplasty, or hospitalization for angina pectoris)'. No significant reduction of these end-points occurred (RR 1.04) compared to the controls over a two-year period, although quinapril reduced the need for new case angioplasty (P< 0.0 18). The suggestion is that it slows atherogenesis (Pitt, B., O'Neill, B. et al. (2001) *The QUinapril Ischemic Event Trial (QUIET): evaluation of chronic ACE inhibitor therapy in patients with ischemic heart disease and preserved left ventricular function.* Am J Cardiol. 87(9):1058-63).

Subsequent trials were undertaken, such as CAMELOT (Comparison of Amlodipine vs Enalapril to Limit Occurrences of Thrombosis), QUIET (Quinapril Ischemic Event Trial).

The CAMELOT trial (Nissen, S.E., Tuzcu, E.M., Libby, P. 2004) concluded:'Administration of amlodipine to patients with CAD and normal blood pressure, resulted in reduced adverse cardiovascular events (RR: 0.69). Similar, but smaller and nonsignificant treatment effects, occurred with enalapril (RR: 0.85).'

Surprisingly, it also stated: *'For amlodipine, IVUS showed evidence of slowing of atherosclerosis progression.'* This reached only borderline statistical significance.(Nissen, S.E., Tuzcu, E.M., Libby, P. (2004) *Effect of Antihypertensive Agents on Cardiovascular Events in Patients With Coronary Disease and Normal Blood Pressure. The CAMELOT Study: A Randomized Controlled Trial. JAMA.* 2004;292(18):2217-2225).

For an ARB, the VALUE trial (*Valsartan Antihypertensive Long-Term Use Evaluation*) showed a 19% increase in MI in those that took valsartan, rather than amlodipine. This was thought caused by ARB activating potentially deleterious angiotensin II type 2 receptors. It became referred to as the ARB paradox.

Any increase in myocardial infarction from an ARB, or any difference between ARBs and ACEs, has now been refuted by head-to-head analyses. ARBs improve stroke risk, but do not improve the secondary occurrence of myocardial infarction (RR 1.05) or deaths.

See: Masserli, F.H., Bangalore, S.(2017). *Angiotensin Receptor Blockers Reduce Cardiovascular Events, Including the Risk of Myocardial Infarction*.Commentary. Circulation; 135 (22).

Also, Strauss, M.H., Hall, A.S. (2017). *Angiotensin Receptor Blockers Do Not Reduce Risk of Myocardial Infarction, Cardiovascular Death, or Total Mortality: Further Evidence for the ARB-MI Paradox.* Circulation Commentary. Vol 135; 22.

Anticoagulation

When I was a junior doctor, the initial treatment for acute cardiac infarction was warfarin. A subsequent primary prevention trial compared it to aspirin for secondary prevention. (MRC, GP Research Framework(1998).*Thrombosis prevention trial: randomised trial of low-intensity oral anticoagulation with warfarin and low-dose aspirin in the primary prevention of ischaemic heart disease in men at increased risk.* Lancet; 351 (9098):233-242).The result of taking warfarin (average INR=1.49), was a 39% reduction (p=0·003) of fatal events. Warfarin reduced the death rate from all causes by 17% (p=0·04). Aspirin caused by a 32% reduction in non-fatal events (p=0·004). Using both agents had a better outcome than either alone.

'No benefit of warfarin alone or warfarin+aspirin compared with aspirin alone, could be documented for the secondary prevention of ischemic events'(Huynh, T., Théroux, P., et al. (2001) *Aspirin, Warfarin, or the Combination for Secondary Prevention of Coronary Events in Patients With Acute Coronary Syndromes and Prior Coronary Artery Bypass Surgery.* Circulation; 103(25):3069).In this trial INR was controlled between 2.0 and 2.5.

In a similar trial (Hurlen, M., Abdelnoor, M., et al. 2002), using warfarin (INR 2.8 - 4.2, or 2.0 to 2.5) and aspirin (75mgs or 160mgs), a composite of death, nonfatal reinfarction, or thromboembolic cerebral stroke, occurred in 20% of those on aspirin alone, and in 16.7% of those on warfarin alone. For those taking both, there was a 15% reduction.

Episodes of major, nonfatal bleeding, were observed in 0.62 percent of patients receiving warfarin and in 0.17 percent of patients receiving aspirin (P< 0.001).

Their conclusions stated: 'Warfarin, in combination with aspirin or given alone, was superior to aspirin alone in reducing the incidence of composite events after an acute myocardial infarction but was associated with a higher risk of bleeding.' (Hurlen, M., Abdelnoor, M., et al. (2002). *Warfarin, Aspirin, or Both after Myocardial Infarction.* NEJM 347(13):969-974.

Most approaches have used antiplatelet therapies to reduce the risk of ischemic recurrences. Aspirin, when used as a secondary prevention agent, lowers the risk of adverse cardiovascular events, while also decreasing the risk of CV death, compared with a placebo. The addition of a $P2Y_{12}$ inhibitor (clopidogrel, prasugrel, and ticagrelor) to aspirin therapy, is known as dual antiplatelet therapy (DAPT). Cardiovascular recurrences occur despite DAPT, although its benefits beyond one year are accompanied by an increased risk of bleeding and no difference in mortality. The addition of a NOAC was subsequently considered.

The (COMPASS) trial (Cardiovascular Outcomes for People Using the Anticoagulation Strategies) undertook to compare:

Rivaroxaban 2.5mgs bd + aspirin 100mgs daily (the DPI group)
versus - rivaroxaban 5mgs bd alone,
versus - aspirin 100mgs daily.

Cho, S.W., et al. (2019), reviewed 24,824 patients with stable CAD or PAD, of mean age 68.2 years. A history of MI was present in 69% (17028 patients).

A summary of their results is:

- Although there was a reduction in myocardial infarction in the doubly treated (DPI group) compared with aspirin. This did not reach statistical significance.

- With DPI, the occurrence of fatal and disabling stroke decreased (HR: 0.58).

- Fewer CV deaths, strokes, and MIs, occurred in the DPI group than in the aspirin-only group (HR: 0.76).

- In peripheral vascular disease, rivaroxaban 5mg bd., resulted in a decrease in major events, including major amputation (HR: 0.67)

- Mortality was lower in the DPI group than in the aspirin alone group (HR: 0.82).

- Compared to aspirin alone, the DPI group showed more major bleeding (HR: 1.66; $P < 0.0001$).

- There were more major bleeds with rivaroxaban alone.

- No differences occurred in the primary outcome of graft failure, diagnosed oneyear after surgery.

- No significant difference was noted in the occurrence of stroke in the rivaroxaban alone, compared to aspirin alone.

The net-clinical-benefit of rivaroxaban was almost the same as aspirin alone (HR: 0.94). Cho, S.W., Franchi, F., Angiolillo, D.J. ((2019). *Role of oral anticoagulant therapy for secondary prevention in patients with stable atherothrombotic disease manifestations.* Ther. Adv. Haematol.:10: 2040620719861475.

Anticoagulation, Non-Valvular AF and Stroke Reduction

For stroke reduction in AF, other trials must be considered, but it is important to distinguish between those with and without valve abnormalities.

It is now established that the CHA2DS2-VASc score accurately predicts ischaemic stroke risk for those with AF. When the result is > 2 in men, and > 3 in women, the risk is significantly increased (Allan, V., Banerjee, A., et al., (2024) *Net clinical benefit of warfarin in individuals with atrial fibrillation across stroke risk and across primary and secondary care.* Heart; 103(3).)

For older adults with non-valvular AF, warfarin (INR of 1.5 to 2.1 or 2.2 to 3.5), there was a higher bleeding risk associated with higher INRs (mean 2.8) and older age (> 74 years).(Yamaguchi, T. (2000) Optimal Intensity of Warfarin Therapy for *Secondary Prevention of Stroke in Patients with Nonvalvular Atrial Fibrillation : A Multicenter, Prospective, Randomized Trial.* Stroke; 31(4):817.

In patients with paroxysmal, or sub-clinical AF (found on 24hr ECG recordings), the risk of ischaemic stroke becomes 2.5-fold more than those without AF. In older adults of mean age 76.8, the conclusion was that 'Among patients with subclinical atrial fibrillation, apixaban (2.5mgs bd) resulted in a lower risk of stroke or systemic embolism (RR: 0.78) than aspirin (RR: 1.24), but a higher risk of major bleeding.' (Healey, J.S., Lopes, R.D., et al. (2023).*Apixaban for Stroke Prevention in Subclinical Atrial Fibrillation.* NEJM; 390:107-117).

Thrombotic material in atrial fibrillation (AF) can develop in the left atrial appendage. The cause relates to: stasis, endothelial dysfunction and a hypercoagulable state indicated by increased fibrinogen, D-dimer, thromboglobulin and platelet factor 4 levels. In some cases, left atrial appendage excision is now being advised.

According to ESC (2012) Guideline Update on Atrial Fibrillation: Prevention of Thromboembolism in Non-valvular Atrial Fibrillation, one should advise antithrombotic therapy for all patients in AF, except for:

- Those at low risk—CHA2DS2-VASc score < 2.

- Those with lone AF.

- Those aged < 65 years,

- Those with contraindications.

Also note that:

- Patients aged > 65 years, especially women, are at high risk of ischaemic stroke.

- Adjusted-dose warfarin and antiplatelet agents, can reduce the risk of stroke by up to 60%, in patients with AF.

- Aspirin plus clopidogrel, or aspirin only, should be considered in patients who refuse a NOAC, or cannot tolerate anticoagulants for reasons unrelated to bleeding.

- When vitamin K antagonists (VKAs) are unacceptable, use a direct thrombin inhibitor (dabigatran), or an oral factor Xa inhibitor (rivaroxaban, apixaban).

In a major study (Camm. A.J, et al., 2021), the outcomes of using a VKA or a NOAC for anticoagulation (in 25,551 patients with non-valvular AF), was observed for two years. For NOAC use, the results showed:

- Fewer all-cause deaths and ischaemic strokes,

- but the risk of major bleeding was much higher.

- All-cause deaths and major bleeds were significantly lower with a NOAC than with a vitamin K antagonist.

Camm, A.J., Fox, K.A.A, Virdone S for the GARFIELD-AF investigators, *et al., Comparative effectiveness of oral anticoagulants in everyday practice.* Heart 2021;**107:**962-970.

Anticoagulation: Valvular AF and Stroke Reduction

Rheumatic heart disease (RHD) as a component of valvular AF (VAF), is seen mainly in developing countries (Blustin, J.M., McBane, R.D., et al. (2014) *Distribution of thromboembolism in valvular versus non-valvular atrial fibrillation.* Expert Review of Cardiovascular Therapy, 12(10), 1129–1132. https://doi.org/10.1586/14779072.2014.960851.

The risk of AF and thromboembolism from valvular heart disease, is approximately three times greater than for those without valvular heart disease. Evidence from the Framingham study shows that:

Patients with rheumatic heart disease and AF, have an 18-fold higher risk of stroke than age and blood pressure matched controls.

Non-valvular AF is associated with a five to six-fold stroke risk, compared to controls.

(see: Kannel, W.B., Abbott, R.D., Savage, D.D., McNamara, P.M. *Epidemiologic features of chronic atrial fibrillation: the Framingham study.* N Engl J Med 1982;306:1018-22).

Prosthetic Heart Valve Risk

There are no data to support the use of non-vitamin K oral anticoagulants (NOACs) in patients with mechanical heart valves (MHV). Dabigatran, an oral direct thrombin inhibitor, was associated with a higher rate of valve thrombosis and bleeding complications in patients, shortly after placement of an MHV, when compared to warfarin in the RE-ALIGN study.

All mechanical valves require lifelong oral anticoagulation using a vitamin K antagonist. Target INR depends on the type and location of the mechanical heart valve and the coexistence of risk factors for thromboembolism (atrial fibrillation, hypercoagulable state, previous thromboembolic event or LV dysfunction).

For MHVs, there is no longer a general recommendation for concomitant antiplatelet therapy in combination with anticoagulation.

Compared to MHVs, the long-term risk of thromboembolism in patients with bio-prosthetic valves (BPV) and sinus rhythm is low. The risk slightly higher for mitral rather than aortic BPVs. The first 90 to 180 postoperative days are a high-risk period.

After surgical valve implantation, the current guidelines recommend short term (3 to 6 months) anticoagulation using a VKA (INR target of 2.5), in the absence of any indication for chronic anticoagulation. This is based on the small risk of thromboembolism before the BPV is fully endothelialised. Alternatively, low-dose aspirin (75 to 100 mg daily) can be given for the first three months following aortic BPV implantation or for life.

After TAVI, 2017 European guidelines favour dual antiplatelet therapy for the first 3 to 6 months, followed by lifelong single antiplatelet therapy . (Brandt, R.R., Pibarot, P. (2021).Prosthetic heart valves: Part 2 - Antithrombotic management, European Society of Cardiology; e-journal of cardiological practice; 20(2)).

Diet

NICE has issued advice (NG238, 2023), for those with established CHD, and those with a high risk of CVD. It is 'to eat a diet in which total fat intake is 30% or less of total energy intake, saturated fats are 7% or less of total energy intake, and where possible saturated fats are replaced by mono-unsaturated and polyunsaturated fats'.

All the dietary nutrient considerations mentioned under primary prevention, might also apply to secondary prevention. Although substantial evidence of benefit may be lacking, at least most are harmless (except for vitamins A, E, K, and an excess of D).

We usually advise patients with hypertension and heart failure to reduce their sodium intake and increase their potassium intake. In naturally hypotensive patients, this can lead to vasomotor syncope, especially in hot weather. Those on a low salt diet who are given a diuretic, an ACE inhibitor, an ARB or a peripheral dilator, might also risk vasomotor syncope.

# Figures & Illustrations

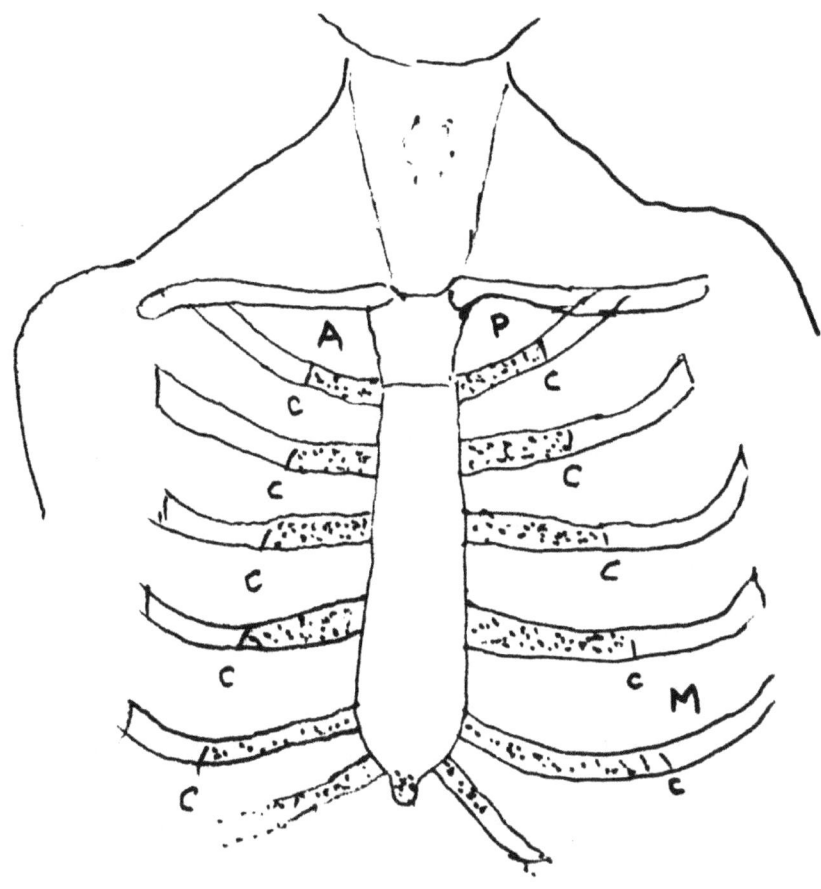

Fig 1: Anterior Chest Wall. Costochondral Cartilages (dotted Areas). A= Aortic Area; P=Pulmonary Area; M= Mitral Area

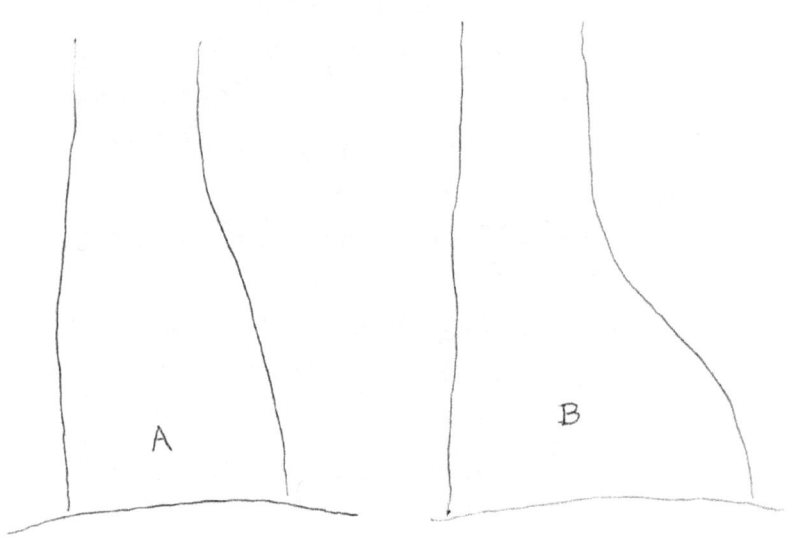

Fig 2: A: Normal Heart Shape on PA CXR.   B: Heart Shape with LVH

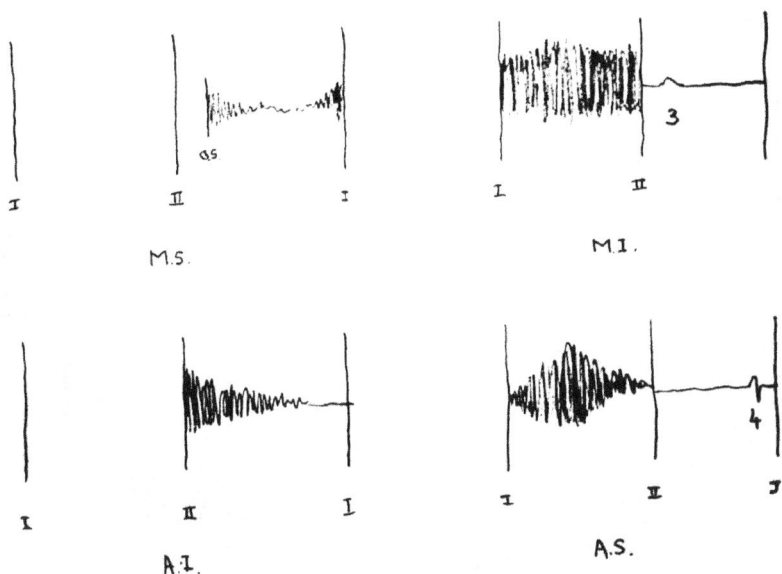

Fig 3: Stylised Phonocardiograms. MS=Mitral Stenosis; MI= Mitral Incompetence: AS=Aortic Stenosis; AI=Aortic Incompetence. I=1st Heart Sound; II=2nd Heart Sound.

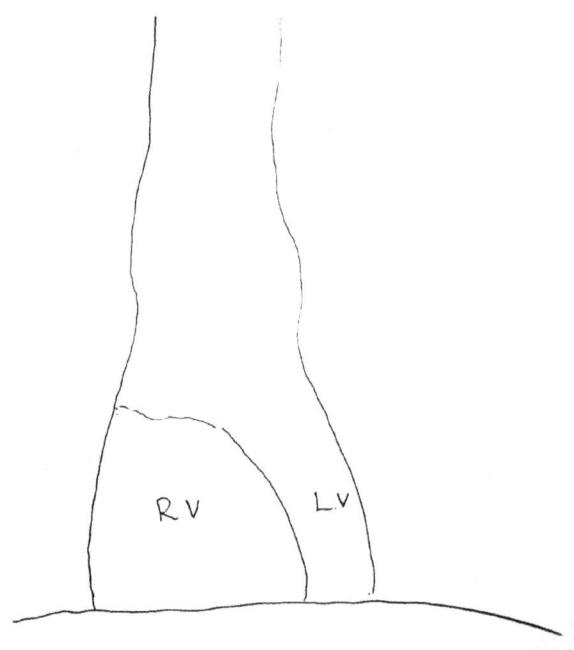

Fig 4: RV overlapping LV as seen on a PA CXR

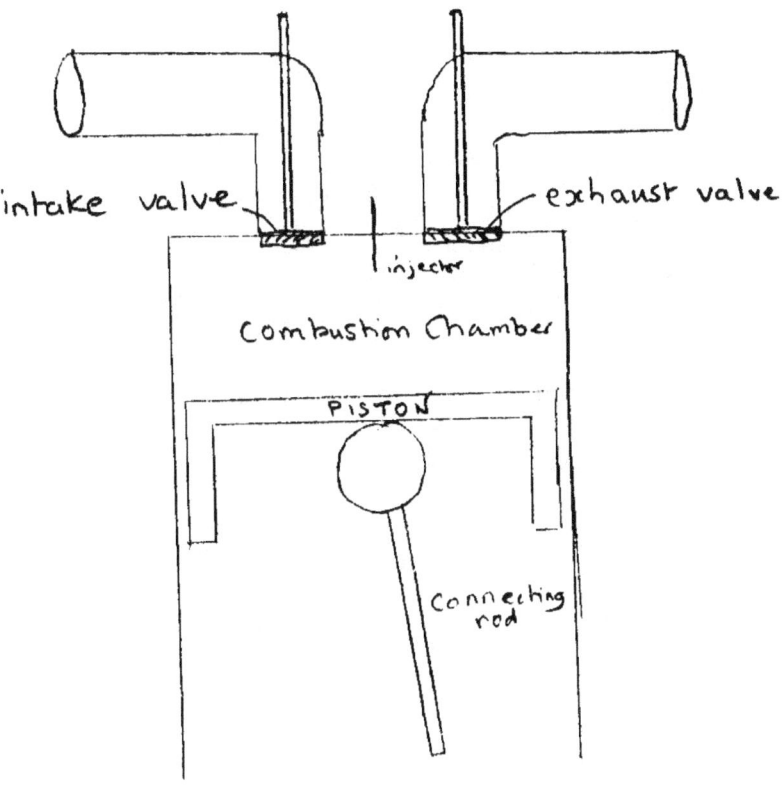

Fig 5: Combustion Engine Cylinder with Intake and Outlet Valves. Compare to LV in Fig 6.

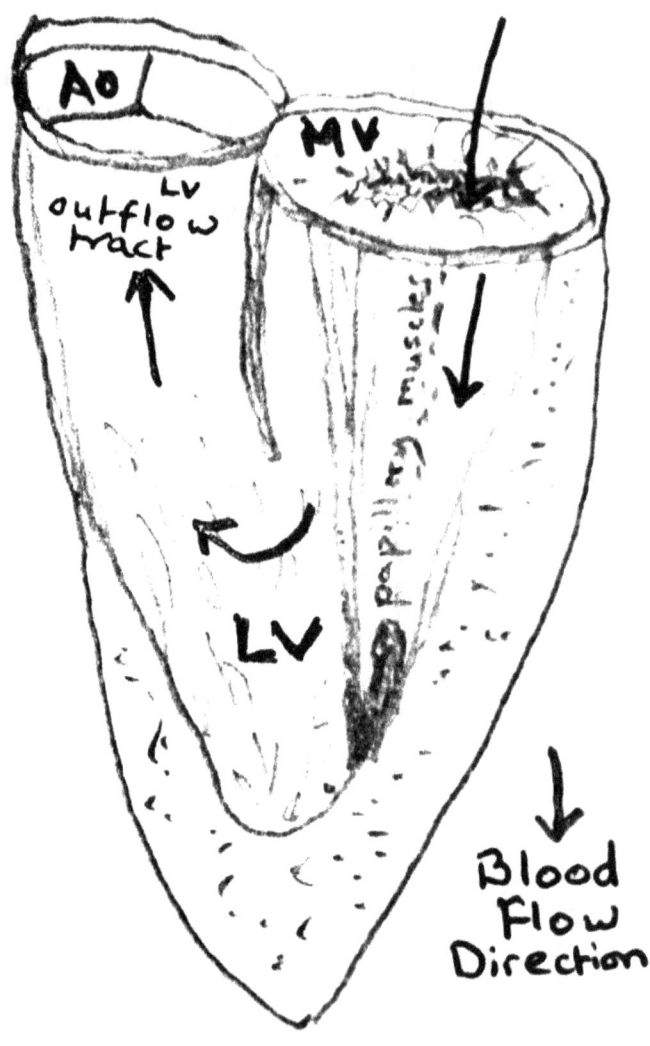

Fig 6: LV with Mitral Inlet Valve (MV) and Aortic Outlet Valve (Ao). Compare to Engine Cylinder Fig 5.

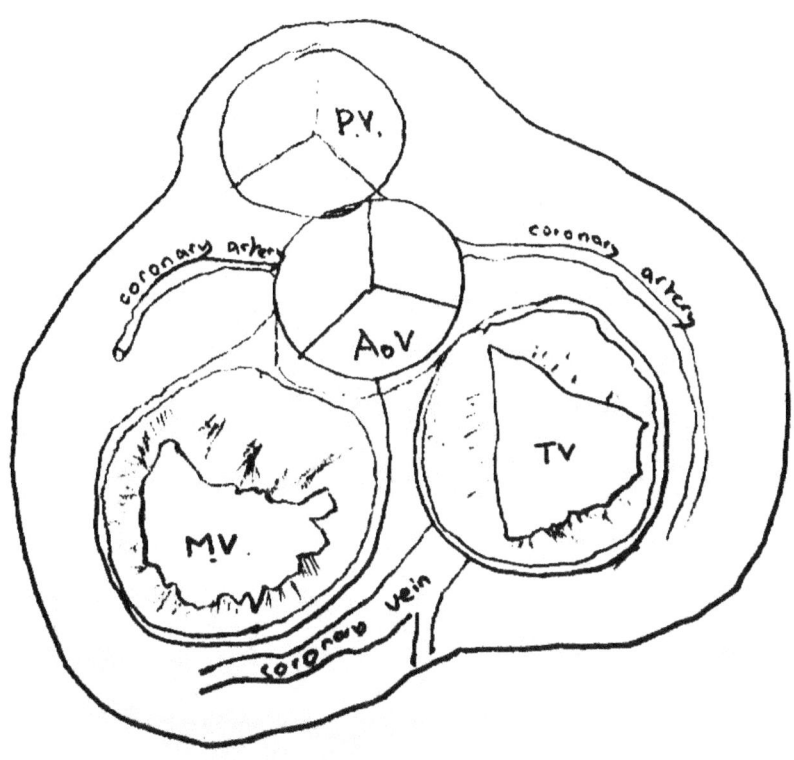

Fig 7: Proximity of Heart Valves in the same Plane. Ao =Aortic; MV=Mitral; PV= Pulmonary; TV= Tricuspid Valve

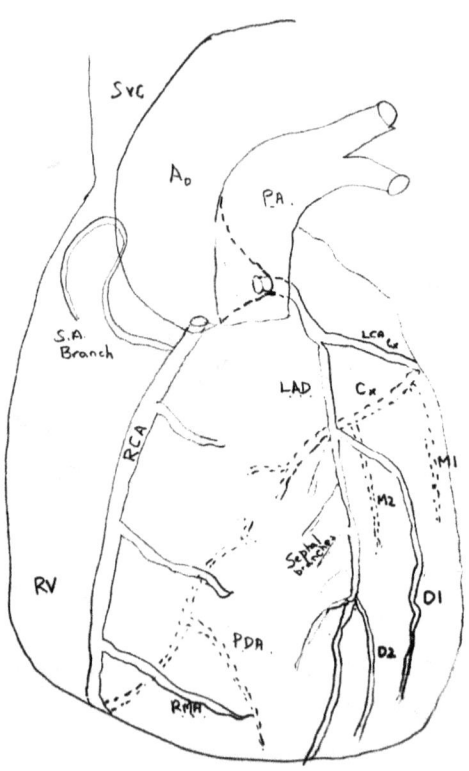

Fig 8: Coronary Artery Tree. RCA=Right Coronary; LAD=Left-Anterior Descending; Cx=Circumflex; D1 & 2=Diagonals; M1&2 Marginals

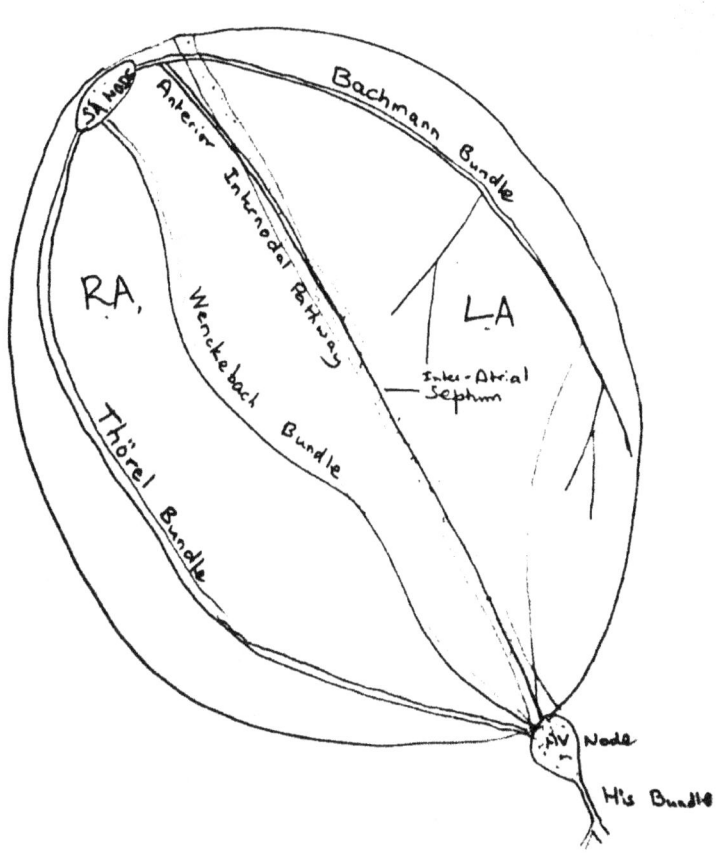

Fig 9: Preferential Atrial Conduction Pathways.

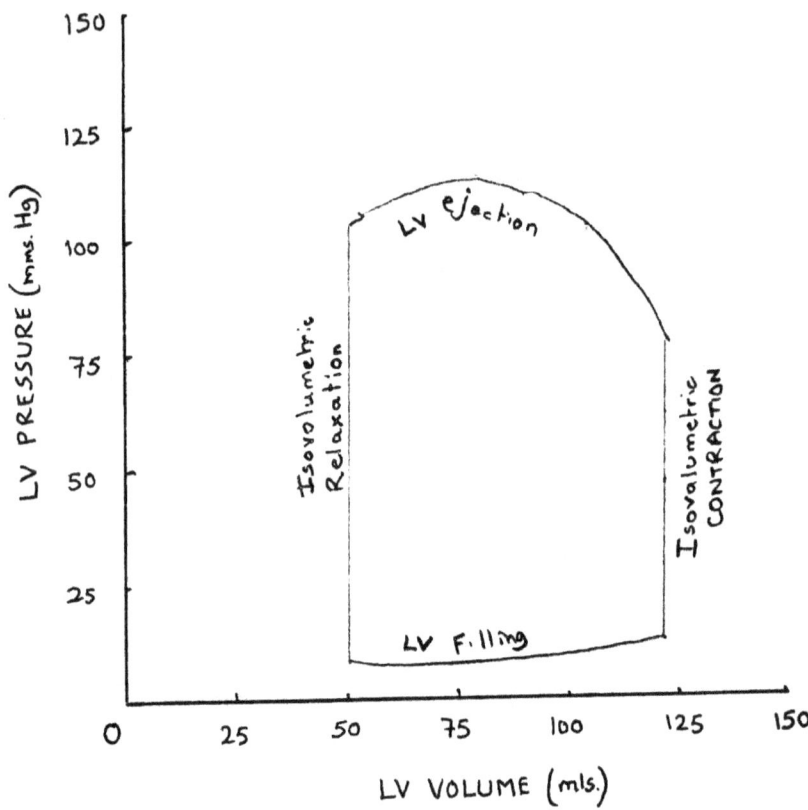

Fig !0: Four Phases of the LV Pressure – Volume Cycle

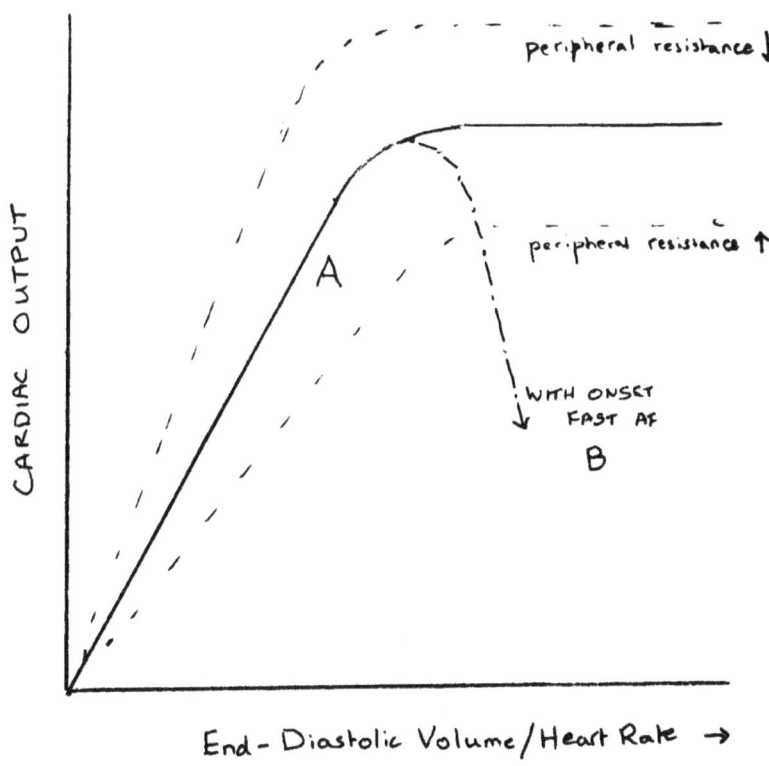

Fig 11: The Frank-Starling Relationship: A: Steadily rising Cardiac Output versus Heart Rate and / or End Diastolic Volume

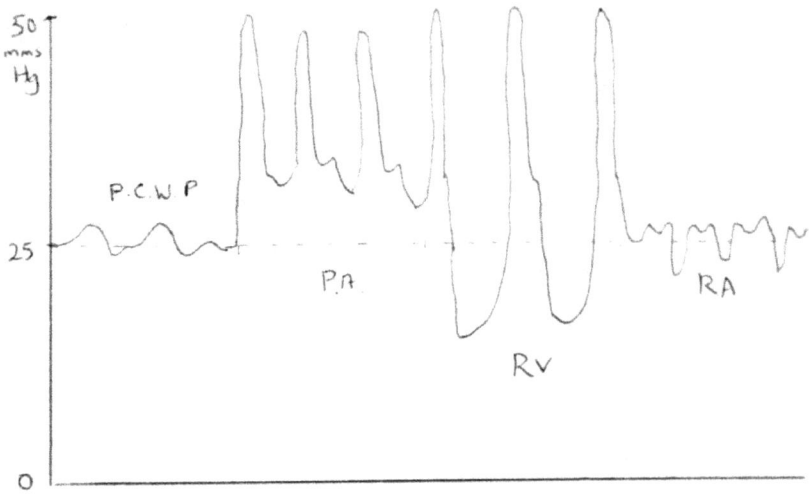

Fig 12: Pull-Back Pressures from Pulmonary Wedge Pressure (PCWP) back to Pulmonary Artery (PA), Right Ventricle (RV) and Right Atrium (RA).

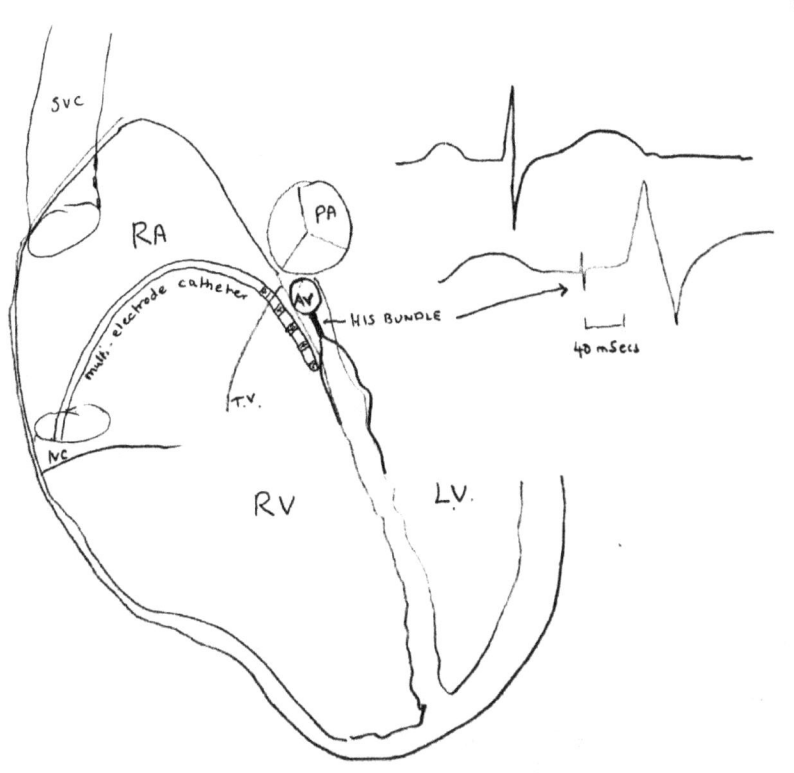

Fig 13: Electrode Catheter in RA with tip near His Bundle. His Deflection between P-wave and QRS

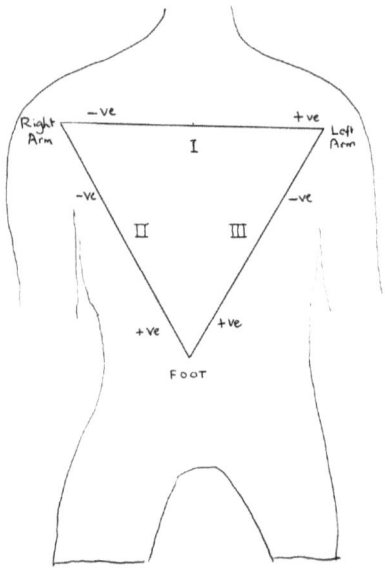

Fig 14: Einthoven's Triangle for Frontal Plane ECG leads. Note which vector directions are +ve and -ve.
Equivalent diagrams can be drawn for the horizontal and sagittal planes

Fig 15: Intracellular and Extracellular Electrograms

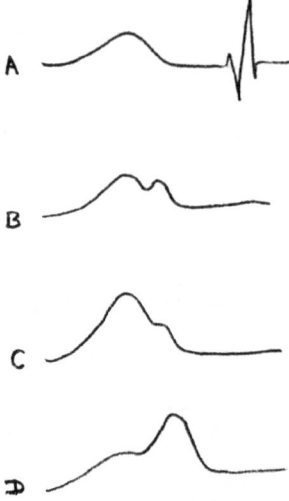

Fig 16: P-wave Morphologies. A=Normal; B=Bifid; C=P-Pulmonale; D=P-Mitrale

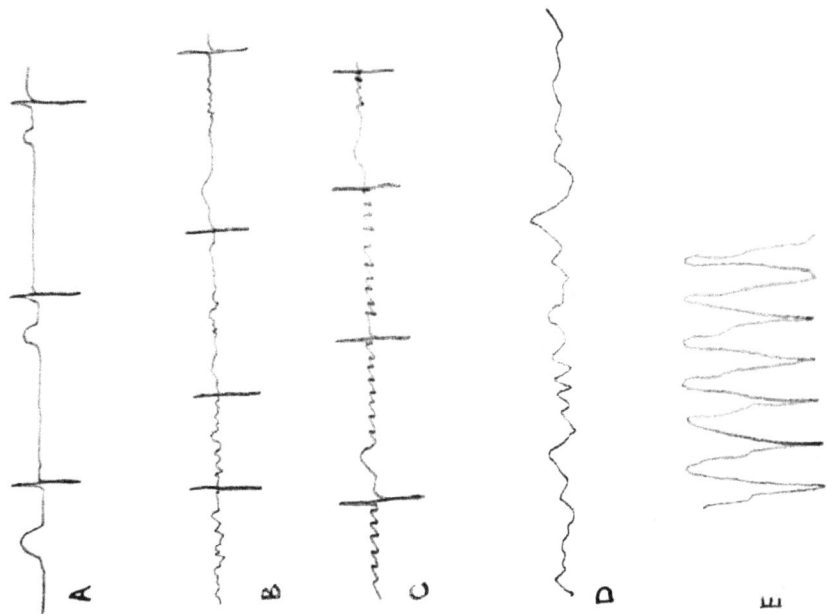

Fig 17: A=Sinus Rhythm; B=Atrial Fibrillation; C=Atrial Flutter; D=Ventricular Fibrillation; E=Ventricular Tachycardia

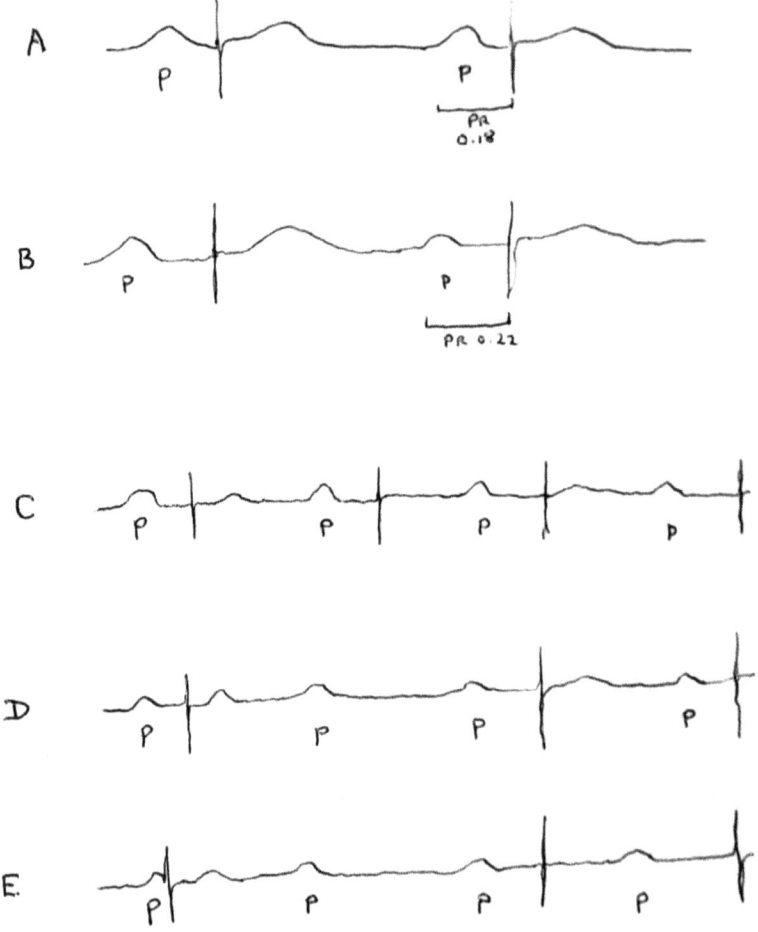

Fig 18: Various Degrees of AV Block: A=Normal; B: First Degree Block; C=Wenckebach Block; D=2:1 Block; E=Complete Block

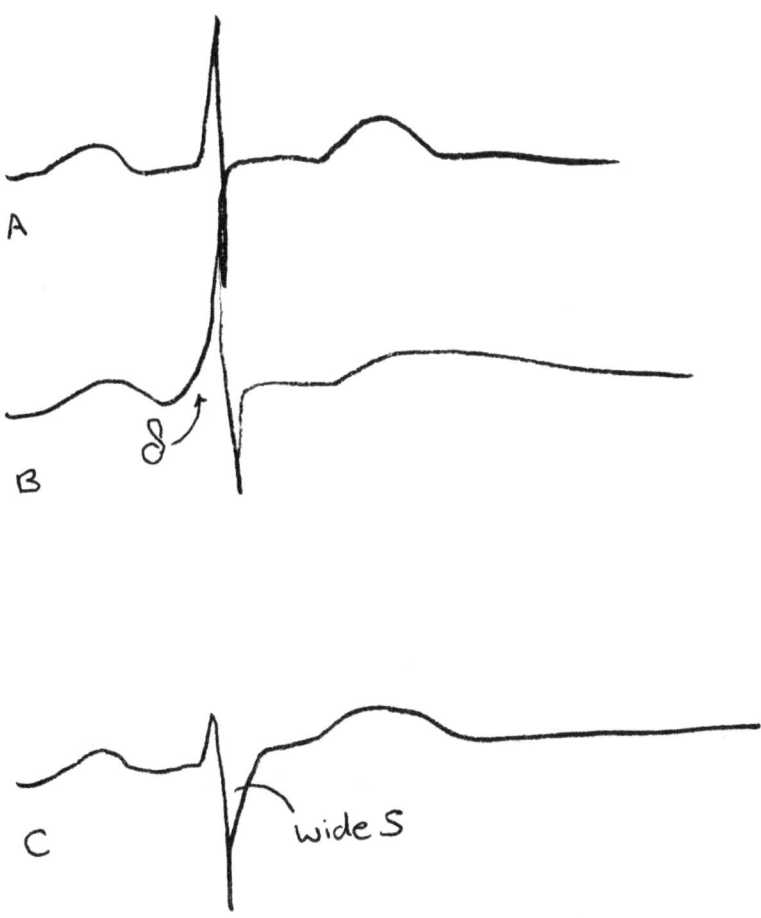

Fig19: Normal ECG; B=Delta Wave of pre-excitation; C= wide S wave from RBBB

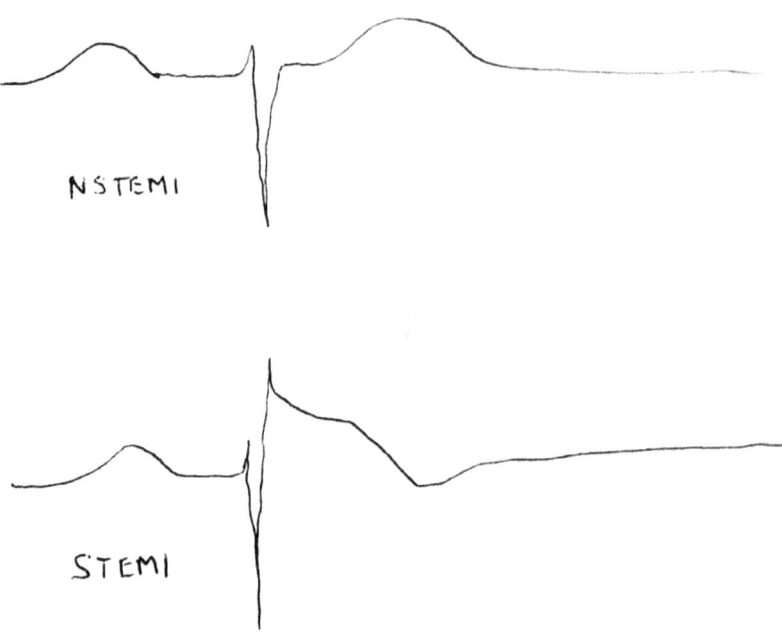

Fig 20: Tio Panel: Non ST Elevation in Myocardial Infarction; Lower Panel ST Elevated in Myocardial Infarction

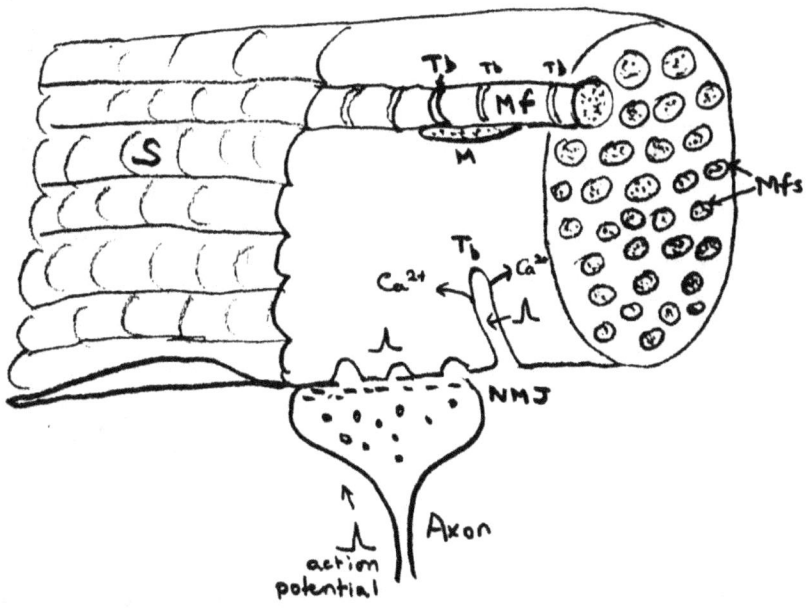

Fig 21: Diagram of Microscopic Myofibril. NMJ=Neuromuscular Junction; Tb=Transverse Tubules conducting action potentials; S=Sarcolemma

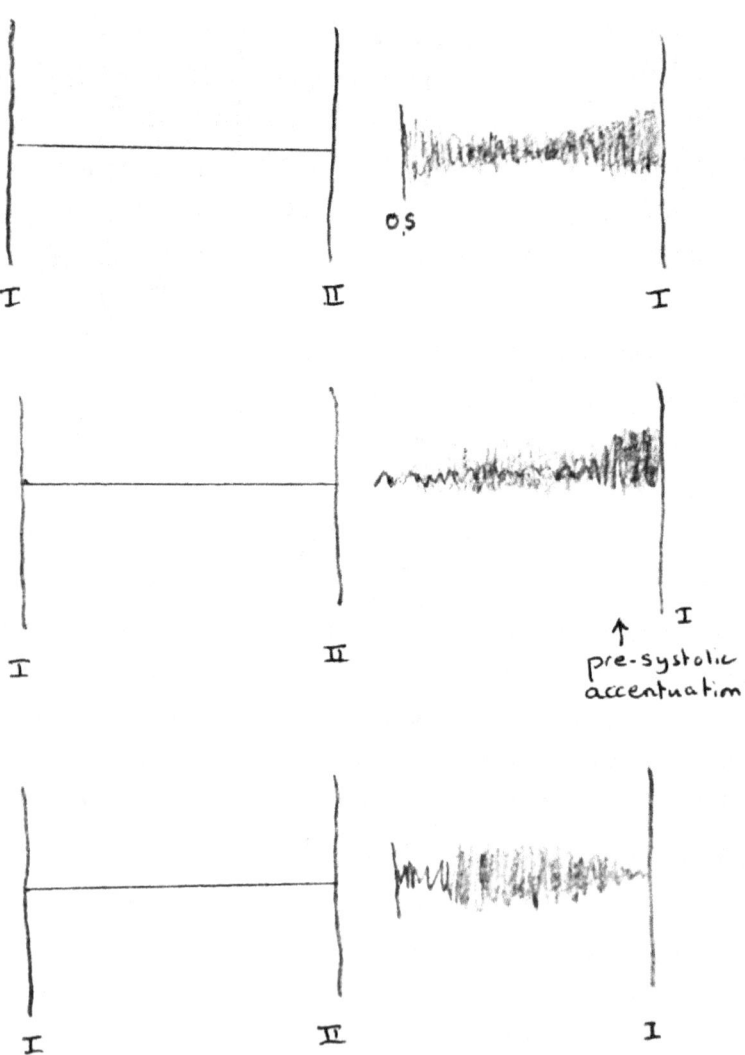

Fig 22: Stylised Murmurs of MS. Top & Bottom have an opening snap (pliable valve); Top with pre-accentuation;.Middle pane: Stiff MV no OS.

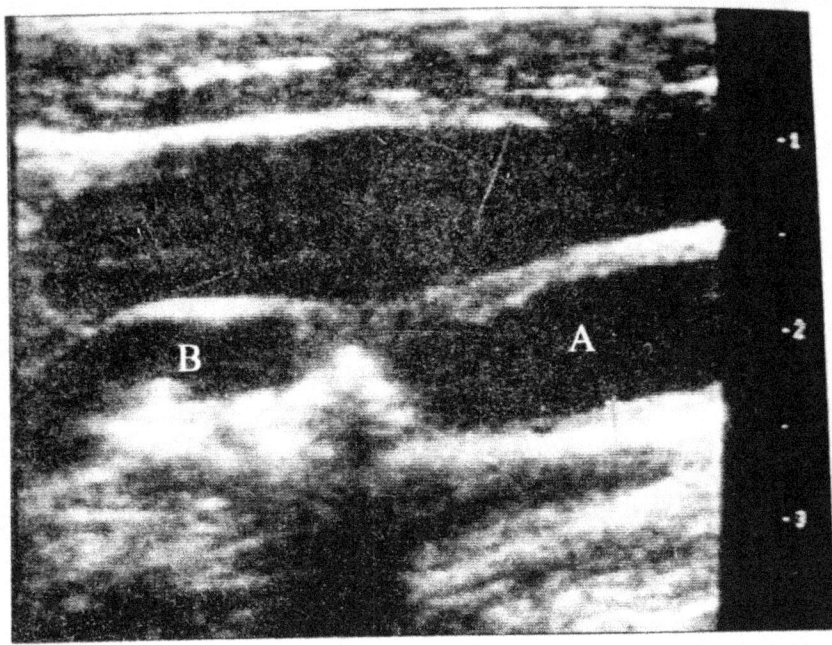

Fig 23: Carotid Atheroma. Ultrasound Image. A= Carotid Artery Lumen with thickened intima; B= Large Obstructive Calcified Plaque

Abbreviations Used

AF: Atrial Fibrillation or Flutter.

AI: Aortic Incompetence.

AS: Aortic Stenosis.

ASD: Atrial Septal Defect.

ATP: Adenosine TriPhosphate.

AV Node: Atrioventricular Node.

CABG: Coronary Artery Bypass Graft.

CAD: Coronary Artery Disease.

CCU: Coronary Care Unit

CHD: Coronary Heart Disease.

CRP: C-Reactive Protein. Blood test for inflammation.

CT: Computed (X-Ray) Tomography.

CVS: Cardiovascular System.

CXR: Chest X-ray.

DOAC: Direct Oral AntiCoagulants (Factor Xa inhibitors and dabigatran).

DNA: Deoxy-Ribonucleic Acid.

ESR: Erythrocyte Sedimentation Rate. Blood test for inflammation.

GWAS: Genome-Wide Association Studies.

HCM: Hypertrophic Cardiomyopathy.

HDL: High Density Lipoprotein.

IHD: Ischaemic Heart Disease.

INR: International Normalised Ratio (for Vitamin K anticoagulant control).

IVUS: Intravascular Ultrasound Study.

LDL: Low Density Lipoprotein.

LMWH: Low Molecular Weight Heparin.

LVH: Left Ventricular Hypertrophy

MHV: Mechanical Heart Valve.

MI: Myocardial Infarction and Mitral Incompetence.

MR: Mitral Valve Regurgitation.

MRI: Magnetic Resonance Imaging

MS: Mitral Stenosis or Multiple Sclerosis.

MVP: Mitral Valve Prolapse.

NICE: National Institute for Health and Care Excellence.

NO: Nitric Oxide.

NOAC: Novel Oral AntiCoagulants. (Factor Xa inhibitors and dabigatran).

PAD: Peripheral Artery Disease.

PCI: Percutaneous Coronary Intervention (angioplasty, TAVI).

PVD: Peripheral Vascular Disease.

PUO: Pyrexia of Unknown Origin.

RAAS: Renin Angiotensin Aldosterone System.

RNA: Ribonucleic Acid

SOB: Shortness of Breath.

SNP: Single Nucleotide Polymorphism (SNP). ('SNIP').

TOE: Trans-Oesophageal Echocardiogram.

TAVI: Transcatheter Aortic Valve Implantation.

URTI: Upper Respiratory Tract Infection.

VKA: Vitamin K Antagonist (warfarin etc.).

VSD: Ventricular Septal Defect.

Bibliography

AFFIRM study: Wyse, D.G., Waldo, J.P and AFFIRM writing group (2002).*A Comparison of Rate Control and Rhythm Control in Patients with Atrial Fibrillation. NEJM; 347:1825-1833.*

Afilalo J, Duque G, et al.. *Statins for secondary prevention in elderly patients: a hierarchical Bayesian meta-analysis.* J Am Coll Cardiol. 2008;51:37–45.).

Agatston, A.S., Janowitz, W. R., et al. *Quantification of coronary artery calcium using ultrafast computed tomography.*Journal of the American College of Cardiology. 1990.(15): 827-832).

Ahrén, B., Bengstsson, H.I., Hedner, P. (1986) *Effects of norepinephrine on basal and thyrotropin-stimulated thyroid hormone secretion in the mouse.* Endocrinology; 119(3): 1058-62).

Albert, C.M., Hennekens, C.H., et al. (1998) Fish Consumption and Risk of Sudden Death. JAMA; 279:23-28.

Allam, A.H., Thompson, R.C. et al.)(2011). *Atherosclerosis in ancient Egyptian mummies: the Horus study. JACC Cardiovascular Imaging.* Apr;4(4):315-27.

Allan, V., Banerjee, A., et al., (2024) *Net clinical benefit of warfarin in individuals with atrial fibrillation across stroke risk and across primary and secondary care.* Heart 103(3).)

Allessie M. A. L. W., Bonke F. I. M., Hollen S. (1985). J. *Experimental Evaluation of Moe's Multiple Wavelet Hypothesis of Atrial Fibrillation.* New York, NY, USA: Grune & Stratton).

ALLHAT study: *Major Outcomes in High-Risk Hypertensive Patients Randomized to Angiotensin-Converting Enzyme Inhibitor or Calcium Channel Blocker vs Diuretic.* JAMA.2002 ; 288(23):2981-2997. doi:10.1001/jama.288.23.2981).

ALLHAT Study. *Major Cardiovascular Events in Hypertensive Patients Randomized to Doxazosin vs Chlorthalidone. The Antihypertensive and Lipid-Lowering Treatment to Prevent Heart Attack Trial*

ALLHAT *JAMA.* 2000;283(15):1967-1975. doi:10.1001/jama.28 3.15.1967.Antithrombotic Trialists' Secretariat, Clinical Trial Service Unit, Radcliffe Infirmary, Oxford. *Collaborative meta-analysis of randomised trials of antiplatelet therapy for prevention of death, myocardial infarction, and stroke in high risk patients.* BMJ 2002; 324:71.

ARIC Study (1985 ongoing). Atherosclerosis Risk in Communities. https://www2.cscc.unc.edu/aric/external link. ALSO: The ARIC (*Atherosclerosis Risk In Communities) Study: JACC* Focus Seminar 3/8. JACC. 2021 Jun, 77 (23) 2939–2959.

ARISTOTLE Trial. Granger, C.B., Alexander, J.H. et al. (2011) *Apixaban versus Warfarin in Patients with Atrial Fibrillation.* N Engl. J. Med;365: 981-992.

ARNI trial for the use of the angiotensin-receptor neprilysin inhibitor in heart failure. (see also the PARADIGM-HF trial).

Aune, D., Sen, A. (2016). *BMI and all-cause mortality: systematic review and non-linear dose-response meta-analysis of 230 cohort studies with 3.74 million deaths among 30.3 million participants. BMJ*;353:i2156.

Bachar, J.B., Manna, B. (2023).*Coronary Artery Bypass Graft.* StatPearls. National Library of Medicine.

B.J.C. Staff (2016) *Direct current cardioversion and thromboprophylaxis in atrial fibrillation.* Br J Cardiol;23(suppl 2):S1–S12).

Bachar, J.B., Manna, B. (2023).*Coronary Artery Bypass Graft.* StatPearls. National Library of Medicine.

Badskjaer. L.B., Soerensen OH. Et al.(1978). Vitamin D and ischaemic heart disease. Metab Res. 1978;10:553–556. doi: 10.1055/s-0028-10933 90.

Baigent, C., Blackwell, L. et al (2009) *Aspirin in the primary and secondary prevention of vascular disease.* Lancet 373; 1849-60.

Balasuriya N., Rupasinghe H.P. (2012). *Antihypertensive properties of flavonoid-rich apple peel extract.* Food Chem. 2012;135:2320–2325).

Barton, Marc. *Willem Einthoven and the Electrocardiogram.* Past Medical History. https://www.pastmedicalhistory.co.uk/willem-einthoven-and-the-electrocardiogram/

Beall, G., Doniach, D. et al. 1969. *Inhibition of the long-acting thyroid stimulator (LATS) by soluble thyroid fractions.* J Lab Clin Med; 73(6):988-99).

Bernier R, Al-Shehri M, et al., (2018). *A Population-Based Study of Adherence to Guideline Recommendations and Appropriate-Use Criteria for Implantable Cardioverter Defibrillators.* Can J Cardiol.; 34(12):1677-1681. PubMed.

Besterman, E., Creese, R. *Waller – Pioneer of Electrocardiography.* British Heart Journal, 1979, 42, 61-61

Birtwhistle, Mike (September 2014): *Saving lives and averting costs? The case for earlier diagnosis just got stronger.* Cancer News.

Blustin, J.M., McBane, R.D., et al. (2014) *Distribution of thromboembolism in valvular versus non-valvular atrial fibrillation.* Expert Review of Cardiovascular Therapy, 12(10), 1129–1132. https://doi.org/10.1586/14779072.2014.960851.

Boden, W.E. et al. (COURAGE TRIAL)(2007).*Optimal Medical Therapy with or without PCI for Stable Coronary Disease.* NEJM; 356 (15): 1503-1516.

Bowditch, H.P. (1871) *Über die Eigenthümlichkeiten der Reizbarkeit, welche die Muskelfasern des Herzens zeigen. Berichte über die Verhandlungen der Königlich Sächsischen Gesellschaft zu Leipzig.* Mathematisch-Physische Classe 23: 652689, 1871. 2

Brandt, R.R., Pibarot, P. (2021).*Prosthetic heart valves: Part 2 - Antithrombotic management,* European Society of Cardiology; e-journal of cardiological practice; 20(2).

Camm, A.J, Fox, K.A.A, Virdone, S. for the GARFIELD-AF investigators, *et al.*, (2021). *Comparative effectiveness of oral anticoagulants in everyday practice.* Heart 107(12):962-970.

Camm, John. *'Why is rhythm control for atrial fibrillation becoming more popular?'* European Soc. Cardiol. (2023);1(2).

Canobbio, MM., Aboulhosn, J. et al. (2017). *The management of pregnancy in patients with complex congenital heart disease.* Circulation; 135(8): 85-87.

CAST: *Randomised placebo-controlled trial of early aspirin use in 20,000 patients with acute ischaemic stroke.* (Chinese Acute Stroke Trial) Collaborative Group. Lancet, 1997; 349:1641.

Castellano, J.M., Popcock, S.J., et al. (2022). *Polypill Strategy in Secondary Cardiovascular Prevention.* N Engl. J. Med. 2022;387:967-977.

Cavero, I., Holzgrefe, H. et al. (2023). Internodal Conduction Pathways: revisiting a century-long debate on their existence, morphology and location in the context of 2023 best science. Advances in Physiology Education; 47(4). Doi.org/10.1152/advan.00029.2023.

CHA_2DS_2-VASc stroke risk score: see Clincalc.com and Medcalc.

CHARM trial for the use of candesartan in heart failure.

Cho, S.W., Franchi, F., Angiolillo, D.J. ((2019). *Role of oral anticoagulant therapy for secondary prevention in patients with stable atherothrombotic disease manifestations.* Ther. Adv. Haematol.:10 : 2040620719861475.

Choi Y.M., Shim K.S., et al. (2012). *Transforming growth factor beta receptor II polymorphisms are associated with Kawasaki disease.* Korean J. Pediatr. 2012;55:18–23. doi: 10.3345/kjp.2012.55.1.18.

CIBIS-II trial for bisoprolol use in heart failure.

CLAIM Trial. Oparil S, Williams D, et al. (2001). *Comparative efficacy of olmesartan, losartan, valsartan, and irbesartan in the control of essential hypertension.* J Clin. Hypertension (Greenwich); 3 :283–91.

CLAS Study: Azen, S.P., Mack, W.J., et al. (1996). *Progression of coronary artery disease predicts coronary artery events: long-term follow-up from the Cholesterol Lowering Atherosclerosis Study (CLAS).* Circulation; 93: 34-41.

Clement, D.L., de Boyzere, M.L. et al. *Prognostic Value of Ambulatory Blood-Pressure Recordings in Patients with Treated Hypertension.* N Engl J Med 2003;348:2407-2415).

Cleveland Clinic Cardiology Board Review. 3rd Edition. 2022. p543. *Comparative effectiveness of oral anticoagulants in everyday practice*

Coats J. (1869) *Wie ändern sich durch die Erregung des n. vagus die Arbeit und die innern Reize des Herzens? Berichte über die Verhandlungen der Königlich Sächsischen Gesellschaft zu Leipzig.* Mathematisch-Physische Classe 21: 360391, 1869

CONSENSUS and SOLVD trials for the use of enalapril in heart failure.

COPERNICUS trial for the use of carvedilol in heart failure.

COURAGE TRIAL: *Optimal Medical Therapy with or without PCI for Stable Coronary Disease.* Boden W.E. et al. (2007). NEJM; 356 (15): 1503-1516).

Cowley, A.W. (2006) *The genetic dissection of essential hypertension.* Nature Reviews Genetics: 7; 829-840.

Criteria Committee, New York Heart Association, Inc. *Diseases of the Heart and Blood Vessels. Nomenclature and Criteria for diagnosis.* Sixth edition Boston, Little, Brown and Co. 1964, p 114.

Crocini, C., , Coppini, R., et al. (2014). *Defects in T-tubular electrical activity underlie local alterations of calcium release in heart failure.* Proc. Nat. Aad. Sciences of the USA;111 (42):15196-15201

Cutler E.C, Levine S.A. (1923). *Cardiotomy and valvulotomy for mitral stenosis; experimental observations and clinical notes concerning an operated case with recovery.* Bost. Med. Surg. J. ;188(26):1023–7.

Cuzick, J., Thorat, M.A., et al. Estimates of benfits and harms of prophylactic use os aspirin in the general population. Annal of Oncology 28: 47-57.

da Luz, P.L., Serrano Jr, C.V., et al.(1999). *The effect of red wine on experimental atherosclerosis: lipid-independent protection.* Experimental and Molecular Pathology, 65(3), pp.150-159).

Dahlöf, B., Devereux, R.B., et al (2002) *Cardiovascular morbidity and mortality in the Losartan Intervention For Endpoint reduction in hypertension study (LIFE): a randomised trial against atenolol.* Lancet; 359(9311):995-1003.

Danchin, N., Cucherat, M., Thuillez, C. (2006). *Angiotensin-Converting Enzyme Inhibitors in Patients With Coronary Artery Disease and Absence of Heart Failure or Left Ventricular Systolic Dysfunction.*

*An Overview of Long-term Randomized Controlled Trials.*Arch Intern Med. 2006;166(7):787-796).

Darden, D., Richardson, C., Jackson, E.A. (2013) *Physical Activity and Exercise for Secondary Prevention among Patients with Cardiovascular Disease.* Curr. Cardiovasc. Risk Rep 7(6):10.1007.

DART Study: Burr, M.l., Fehily, A.M. et al., (1998). *Effects of changes in fat, fish and fibre intakes on death and myocardial re-infarction: diet and re-infarction trial.* Lancet; 2 (8666): 757-761.

Davis, K. et al. *Complications of coronary arteriography from the Collaborative Study of Coronary Artery Surgery (CASS).* Circulation.1979; 59:1105–1112).

Davies, Michael J. *Atlas of Coronary Artery Disease* (1998). Lippincott Raven.

de Feyter, P., Van Eenige, M.J., Dighton., D.H. et al (1982). *Prognostic Value of Exercise Testing. Coronary Arteriography and Left Ventriculography, 6-8 Weeks after Cardiac Infarction.* Circulation 66 (3): 527-536).

de Feyter, P.J., van Eenige, M.J., et al (1982). *Experience with intracoronary streptokinase in 36 patients with acute evolving myocardial infarction.* European Heart Journal; (3, 15): 441–448.

de Logeril, M, et al. (1999). *The Lyon Diet.* Circulation; 99(6): 779-85).

Dighton, D. H. (2006). *Heart Sense. How to look after your heart.* HeartShield Publications. ISBN 0-9551072-1-0.

Dighton, D.H. (2005). *Eat to Your Heart's Content.* HeartShield Publications. ISBN 0-9551072-0-2.

Dighton, D.H. (2024). *The Art and Science of Medical Practice.* Medicause. ISBN (hardback: 978-1-7385207-7-0; Paperback: 978-1-7385207-3-2; ebook: 978-1-7385207-4-9

Dolgin, M. Association NYH, Fox AC, Gorlin R, Levin RI, New York Heart Association. Criteria Committee. *Nomenclature and criteria for diagnosis of diseases of the heart and great vessels.* 9th ed. Boston, MA: Lippincott Williams and Wilkins; March 1, 1994.

Doll, R., Bradford-Hill, A. (1950). *Smoking and Carcinoma of the Lung.* BMJ; 2:739.

Dutch Family Cohort. 2021. Lessons learned from 25 years of research into long-term consequences of prenatal exposure to the 1944-1945 Dutch famine. International Journal of Environmental Health Research. May 2021.

Elwood, P.C., Cochrane, A.L., et al. (1974). A randomised controlled trial of acetyl salicylic acid in the secondary prevention of mortality from myocardial infarction. BMJ; 1:436.

EMPHASID-HF Trial. Zannad, F., MMurray, J.V.J., et al. (2011) *Eplerenone in Patients with Systolic Heart Failure and Mild Symptoms.* N Engl. J. Med.; 364:11-21.

ENGAGE and AF-TIMI Trials. Giugliano, R.P., Ruff, C.T., et al. (2013) *Edoxaban versus Warfarin in Patients with Atrial Fibrillation.* N Engl J Med 2013;369:2093-2104.

Ferrari, R., Ford. I., et al. (2008) (The BEAUTIFUL study): *Randomized trial of ivabradine in patients with stable coronary artery disease and left ventricular systolic dysfunction - baseline characteristics of the study population.* Cardiology;110(4):271-82.

Filippo. TeslerUgo (2020). *A History of Cardiac Surgery: An Adventurous Voyage from Antiquity to the Artificial Heart.* Cambridge Scholars Publishing; ISBN: 978-1-527-54480-2.

Fox K, Ford I, et al., (2014) (The SIGNIFY Study). *Ivabradine in stable coronary artery disease without clinical heart failure.* N Engl J Med. 2014 Sep 18;371(12):1091-9.

Frank, O. 1926. *Zur Dynamik des Herzmuskels. Z Biol 32: 370–437, 1895).* Starling contributed later (Starling EH and Visscher MB. *The regulation of the energy output of the heart.* J. Physiol. *62*: 243–261, 1926.

Frustaci A, Russo MA, Chimenti C. (2009). *Randomized study on the efficacy of immunosuppressive therapy in patients with virus-negative inflammatory cardiomyopathy: the TIMIC study.* Eur Heart J: 1995-2002.

Gaetano, G. de., Cerletti, C. et al. (2001) European Project. FAIR CT 97 3261 Project participants, in Nutrition, Metabolism and Cardiovascular Diseases; 11:47-50.

Gallucci, G., Tartarone, A., et al. (2020). *Cardiovascular risk of smoking and benefits of smoking cessation.* J. Thoracic Dis.; 12(7): 3866-3876.

GBD 2017 Causes of Death Collaborators. Global, regional, and national age-sex-specific mortality for 282 causes of death in 195 countries and territories, 1980-2017: a systematic analysis for the Global Burden of Disease Study 2017. Lancet. 2018 Nov 10;392(10159):1736-1788.

Geleijnse J.M., Launer L.J., et al.(2002). *Inverse association of tea and flavonoid intakes with incident myocardial infarction: The Rotterdam Study.* Am. J. Clin. Nutr. 2002;75:880–886.

Gerard Manley Hopkins. Poet and Jesuit priest:1844-89). John Duns Scotus (1266-1308). He referred to haecceity.

GISSI Study (1999): Gruppo Italiano per lo studio della sopravivenza nell'infarto miocardio. *Dietary supplementation with n-3 polyunsaturated fatty acids and vitamin E after myocardial infarction*: results of the GISSI Prevenzione trial. Lancet; 335: 447-455.

Griep, L.M.O., Verschuren, W.M.M., et al. (2011). *Colors of fruit and vegetables and 10-year incidence of stroke.* Stroke; 42(11):3190-5.

Guadina, M., Sander, S., et al (2023) *Graft Failure After Coronary Artery Bypass Grafting and Its Association With Patient Characteristics and Clinical Events: A Pooled Individual Patient Data Analysis of Clinical Trials With Imaging Follow-Up.* Circulation: 148(17):1305-1315.

Guinness Book of Records. MLA. Guinness World Records. [London] : Guinness World Records, 2000.

Haider. F., Ghafoor, H., et al. Vitamin D and Cardiovascular Diseases: An Update. Cureus. 2023 Nov 30;15(11).

Hanaffy, D.A., Erdianto, W.P (2023) *Three Ablation Techniques for Atrial Fibrillation during Concomitant Cardiac Surgery: A Systematic Review and Network Meta-Analysis. J. Clin. Med; 12(7): 5716.* doi: 10.3390/jcm12175716.

Hare, D.L., Toukhsati, S.R., et al. (2014). *Depression and cardiovascular disease: a clinical review.* Eur. Heart J. 2014;35:1365–1372).

Healey, J.S., Lopes, R.D., et al. (2023). *Apixaban for Stroke Prevention in Subclinical Atrial Fibrillation.* NEJM; 390:107-117. *Heart* 2021;**107:**962-970.

Hertog M.G., Feskens E.J., et al. *Dietary antioxidant flavonoids and risk of coronary heart disease: The Zutphen Elderly Study.* Lancet. 1993;342:1007–1011.
Hierarchical Phase Contrast Tomography (HiP-CT) scanning. Link: https://doi.org/10.1148/radiol.232731.

Hirsch, A.T., Criqui, M.H., et al. (2001). *Peripheral Arterial Disease Detection, Awareness, and Treatment in Primary Care* (PARTNERS STUDY). JAMA; 286(11):1317-1324.

Ho, M.J., Bellusci A., Wright J.M. (2009). *Blood pressure lowering efficacy of coenzyme Q10 for primary hypertension.* Cochrane Database Syst. Rev. 2009;2009(4):CD007435.

Hodgkin, A.L., Huxley, A.F. *A quantitative description of membrane current and its application to conduction and excitation in nerve.* J. Physiology. 1952 Aug 28; 117(4): 500–544).

Hohnloser, S.H., Fudim, M., et al. *Efficacy and Safety of Apixaban Versus Warfarin in Patients With Atrial Fibrillation and Extremes in Body Weight: Insights From the ARISTOTLE Trial.* Circulation: 139)20). (https://doi.org/10.1161/ CIRCULATIONAHA. 118.037955.

Holmes, T.H, Rahe, R.H. *The social readjustment rating scale.* Journal of Psychosomatic Research. 1967;11:213–218.

Hori, M., Kitakaze, M. (1991) *Adenosine, the heart and coronary circulation.* Hypertension; 18 (5):565-74).

Huang, J., Khatan, P., et al. (2023) *Intakes of folate, vitamin B6, and vitamin B12 and cardiovascular disease risk: a national population-based cross-sectional study.* Cardiovasc. Med. 14;10:1237103.

Huynh, T., Théroux, P., et al. (2001) *Aspirin, Warfarin, or the Combination for Secondary Prevention of Coronary Events in Patients With Acute Coronary Syndromes and Prior Coronary Artery Bypass Surgery.* Circulation; 103(25):3069.

Ikeda, Y., Shimada, K., et al., (2014). *Low-dose aspirin for primary prevention of cardiovascular events in Japanese patients 60 years or older with atherosclerotic risk factors: a randomized clinical trial.* JAMA, 17;312(23):2510-20.

Iori E, Bendinelli S, et al. (2003). *Coronary artery calcium identified by multislice TC as marker of early coronary artery disease.* Monaldi Arch Chest Dis. 2003; 60(1):63-72).

ISIS-2 Trial. *Randomised trial of intravenous streptokinase, oral aspirin, both, or neither among 17,187 cases of suspected acute myocardial infarction:* ISIS-2. ISIS-2 (Second International Study of Infarct Survival) Collaborative Group. Lancet; 2(8607):349-60).

Ito, M., Toda, T., et al. (1986). *Effect of magnesium deficiency on ultrastructural changes in coronary arteries of swine.* Acta Pathologica Japonica; 36: 225-234.

Iversen, K., Ihlemann, N., Gill S.U., et al.(2019). *Partial oral versus intravenous antibiotic treatment of endocarditis.* N. Engl. J. Med., 380: 415-424).

Jahangiri. R, Rezapour. A, et al. (2022). *Cost-effectiveness of fixed-dose combination pill (Polypill) in primary and secondary prevention of cardiovascular disease: A systematic literature review.* Plos One. 2022;17(7):e0271908).

Jones, W.S., Mulder, H. et al. (2021). *Comparative Effectiveness of Aspirin Dosing in Cardiovascular Disease.* NEJM; 384:1981-1990.

Kandil, O., Motawea, K.R., et al., (2022). *Polypills for Primary Prevention of Cardiovascular Disease: A Systematic Review and Meta-Analysis.* Frontiers Cardiovasc. Med.; 14: 9:880054).

Kanugala, A.K., Kaur, J. et al (2023)*Renin-Angiotensin System: Updated Understanding and Role in Physiological and Pathophysiological States.* Cureus; 15(6): e40725.

Kawasoe, S., Ohishi, M. (2024). *Regression of left ventricular hypertrophy.* Hypertension Research; 47: 1225-6.

Kent, A.F.S. (1913): *The structure of the cardiac tissue at the auriculoventricular junction.* J Exp Physiol 47:193, 1913-1914

Kwiecinski, J., et al. *Vulnerable plaque imaging using ^{18}F-sodium fluoride positron emission tomography.* Br. J. Radiology2020. 93 (1113): 20190797).

Laplaces' Law. (Pierre-Simon Laplace, 1806). *Laplace's Law and Tension.* Gilbert-Kawai, E.T., Wittenberg, M.D. (2014). Cambridge University Press.

Laurence, D.H. (1963). *Clinical Pharmacology.* (J.&A. Churchill Ltd).

Levey, G.S. (1971).*Catecholamine Sensitivity, Thyroid Hormone and the Heart A Reevaluation.* American J. of Medicine; 50: 413-420).

Lindlahr, V.H. (1942). *You are What You Eat,* (National Nutrition Society, Inc.).

Maclagen, T (1876). *The treatment of acute rheumatism by salicin.* Lancet 1:383-84.

Mahmood, S.S. et al. *The Framingham Heart Study and the Epidemiology of Cardiovascular Diseases: A Historical Perspective.* Lancet. 2014; 383 (9921):999-1008).

Majeed, M.L., Ghafil, F.A., et al. (2019) *Anti-Atherosclerotic and Anti-Inflammatory Effects of Curcumin on Hypercholesterolemic Male Rabbits.* Indian J Clin Biochem.; 36(1): 74-80).

Mancia, G., Facchetti, R., et al., (2015) *Adverse prognostic value of persistent office blood pressure elevation in white coat hypertension.* Hypertension. 66(2):437-44).

Manda, Y.R., Baradhi, K.M. et al (2023).*Cardiac Catheterization Risks and Complications.* National Library of Medicine.

Marenberg M.E., Risch N., Berkman L.F., et al. (1994). *Genetic susceptibility to death from coronary heart disease in a study of twins.* N. Engl. J. Med.;330:1041–1046. doi: 10.1056/NEJM199404143301503).

Marino, T., Lange, R.A., et al. (2022). *Anticoagulants and Thrombolytics in Pregnancy.* https://emedicine.medscape.com/article/164069

Marmot, M.G., Smith, G.D., et al. (1991). *Health inequalities among British civil servants: the Whitehall II study.* Lancet;337(8754):1387-93.

Manson JE, Cook NR, Lee IM, et al. *Vitamin D supplements and prevention of cancer and cardiovascular disease.* N Engl J Med. 2019;380:33–44.

Masserli, F.H., Bangalore, S. (2017).*Angiotensin Receptor Blockers Reduce Cardiovascular Events, Including the Risk of Myocardial Infarction.* Commentary. Circulation; 135 (22).

McAlister, F. A., Wiebe, N., et al.,(2009). *Meta-analysis: beta-blocker dose, heart rate reduction, and death in patients with heart failure.* Ann. Int. Med.150(11):784-94).

McDiarmid AK, Swoboda PP, et al. *Myocardial Effects of Aldosterone Antagonism in Heart Failure With Preserved Ejection Fraction.* J. Am. Heart Assoc. 2020 Jan 07; 9(1):e011521)

McNeil, J.J., Nelson, M.R., et al (2018). *Effect of Aspirin on All-Cause Mortality in the Healthy Elderly.* New Engl. J. Med. 379:1519-1528.

Mehta, A., Kondamudi, N., et al. (2020). *Running away from cardiovascular disease at the right speed: The impact of aerobic physical activity and cardiorespiratory fitness on cardiovascular disease risk and associated subclinical phenotypes.* Prog. Cardiovasc. Dis. 63(6):762-774.

MERIT-HF for metoprolol use in heart failure.

Münzel T, Steven S, Daiber A. (2014). *Organic nitrates: update on mechanisms underlying vasodilation, tolerance and endothelial dysfunction.* Vascul. Pharmacol. 2014 Dec; 63(3):105-13.

Murabito J.M., Pencina et al. (2005) *Sibling cardiovascular disease as a risk factor for cardiovascular disease in middle-aged adults.* JAMA. 294:3117–3123. doi: 10.1001/JAMA: 294.24.3117.

Murphy, M. (Feb. 2017) *Anticoagulation During Pregnancy*. Ash Clinical News.

National Literacy Trust (2020). www.literacytrust.org.uk

Nickenig, G., Weber, M. et al. (2019) *Transcatheter edge-to-edge repair for reduction of tricuspid regurgitation: 6-month outcomes of the TRILUMINATE single-arm study.* The Lancet;394:2002–11.

Nissen, S.E., Tuzcu, E.M., Libby, P. (2004) *Effect of Antihypertensive Agents on Cardiovascular Events in Patients With Coronary Disease and Normal Blood Pressure. The CAMELOT Study: A Randomized Controlled Trial. JAMA.* 2004;292(18):2217-2225).

O'Brien EC, Simon, D.N., et al. *The ORBIT bleeding score: a simple bedside score to assess bleeding risk in atrial fibrillation.* Eur Heart J. 2015 Dec 7; 36(46): 3258-3264.

Oparil S, Williams D, et al. (2001).*Comparative efficacy of olmesartan, losartan, valsartan, and irbesartan in the control of essential hypertension. J Clin Hypertens (Greenwich)*; 3: 283–91.

Palomaki G.E., Melillo S., Bradley L.A. (2010). *Association between 9p21 genomic markers and heart disease: A meta-analysis.* JAMA. 2010;303:648–656. doi: 10.1001/jama.2010.118.

Parker, W., Iqbal, J. (2020)*Comparison of Contemporary Drug-eluting Coronary Stents – Is Any Stent Better than the Others?* Heart International. 2020;14(1):34-42).

Patel, P. A.(2021). *Heart Failure Review*. 26(2):217-226.

Perera, D., Clayton.T. et al. REVIVED-BCIS-2 Trial (2022) *Percutaneous Revascularization for Ischemic Left Ventricular Dysfunction.* N Engl J Med.;387:1351-1360.

Piano, M.R. (2017)*Alcohol's Effects on the Cardiovascular System*. Alcohol Res; 38(2):210-241.

Pitt, B., O'Neill, B. et al. (2001) *The QUinapril Ischemic Event Trial (QUIET): evaluation of chronic ACE inhibitor therapy in patients with ischemic heart disease and preserved left ventricular function*. Am J Cardiol. 87(9):1058-63).

Pitt, B., Zannad, F., et al. (1999). *The Effect of Spironolactone on Morbidity and Mortality in Patients with Severe Heart Failure*. (The RALES trial): New Engl. J. Med. 1999; 341:709-717).

Plichart, M. et al. *Carotid intima-media thickness in plaque-free site, carotid plaques and coronary heart disease risk prediction in older adults. The Three-City Study*. Atherosclerosis (2011), (2); 917-924).

RACE II Trial. Isabelle C. Van Gelder, M.D., Hessel F. Groenveld, M.D et al. N Engl J Med 2010; 362:1363-1373.

RALES trial. Pitt, B., Zannad, F., et al. *The Effect of Spironolactone on Morbidity and Mortality in Patients with Severe Heart Failure.* New Engl. J. Med. 1999; 341:709-717

Rao, S., Siddiqi, T.J., et al. (2022). *Association of polypill therapy with cardiovascular outcomes, mortality, and adherence: A systematic review and meta-analysis of randomized controlled trials.* Prog. Cardiovasc. Dis.; 73:48-55).

RE-LY Trial. Connolly, S.J., Ezekowitz, M.D., et al. (2009) *Dabigatran versus Warfarin in Patients with Atrial Fibrillation.* N Engl J Med 2009; 361:1139-1151

Renda, G., de Catarina, R. (2011). *The AVERROES Trial – Clinical Implications.* European Society of Cardiology. Vol. 9, N° 27 - 12 Apr 2011. (oral anti-Xa inhibitor apixaban is superior to aspirin).

Ridker, P.M., Manson, J.E., et al. (1991). *Low-dose aspirin therapy for chronic stable angina. A randomized, placebo-controlled clinical trial.* Ann. Int. Med. 15;114(10):835-9).

ROCKET AF Trial. Patel, M.R., Mahaffey, K.W., et al (2011) *Rivaroxaban versus Warfarin in Nonvalvular Atrial Fibrillation.* N Engl J Med 2011; 365:883-891.

Rodriguez F., Maron D.J., et al., (2017) *Association between intensity of statin therapy and mortality in patients with atherosclerotic cardiovascular disease.* JAMA Cardiol 2017; **2**: 47.

Ronksley P.E, Brien S.E, Turner B.J, et al. *Association of alcohol consumption with selected cardiovascular disease outcomes: A systematic review and meta-analysis.* BMJ. 2011;342:d671. doi: 10.1136/bmj.d671.

Rosenman, R.H., Friedman, M. (1974) *Type A Behaviour and Your Heart.* Knopf Doubleday Publishing Group.

Rymer, J., Anand, S.S. et al. (2023). *Rivaroxaban Plus Aspirin Versus Aspirin Alone After Endovascular Revascularization for Symptomatic PAD.* (Insights From VOYAGER PAD).Circulation

Saha, S.A., Molmar, J. Arora, R.R. (2007). *Tissue ACE inhibitors for secondary prevention of cardiovascular disease in patients with preserved left ventricular function: a pooled meta-analysis of randomized placebo-controlled trials.* Database of Abstracts of Reviews of Effects (DARE): Quality-assessed Reviews. Bookshelf ID: NBK73586).

Salonen, J.T., Ylä-Herrtuala, S., et al. (1992). *Auto-antibody against oxidised and progression of carotid atherosclerosis.* Lancet; 339: 883-6.

Saris, J.J., 't Hoen, P.A., et al. *Prorenin induces intracellular signalling in cardiomyocytes independently of angiotensin II.* Hypertension. 2006; 48:564–571.

Schneider, T., Martens, P.R., et al (2000). *Multicenter, randomized, controlled trial of 150-J biphasic shocks compared with 200-J to 360-J monophasic shocks in the resuscitation of out-of-hospital cardiac arrest victims.* Circulation 102:1780–1787.

Serruys, P.W., Morice, Marie-Claude, et al. (2009). *Percutaneous Coronary Intervention versus Coronary-Artery Bypass Grafting for Severe Coronary Artery Disease.* N Engl J Med; 360:961-972).

Shakespeare, William. King Henry the Fifth. Act II, Scene III.

Shams, P Ahmed, I. (2023). Cardiac Amyloidosis. StatPearls (Internet). National Library of Medicine.

SHIFT trial for the use of ivabradine in heart failure.

Silvain, J., Cayla, G., et al. (2024). *Beta-Blocker Interruption or Continuation after Myocardial Infarction.* N Engl J Med 391:1277-1286.)

Sorojja, P. et al. *Transcatheter Repair for Patients with Tricuspid Incompetence.* N Engl J Med 2023; 388:1833-1842 DOI: 10.1056/NEJMoa2300525).

Starling, EH and Visscher MB. (1926) The regulation of the energy output of the heart. *J Physiol 62*: 243–261.

Steering Committee of the Physicians' Health Study Research Group (no authors listed). 1989. *Final report on the aspirin component of the ongoing Physicians' Health Study.* NEJM.; 321(3):129-35.

Steiner, M., Kahn, A.H., et al. (1996). *A double-blind crossover study in moderately hypercholesterolemic men that compared the effect of aged garlic extract and placebo administration on blood lipids.* Am. J. Clin. Nutrition;64(6):866-70.

Strauss, M.H., Hall, A.S. (2017). *Angiotensin Receptor Blockers Do Not Reduce Risk of Myocardial Infarction, Cardiovascular Death, or Total Mortality: Further Evidence for the ARB-MI Paradox.* Circulation Commentary. Vol 135; 22.

SYNTAX trial. Serruys, P.W., Morice, Marie-Claude, et al. (2009). *Percutaneous Coronary Intervention versus Coronary-Artery Bypass Grafting for Severe Coronary Artery Disease.* N Engl J Med; 360:961-972).

Sun, W., Zhang, H., et al (2016). *Comparison of the Efficacy and Safety of Different ACE Inhibitors in Patients With Chronic Heart Failure: A PRISMA-Compliant Network Meta-Analysis.* Medicine (Baltimore)Feb; 95(6):e2554. doi: 10.1097/MD.0000000000002554).

Sutcliffe, P., Connock, M., et al. (2013).*Aspirin for prophylactic use in the primary prevention of cardiovascular disease and cancer: a systematic review and overview of reviews.* Health Technol. Assess.;17(43):1-253.

Swedberg, K., Komajda, M., et al. (2010). *Ivabradine and outcomes in chronic heart failure (SHIFT): a randomised placebo-controlled study.* Lancet. 376: 875-885.

Teppo, K., Lip, G.Y.H., et al. *Comparing CHA2DS2-VA and CHA2DS2-VASc scores for stroke risk stratification in patients with atrial fibrillation: a temporal trends analysis from the retrospective Finnish AntiCoagulation in Atrial Fibrillation (FinACAF) cohort.* Lancet, Regional Health Europe. 2024; 43: 100967).

Thompson, B., Waterhouse, M., et al. (2023). *Vitamin D supplementation and major cardiovascular events: D-Health randomised controlled trial.* BMJ. 2023;381:e075230.)

Thorogood, M., Mann, J., McPherson, K. (1994).*Risk of death from cancer and ischaemic heart disease in meat and non-meat eaters.* BMJ; 308(6945):1667-70.

US Preventive Services Task Force (2022). *Statin Use for the Primary Prevention of Cardiovascular Disease in Adults. US Preventive Services Task Force Recommendation Statement.JAMA.* 2022;328(8):746-753. doi:10.1001/jama.2022.13044.

Val-HeFT trial for the use of valsartan in heart failure.

Van Gelder, I.C., Groenveld, H.F., et al. RACE II Trial. N Engl J Med 2010; 362:1363-137

Vane, J.R (1971) *Inhibition of Prostaglandin Synthesis as a Mechanism of Action for Aspirin-like Drugs.* Nature New Biology; 231: 232.

Wald, N.J., Law, M.R. (2003) *A strategy to reduce cardiovascular disease by more than 80%.* BMJ (Clinical research ed.). (2003) 326:1419.

Watson, James D., Crick Francis H.C. (1953) *Molecular Structure of Nucleic Acids.* Nature 171:737-8.

Watson, James D., Crick. Francis H.C. (1953). *Genetical Implications of the Structure of Deoxyribonucleic Acid.* Nature: 30;171(4361):964-7.

Wersching, H. (2011) An apple a day keeps stroke away? Consumption of white fruits and vegetables is associated with lower risk of stroke. Stroke. 2011;42:3001–3002. doi: 10.1161/STROKEAHA.111.626754).

Wheeler, M.T. et al. (2010). *Cost effectiveness of pre-participation screening for prevention of sudden cardiac death in young athletes.* Ann. Int. Med: 152(5): 276-286)doi: 10.1059/0003-4819-152-5-201003020-00005.

Wilson, W. et al. (2007) *Prevention of infective endocarditis*. Circulation; 116: 1736-1754.

Withering, William (1785). *An Account of the Foxglove and some of its Medical Uses With Practical Remarks on Dropsy and Other Diseases.*

Wolff L, Parkinson J, and White P.D. *Bundle-branch block with short P-R interval in healthy young people prone to paroxysmal tachycardia.* Am Heart J. 1930; 5:685–704. doi: 10.1016/S0002-8703(30)90086-5).

Won H.H., Natarajan P., et al. (2015) *Disproportionate Contributions of Select Genomic Compartments and Cell Types to Genetic Risk for Coronary Artery Disease.* PLoS Genet. 2015;11:e1005622.

Wu, A.D., Lindson, N., et al. (2022). *Smoking Cessation for Secondary Prevention of Cardiovascular Disease.* Cochrane Database Syst. Rev. 2022(8) CD014936).

Wyse, D.G., Waldo, J.P and AFFIRM writing group (2002). *A Comparison of Rate Control and Rhythm Control in Patients with Atrial Fibrillation.* NEJM; 347:1825-1833.

Yamaguchi, T. (2000) *Optimal Intensity of Warfarin Therapy forSecondary Prevention of Stroke in Patients with Nonvalvular Atrial Fibrillation : A Multicenter, Prospective, Randomized Trial.* Stroke; 31(4):817).

Zahoor, M.M., Mazhar, S., et al (2024) *Factor Xa inhibitors versus warfarin in patients with non-valvular atrial fibrillation and diabetes mellitus: a systematic review and meta-analysis of randomized controlled trials.* Annals of Medicine & Surgery 86(2): 986-993.

INDEX

'a' wave, 37, 97, 154
   'v' wave, 37, 97, 98, 143, 153
   'y'-descent', 98

2

24-hour ECG, 14
2nd sound splitting, 39

3

3-D images, 87
3rd heart sound, 78, 79, 142, 213

4

4th heart sound, 78, 79, 214

9

9p21, 168, 301, 386

A

A&E, 51
A_2, 39, 81, 149, 157, 160, 250
A_2P_2, 81
AAI pacing, 278
ablation, 14, 102, 127, 217, 132, 135, 255, 256, 257, 380
abnormal results, 56
abscess formation, 205
academic, 52, 283, 286, 288
accidents, 68
accuracy, 24, 26, 34, 39, 44, 47, 49
ACE, 48, 88, 152, 189, 190, 197, 210, 215, 227, 233, 242, 257, 258, 259, 260, 261, 282, 317, 319, 331, 332, 333, 341, 387, 389, 391
ACE inhibition, 48, 190
ACE inhibitors, 152, 210, 215, 227, 233, 257, 260, 261, 282, 331, 332, 389
acetylcholine, 21, 78, 118, 241
acetylsalicylic acid, 237
Ach, 119, 128, 131
acid regurgitation, 10
acid/base, 185
acromegaly, 32, 208
acronym, 112, 125, 194, 278
actin filaments, 258
action potential, 114, 122, 227
actor, 20
acute coronary syndrome, 238, 263, 264
Adam-Stokes, 13, 79, 120, 121
addictive, 150, 174
Adenosine, 250, 366, 381
adrenergic antagonists, 245
Advanced Life Support, 255
AF, 33, 61, 62, 68, 81, 83, 85, 107, 108, 117, 123, 124, 127, 128, 129, 137, 138, 139, 141, 142, 157, 184, 186, 194, 195, 200, 208, 214, 227, 236, 239, 240, 243, 244, 247, 248, 249, 250, 251, 252, 255, 256, 262, 309, 337, 338, 339, 340, 366, 373, 377, 388

AFFIRM Study, 124
African American, 166
Afro-Caribbean, 260, 261
After-load, 187
against the odds, 67
Agatston score, 93
age, 7, 22, 55, 284
agenda, 30, 54
AI, 35, 39, 40, 48, 69, 85, 145, 149, 150, 151, 152, 158, 177, 184, 210, 298, 366, 395
Alan Gardiner, 65
Alan Gelson, 32
Alan Harris, 39
alcohol, 13, 14, 26, 29, 53, 119, 122, 124, 175, 225, 241, 306, 310, 311, 312, 313, 314, 387, 388
alcoholism, 26
aldosterone, 189, 192, 233, 234, 257, 258, 368, 385
algorithm, 268
algorithmic, 8
alien life, 69
ALLHAT Study, 246, 370
alpha stimulation, 245
alpha-1 blockers, 245
alpha-2 adrenergic receptors, 245
alpha-blockade, 178
altered consciousness, 12
altruistic, 53
ambrisentan, 265
American Civil War, 10
amiloride, 232
amino acids, 28, 295, 296
amiodarone, 128, 134, 186, 197, 217, 247, 248, 250, 273
amlodipine, 261, 262, 263, 333
amoxicillin, 203
amphetamine-based, 150
amputation, 182, 336
Amsterdam, 55

amyloid, 140, 196, 197
amyloidosis, 193, 196, 197, 390
anachronistic, 27
anaemia, 11, 61, 126
anaerobic agent, 312
anaesthesia, 225, 252
analgesia, 120, 252
anatomical, 45, 50, 58
anatomy, 3, 16, 44, 46, 72, 75, 77, 110, 170, 173, 183, 281, 291
Andreas Gruentzig, 271
Andrew Joseph, 179
androgen blocking effect, 234
anecdotal, 46, 236, 247, 294
aneurysm, 176, 179, 180, 181, 205
angina, 3, 8, 9, 10, 12, 60, 82, 101, 145, 167, 169, 172, 179, 182, 184, 186, 196, 207, 224, 238, 244, 262, 263, 264, 265, 274, 275, 276, 284, 285, 287, 291, 293, 315, 327, 332, 333, 388
angiogenesis, 75
angiogram, 55, 60, 140, 143, 199, 291
angiography, 60, 77, 98, 101, 147, 172, 173, 178, 182, 270, 272, 284
angioplasty, 183, 217, 263, 269, 271, 272, 274, 275, 276, 282, 284, 333, 368
angiosarcomas, 221
angiotensin II, 192, 258, 259, 333, 389
angiotensin II Receptor Blockers, 261
Angiotensin-Converting Enzyme(ACE) inhibitors, 259
angiotensin-I, 189
angiotensin-II, 88, 189
angle of the jaw, 36, 37
ankle / brachial index (API), 182
ankle arteries, 182
ankle swelling, 9, 12, 42, 60, 236, 263
ankylosing spondylitis, 210
annuloplasty, 144, 153
anonymous, 16
anoxic tissue, 169
antecedent potential, 51

antegrade, 102, 131, 132, 138, 177
anterior chest, 9, 10, 55, 177
anterior descending coronary artery, 12, 74, 75, 76, 86, 162, 171
anterior infarction, 86, 171, 193, 277
anthracycline, 191
anthrocyanidin content, 312
antiarrhythmic drugs, 246
antibiotics, 21, 134, 141, 203, 204, 206, 283
anti-CD3/CD4/20, 198
anticoagulants, 216, 224, 237, 338, 339, 340, 373, 374
anticoagulation, 125, 126, 153, 200, 208, 210, 217, 218, 236, 251, 334, 339, 340, 341
anti-fibrotic, 258
anti-myosin scintigraphy, 200
anti-nuclear antibodies, 199
anti-oxidants, 175, 312, 314
antiphospholipid antibody (APLA), 210
antiphospholipid syndrome, 143
antiplatelet, 126, 238, 303, 327, 335, 338, 340, 341, 370
anti-psychotics, 134
antithrombotic therapy, 338
anxious hypertensive cases, 243
AOO, 278
aortic aneurysm, 9, 160, 179, 180, 206, 271, 298, 303
aortic area, 36, 39
aortic flow, 151
aortic homografts, 271
aortic incompetence, 33, 43, 158, 159
aortic rupture, 178
aortic valve cusp prolapse, 158
aortic valve replacement, 136
aortic valve stenosis, 33
aperture area, 138, 147, 214
apex beat, 37, 73, 153
apixaban, 125, 239, 240, 338, 339, 371, 380, 381, 388
APOE, 168
apoptosis, 190, 191, 301

appendicitis, 42
appetite suppressant, 174, 222
appraisal, 7, 44, 281
aPTT, 218
ARB, 88, 190, 197, 215, 227, 233, 242, 257, 258, 261, 331, 333, 334, 341, 390
arginine, 169, 175, 295, 314
ARIC study, 312, 313
ARISTOTLE trial, 126, 240, 371, 381
ARNI study, 190, 371
arrhythmia, 195, 221
arrhythmogenic RV dysplasia, 298
arrogance, 326
art, 7, 15, 16, 18, 42, 47, 50, 57, 58, 65, 69, 73, 129, 179, 203, 230, 235, 240, 274, 292
art of medicine, 47, 69, 292
arterial disease, 74
arterial grafts, 273
arterial protheses, 269
arterial puncture, 64
arterial stiffness, 89
arterial thrombus, 94, 182
arterial tree, 167, 168
arteries, 275, 284, 291, 326
arteriosclerosis, 3, 87, 89, 165, 294, 304
artery scanning, 286
art-form, 50
arthritis, 32, 33, 58, 284
Agatston, Arthur, 93
AS, 39, 40, 144, 145, 146, 147, 148, 149, 366, 395
asbestos, 221
ascending aorta, 75, 144, 177, 180
ascending aortic dissections, 177, 178
Asclepius, 311
ASD, 100, 155, 156, 157, 159, 162, 163, 215, 216, 298, 366
Asian, 53

aspirin, 21, 22, 125, 237, 238, 282, 318, 319, 327, 334, 335, 338, 371, 381, 382, 385, 389, 391, 392

assessment, 45

assumption, 48, 57, 243

asthma, 11, 61, 62, 129, 209

asthmatics, 250

asymptomatic, 291

asystole, 13, 117

AT_2 receptors, 259

A-T-C-G (amino acids), 295

atenolol, 129, 210, 244, 261, 375

atherogenic nutrients, 295

atheroma, 281, 282, 284, 285, 286

atheroma prevention, 176, 224, 312

atheroprotective, 175, 176

atherosclerosis, 74, 91, 165, 167, 168, 176, 177, 283, 294, 370, 374, 387

athlete, 195, 201

athletic, 306

athletically fit, 194

athleticism, 79

atorvastatin, 266, 320, 321, 328, 329

ATP, 242, 296, 316, 366

ATPase, 114, 186

atrial depolarisation, 78, 107

atrial dilatation, 33, 108, 123

atrial fibrillation, 14, 33, 37, 68, 107-109, 123-125, 127, 128, 217, 225, 227, 239, 244, 251, 273, 337, 338, 339, 340, 369, 370, 371, 372, 377, 380, 381, 386, 388, 391, 393, 394

atrial fibrosis, 33, 79, 107, 108, 123

atrial flutter, 107, 127

atrial myxoma, 154, 220

atrial size, 128

atrial systole, 97, 137, 138, 146, 220

atrial tachycardias, 244, 249

atropine, 53, 120

attached, 23

attention-seeking, 17

attitudes and beliefs, 29
atypical, 57, 182, 194, 281
Aubrey Leatham, 2, 16, 38, 79
auscultation, 38, 39, 41, 159, 395
auscultatory findings, 39, 40
Austin Flint murmur, 149
autoimmunity, 198
automated external defibrillators, 251
autonomic manoeuvres, 131
autonomic nervous system, 192
autonomic responsiveness, 52
autosomal, 25, 154, 297, 298, 303
AV block, 79, 102, 108, 109, 120, 121, 147, 250, 278
AV nodal re-entrant tachycardia, 130, 131
AV node, 77, 78, 79, 102, 107, 108, 109, 121, 127, 130, 131, 132, 133, 147, 153, 186, 256, 262, 395
AV reciprocating tachycardia, 130
availability, 26, 52
AVERROES study, 126, 388
axis deviation, 110, 111, 112
azithromycin, 203

B

B_{12} and B_6, 314
back-pressure, 86
bacterial genome, 203
balloon and stent, 274
balloon valvuloplasty, 149, 214
bare metal stents, 275
Barlow's syndrome, 141
base pairs, 296
Bawa-Garba, 63
Bayes, 45, 46
Bayes' Theorem, 46

BEAUTIFUL trial, 228, 378
behaviour, 7, 70
BehÇet disease, 177
bending, 167, 168
bendroflumethazide, 232
benzyl-penicillin, 21
bereavement, 26
Bernoulli, 82, 147, 171, 285
berries, 175, 317, 318
beta-1 receptors, 228, 242
beta-2 receptors, 242, 244
beta-blockers, 33, 128, 129, 144, 152, 183, 227, 243, 244, 247, 254, 282, 319, 330, 385
bias, 68, 171, 310, 326
bicuspid, 144, 151, 160, 202, 298
bifid P-wave, 107, 138
bifurcation, 143, 167
BioFreedom valve, 275
biology, 48
bio-mechanical valves, 202
biopsy, 53, 54
biphasic, 251, 252, 389
bipolar, 104
birth weight, 213
bisoprolol, 129, 243, 244, 374
bi-stable Situations, 82
blackberries, 318
blackout, 16
blackouts, 79, 289
bleed, 126
BLOB: BLindingly OBvious, 283
blockage, 75, 77, 93, 244
blood, 31, 46, 48, 51, 53, 57, 58, 284, 285, 286, 289, 290, 293, 294, 326
blood cholesterol, 91, 172, 266, 284, 316, 318, 320, 326
blood culture, 62, 204
blood flow, 74, 78, 82, 85, 86, 89, 95, 99, 101, 120, 150, 151, 157, 167, 169, 170, 171, 182, 183, 213, 214, 221, 250, 263, 285

blood HDL, 169
blood lipid, 92, 284, 285
blood pressure, 12, 13, 35, 78, 88, 118, 157, 178, 213, 214, 242, 245, 258, 259, 260, 261, 285, 289, 293, 294, 302, 304, 309, 317, 326, 331, 333, 340, 384
blood sugar, 290
blood vessel, 119
blue toes, 181
BMI, 111, 174, 225, 307, 308, 371
BNF, 225, 234, 247, 249, 260, 261
BNP, 191
body weight, 100, 225, 308
body-centric, 70
bosentan, 265
Bowditch, 81, 372
brachiocephalic vein, 277
bradycardia, 9, 13, 14, 21, 79, 118, 119, 127, 134, 207, 227, 228
bradykinin, 259
breathing, 6, 11, 32, 35, 37, 40, 41, 70, 98, 156, 277
breathless, 326
breathlessness, 7, 11, 12, 122, 123, 129, 185, 194
bretylium, 128
British, 20, 27, 51, 79, 84, 103, 232, 283, 306, 311, 372, 384
British attitudes, 27
Brugada syndrome, 134, 254, 298
bruising, 235, 236, 237
B-type natriuretic peptide, 191
bumetanide, 231, 233
bundle branch block, 80, 81, 87, 110, 112, 134, 207, 227
bundle branches, 77
Bundle of His, 78, 102
Bundle of Kent, 109, 115, 131, 132
bureaucrats, 7, 53, 55, 64, 69, 179, 324
business people, 19, 288
bypass, 8, 16, 55, 109, 172, 183, 269, 270, 273, 281, 291, 331, 333

C

C&S, 49
Ca^{2+} release, 114
CABG, 23, 26, 30, 43, 60, 174, 175, 178, 231, 270, 271, 272, 273, 276, 281, 282, 284, 366, 391
CAC, 93
CAD heritability, 299
CAD patients, 172, 303
CAD-related mortality, 166
Caesarian section, 216
caffeine, 124, 241
calcification, 85, 93, 138, 139, 206, 274
calcified, 21, 40
calcineurin inhibitors, 275
calcium, 9, 60, 82, 88, 93, 99, 106, 107, 113, 114, 134, 152, 161, 168, 172, 186, 191, 197, 215, 242, 247, 248, 262, 263, 264, 266, 284, 291, 3 69, 375, 382
calcium channel blockers, 88, 152, 215, 262
calcium ion (Ca^{2+}), 114
calcium ions, 82, 106
calcium score, 60, 93, 161, 172, 284, 291
calculation, 49, 147
calf muscles, 13, 120, 181, 182
calf tenderness, 42
CAMELOT Study, 333, 386
Camm. A.J, 339
cAMP, 227, 228
cancer, 24, 25, 58, 59, 287
candesartan, 190, 261, 373
candida, 201
capillary electrometer, 103
captopril, 210
carbohydrate metabolism, 300
carcinoid, 136, 152, 222, 242

cardiac, 21, 34, 38, 55, 56, 64, 276, 277, 281, 282, 283, 284, 287, 289, 310, 326
  cardiac anatomy, 72
  cardiac biopsy, 194, 196
  cardiac catheterisation, 2, 84, 85, 97, 98, 99, 134, 143,147, 158, 178, 201, 236, 241, 272, 274
  cardiac cirrhosis, 153
  cardiac contour, 73
  cardiac cycle, 80, 95
  cardiac electrical excitability, 114
  cardiac infarction, 42, 76, 77, 81, 90, 93, 94, 96, 112, 122, 142, 158, 167, 169, 171, 172, 174, 176, 184, 210, 211, 217, 224, 228, 237, 240, 241, 249, 267, 277, 284, 330, 334
  cardiac infection, 122
  cardiac ischaemia (*see ischaemia*), 122
  cardiac management, 1
  cardiac output, 34, 62, 81, 83, 99, 118, 128, 129, 147, 178, 184, 186, 188, 213, 230, 326
  cardiac radiology, 73, 92
  cardiac screening, 3, 281, 282, 283, 284, 285, 287, 288, 290
  cardiac suppressants, 134, 247, 248, 249, 250
  cardiac surgery, 2, 55, 206, 229, 255, 269, 270
  cardiac transplantation, 164, 197, 271
  Cardiac Value™ of food, 176
  cardiac work, 9
  cardiologists, 2, 118, 167, 212, 272
  cardiology, 16, 39, 56
  cardiomyocyte, 190
  cardiomyopathy, 3, 14, 50, 87, 96, 122, 123, 124, 125, 146, 193, 194, 200, 201, 208, 209, 215, 217, 253, 254, 258, 298, 378
  cardio-pulmonary failure, 62
  cardioversion, 251, 252, 371
  career, 44, 53, 54, 55, 56, 64
  caring, 24
  carotenoid, 312
  carotid artery, 36, 40, 52, 60, 172

carotid atheroma, 91, 93, 169, 170, 172, 173, 174, 176, 267, 284, 285, 326
cartilages, 10
carvedilol, 189, 243, 244, 374
Catastrophe theory, 82
catecholamines, 62, 119, 122, 189, 207, 242, 243, 245
catheter techniques, 269
catheter technology, 56
catheterisation, 38, 56, 64
cause of death, 165
cavalier, 56
cell membrane, 113, 115
cephalexin, 203
cerebral abscess, 204
cerebral artery disease, 237
cerebral artery thinning, 89
cerebral emboli, 91, 239
cerebral haemorrhage, 89, 205
cerebral perfusion, 78
cGK-I, 263, 264
cGMP, 228, 263, 264
CHA_2DS_2-VASc score, 125, 337, 338
Chagas' disease, 199, 207, 254
characteristics, 47, 56
Charing Cross Hospital, 284
CHARM trial, 190, 373
CHB (complete heart block), 13, 147
check-list, 7, 287
chemokines, 300
chest pain, 9, 290
chest X-ray, 38, 41, 51, 75, 88, 92, 291, 367
chewing (and SBE), 202
children, 25, 294
chlortalidone, 231, 232
choice, 23, 70
cholesterol, 284, 285, 290, 326
cholinesterase, 21, 119

chorionic gonadotrophin, 46
Christiaan Barnard, 271
chromosome deletions, 297
chromosomes, 295, 296
chronic, 57, 58
chronic bacteraemia, 202
Churchill, 310
CIBIS-II study, 189, 374
cilexetil, 261
cilostazol, 183
cinchona tree bark, 247
circulating blood volume, 189, 214, 230
circulation, 8, 72, 75, 77, 78, 99, 100, 101, 118, 120, 156, 157, 163, 167, 169, 171, 191, 212, 213, 230, 250, 251, 270, 285, 381
circumflex, 12, 75, 170, 171
CLAS Study, 170, 320, 374
classic scenario, 185
classification, 11, 177, 246, 271
claudication, 41, 161, 181, 182, 183, 209
clicking, 140
click-murmur syndrome, 141
clicks, 39
Clifford Wilson, 67
clindamycin, 203
clinical, 1, 3, 6, 7, 15, 16, 17, 18, 20, 21, 22, 23, 24, 25, 26, 27, 31, 33, 34, 37, 38, 40, 41, 43, 44, 45, 46, 48, 49, 50, 51, 52, 54, 55, 56, 57, 58, 59, 60, 63, 64, 65, 66, 68, 69, 70, 72, 84, 85, 86, 93, 94, 108, 111, 112, 120, 123, 124, 125, 134, 136, 139, 141, 157, 158, 168, 178, 179, 191, 196, 198, 199, 203, 204, 207, 229, 236, 238, 258, 267, 275, 281, 282, 283, 284, 286, 289, 291, 292, 297, 299, 301, 302, 319, 327, 328, 337, 370, 375, 378, 380, 381, 388
clinical experience, 56
clinical judgement, 22
clinical management, 7, 45, 51, 52
clinical mastery, 22, 65
clinical relevance, 45, 59
clinical wisdom, 1, 69

clinically justifiable, 52
clock mechanism, 326
clopidogrel, 217, 225, 238, 335, 338
clot formation, 93, 167, 169, 221
clotting, 139, 143, 235, 236, 237, 238, 274, 290
clubbing, 33, 40, 162
CO_2, 192
coarctation, 38, 160, 161, 183, 215, 298
cocaine, 14, 29, 33, 124, 241
Cochrane database, 331
Cockney, 20
Coenzyme Q_{10}, 316
cognitive, 27
cohort analysis, 46
cold cyanosed hands, 230
collagen production, 190
collateral arteries, 75, 76
collateral perfusion, 96
colleagues, 23, 58, 64, 281
colour abnormalities, 34
Columbo, 66
combustion engine, 73
commitment, 30
common-sense, 288
COMPASS trial, 335
compatibility, 29
complete heart block, 13, 21, 79, 108, 118, 121, 147, 164, 210, 277, 278
completeness, 23, 34
complex, 43
compliance, 188, 235, 325
complications, 42, 98, 99, 158, 178, 179, 205, 230, 247, 256, 270, 273, 302, 340
computer aided diagnoses, 69
conclusion, 2, 31, 68, 240, 261, 271, 338, 391
conduction, 21
conduction circuit, 121
conduction fibres, 108

conduction pathway, 132
confidence, 47, 50
Conn's syndrome, 208
connective tissue disease, 152
consciousness, 78, 118, 120, 135
consensus, 324
CONSENSUS trial, 190, 374
consolidation, 41
contraction, 34, 37, 77, 80, 81, 82, 109, 114, 128, 140, 147, 150, 160, 187, 188
controlled trial, 251, 315, 373, 377, 389, 391
conversation, 6, 29
convulsion, 13
Cooley, 270, 271, 272
cooperation, 325
COPD, 11, 34, 61, 62, 107, 250, 273
COPERNICUS trial, 189, 374
coping, 10
corona, 74
coronary angiogram, 55, 56, 178, 291
coronary arteries, 51, 74, 75, 76, 97, 168, 169, 170, 172, 199, 206, 209, 211, 264, 274, 291, 300, 315, 382, 395
coronary arteriography, 99, 161, 172, 274, 376
coronary artery, 2, 14, 24, 43, 55, 56, 74, 75, 76, 77, 82, 88, 93, 94, 96, 99, 134, 136, 162, 167, 169, 170, 171, 172, 178, 207, 210, 211, 213, 217, 231, 240, 243, 250, 262, 269, 270, 271, 272, 273, 276, 281, 284, 285, 290, 291, 294, 299, 303, 326, 331, 333, 369, 374, 378, 382, 391
coronary atheroma, 91, 93, 97, 168, 172, 267, 291
coronary cusp, 158
coronary disease, 326
coronary endarterectomy, 270
coronary heart disease, 16, 24
coronary ostia, 76, 178, 179
coronary sinus, 74
coronary stent, 30, 302
coronary thrombosis, 76
coronary vein function, 74

coroner, 54
corporate, 8, 22, 69, 312
corporation, 240, 306
Corrigan pulse, 150
corroboration, 33, 45, 47, 58
costochondral joint, 10, 60
costochondritis, 10, 60
counselling, 195, 203, 297, 323
courage, 56
COURAGE trial, 275, 276
COVID-19, 198, 200, 296
Coxsackie, 198, 199
CPK, 57
C-reactive protein, 169, 191
creatinine, 61
crescendo – decrescendo, 145
CREST Syndrome, 210
critical, 8, 47, 82, 96, 112, 126, 150, 167, 182, 207, 257, 281, 296
cross-flow, 77, 171
crown, 74
CRP, 57, 290
CT, 2, 9, 49, 51, 58, 60, 72, 84, 93, 99, 161, 172, 178, 180, 181, 182, 205, 284, 286, 291, 313, 366, 378, 380
cultural, 7, 20, 27
culture, 49
curcumin, 317
cure, 28, 289
Cushing's Syndrome, 208
CVA, 29, 118, 134
CVS deaths, 166, 309, 313, 318, 331
CVS events, 90, 267, 290, 314, 328, 331
CVS outcomes, 245, 308, 319, 323
CXR, 62, 73, 92, 93, 110, 138, 143, 147, 151, 153, 157, 158, 159, 161, 163, 193, 206, 291, 367
cyanosis, 32, 33, 35, 40, 157, 162, 164, 186
cyclosporine, 271
CYP2C19, 225

CYP2C9, 225
CYP4F2, 225
cystic changes, 144
cystic degeneration, 176, 177
cytokines, 168, 170, 189, 198, 300, 303
cytotoxic effect, 199
Czechs, 20

D

D-rotation type, 163
Da Costa, 10
dabigatran, 125, 218, 239, 339, 340, 367, 368, 388
Dacron, 161, 271
dangerous, 54
DARE trial, 332, 389
David Lipkin, 281
DC reversion, 85, 128, 129, 139, 244, 250, 251, 255
DDD (double chamber) pacing, 278
DDI mode, 278
de Feyter, 171, 240, 376
de Logeril, M, 176, 294, 376
de Motu Cordis, 8
death, 23, 24, 27, 54, 56, 282, 324
DeBakey, 180, 270, 271, 272
Deborah Doniach, 207
decision tools, 284
decompensated MR, 143
deep leg-vein thrombosis, 15
defibrillation, 134, 217, 252
dehydration, 31, 34, 36, 86, 230
delta wave, 109
demeanour, 20, 32
denial, 30, 285
Deoxyribonucleic Acid, 296, 392

depolarisation, 78, 106, 107, 109, 111, 112, 113, 119, 227, 241
depression, 14, 26, 32, 74, 112, 248, 311, 380
descending aorta, 177, 180
desert island, 38
detached, 16, 24, 30
detached approach, 16
detachment, 7
detective, 24
devils in the detail, 276
diabetes, 25, 283, 290
diabetes mellitus, 239, 331, 394
diabetic, 183, 213, 276, 289
diabetics, 174, 290
diagnosis, 3, 7, 10, 11, 12, 13, 16, 17, 29, 30, 31, 32, 34, 37, 38, 39, 42, 43, 44, 46, 47, 49, 50, 58, 59, 62, 63, 65, 66, 67, 69, 111, 122, 141, 146, 155, 172, 178, 181, 191, 194, 196, 203, 206, 207, 222, 229, 276, 285, 286, 291, 302, 324, 325, 372, 375, 377
diagnostic accuracy, 9, 16, 24, 38, 39, 44, 45, 46, 49, 58, 172
diagnostic relevance, 9, 21
diagnostic uncertainty, 44
diagnostic value, 39, 46
dialect, 20
diaphragm, 36, 72, 73, 76, 111
diastolic murmurs, 39
diastolic pressure, 79, 80, 150, 159, 185
diet, 34, 167, 174, 176, 225, 232, 282, 294, 295, 306, 307, 310, 311, 312, 313, 341, 376
dietary nutrients, 295
diethylpropion, 174
differential diagnoses, 66
difficulty in breathing, 9
digital, 70
digital self, 70
digitalis, 28, 129, 130, 186, 187, 217, 225, 226, 227
digitalis alkaloid, 129
Digitalis lanata, 186
digitoxin, 129, 186, 226

digoxin, 31, 33, 57, 128, 129, 144, 186, 226, 227, 247
dihydropyridines, 262
dilatation, 40, 79, 119, 142, 144, 149, 151, 152, 161, 162, 176, 185, 191, 210, 215, 218, 298
dilated, 73, 124, 129, 140, 145, 150, 152, 153, 161, 193, 200, 201, 253, 298
diltiazem, 178, 227, 247, 262, 321
dimensions, 85, 145, 150, 257
diphtheria, 21, 207
diphtheria toxin, 21
discussion, 15, 17, 59
disease progression, 42
dissection, 9, 56, 150, 161, 176, 177, 178, 180, 184, 211, 243, 263, 274, 298, 303, 304, 375
distortion, 24, 50, 51
diuretics, 31, 144, 214, 226, 227, 229, 230, 231, 232, 233, 235, 260, 325
DNA, 296, 367
DOAC, 126, 239, 367
dobutamine, 86, 96, 290, 291
doctor, 17, 20, 23, 29, 42, 53, 54, 56, 65, 287, 325
doctor-patient relationship, 20
doctors, 1, 2, 6, 7, 11, 15, 16, 18, 19, 20, 22, 23, 24, 30, 38, 43, 45, 46, 53, 54, 55, 56, 57, 58, 63, 64, 66, 67, 69, 92, 118, 150, 179, 183, 237, 241, 252, 260, 273, 276, 277, 282, 285, 286, 292, 325, 326
Doll and Hill, 308
Don Russell, 66
Donald Ross, 270
DOO mode, 278
Doppler, 85, 100, 101, 151, 153, 158, 182
dorsalis pedis, 182
double helix, 296
double-blind controlled trials, 28, 325
doubly treated, 336
Douglas bag, 100
doxazosin, 246
DPI group, 335, 336
dress standard, 32

dronedarone, 250
Drug eluting stents, 275
drug intoxication, 33
drugs, 11, 22, 27, 28, 29, 53, 106, 118, 128, 134, 135, 139, 148, 150, 170, 174, 178, 183, 186, 190, 202, 203, 210, 224, 225, 226, 227, 231, 235, 242, 246, 247, 248, 249, 250, 251, 255, 262, 264, 265, 266, 274, 275, 282, 303, 3 19, 320, 321
Duchenne muscular dystrophy, 25
ductus, 156, 159, 160, 163
DVT, 15, 61, 241
dysfunction, 74, 80, 86, 89, 96, 118, 162, 164, 185, 187, 191, 195, 206, 210, 215, 227, 228, 256, 264, 265, 278, 311, 315, 338, 378, 385
dyslipidaemia, 169, 174, 321
dysphagia, 179, 287
dyspnoea, 11, 145
dysrhythmia, 193

E

early diastolic murmur (EDM), 39, 149
early disease detection, 52, 281,285, 323
early intervention, 285
East London, 66
Ebstein anomaly, 152, 161, 163
ECG, 3, 9, 14, 31, 37, 38, 48, 50, 51, 52, 55, 57, 60, 62, 64, 78, 79, 80, 95, 103, 104, 105, 106, 107, 108, 110, 112, 113, 114, 122, 129, 131, 132, 133, 135, 138, 146, 151, 153, 157, 158, 160, 162, 163, 171, 172, 194, 196, 1 99, 207, 208, 226, 248, 284, 286, 289, 290, 324, 337, 395
ECGs for athletes, 324
echocardiography, 38, 39, 40, 49, 62, 84, 85, 86, 87, 89, 100, 101, 128, 139, 141, 142, 146, 149, 154, 158, 160, 196, 205, 221, 243, 286, 289, 290, 2 91
eclectic functioning, 65
ectopic, 11, 107, 121, 122, 130, 315
edoxaban, 239

educated, 19
education, 27, 325, 326
educational status, 7, 19
EEG, 50
eGFR, 126
Egyptians, 237
Ehlers-Danlos syndrome, 141, 143, 298
Einthoven, 103, 104, 105, 111, 112, 371, 395
Einthoven's Triangle, 104, 395
Eisenmenger, 157, 159, 215
ejection fraction, 49, 80, 81, 86, 87, 95, 140, 148, 152, 188, 196, 227, 234, 253, 260, 330
elasticity, 79, 143
elderly, 31, 329, 330, 369
electrical abnormalities, 290
electrical pathway, 14
electrical phenomena, 50
electrical wave, 77, 80, 105, 106, 114
electrode catheter, 102, 133, 252
electrode position, 277
electrogram, 106, 112, 113, 133
electrolyte, 61, 130, 244, 257
electronics, 51, 79, 85, 106
electrophysiology, 14, 77, 84, 97, 102, 127, 133, 253, 272, 290
embarrassing, 18, 20, 30
embarrassment, 30
embolism, 42, 117, 139, 143, 157, 216, 220, 241, 338
emergency, 7, 9, 23, 53, 55, 64, 134, 171, 178, 182, 263
emotion, 8, 9
EMPHASID-HF trial, 190
emphysema, 37, 41, 61, 62, 185
employees, 288
enalapril, 190, 210, 260, 333, 374
end-diastolic volume, 80, 188
endocarditis, 49, 85, 96, 136, 141, 142, 148, 150, 158, 159, 201, 202, 203, 204, 205, 206, 216, 382, 393
endocardium, 77, 78

endothelial, 89, 169, 192, 199, 264, 300, 302, 303, 315, 316, 338, 385
endothelin receptor antagonists, 265
endothelium, 99, 169, 174, 199, 250, 263, 300, 317
endpoint, 271, 330, 391
ENGAGE trial, 240, 377
English, 20, 186
enoximone, 228
enquiry, 15, 24, 70
enzyme, 21, 57, 88, 125, 189, 227, 250, 322, 332
enzymes, 57
ephedrine, 29
epicardium, 79
epidemiology, 88, 165, 167, 384
epinephrine, 128, 134, 190, 241, 246
eplerenone, 190
equation, 69, 70, 82, 147, 171, 188
Eric Berne, 17
error, 24, 44, 48, 49, 58, 59, 64, 86, 112, 293
erythema marginatum, 199
erythromycin, 21, 321
e-selectin, 300
ESR, 57, 58, 122, 199, 290, 367
essential, 1, 3, 19, 29, 40, 42, 43, 46, 47, 60, 65, 81, 89, 132, 224, 225, 231, 243, 255, 262, 290, 291, 293, 304, 316, 374, 375, 386
establishment, 27
ethyl alcohol, 175
eureka, 66
Europa space project, 324
European, 20, 21, 124, 313, 332, 341, 372, 373, 376, 378, 388
everolimis, 275
evidence, 1, 9, 14, 24, 27, 28, 33, 34, 36, 37, 41, 43, 44, 52, 55, 59, 62, 63, 65, 66, 69, 78, 112, 124, 160, 162, 168, 169, 173, 230, 234, 235, 275, 282, 283, 284, 288, 291, 312, 315, 320, 331, 333, 341
examination, 32, 34, 41, 42, 44, 68
examination routine, 32
excuse, 64
executive, 15, 16, 17, 18, 23

exercise, 8, 9, 10, 11, 14, 31, 36, 51, 60, 61, 81, 86, 87, 96, 122, 139, 148, 163, 171, 172, 182, 183, 184, 185, 192, 200, 213, 229, 231, 250, 263, 265, 273, 275, 276, 278, 284, 286, 289, 290, 291, 294, 306, 309, 311, 325, 376
   Exercise and Heart Disease, 309
exome, 297
exophthalmos, 34
expedient, 8, 325
experience, 22, 25, 27, 42, 47, 50, 54, 276, 325
extra conduction pathways, 109, 130, 133, 250
extracorporeal pumps, 269
extrasystole, 14, 33

F

Fabry Disease, 194, 297
facial features, 32
factor Xa inhibitors, 239
factor Xa inhibitors, 125, 239, 367, 368, 394
fainting, 12, 13, 78, 119, 120
faintness, 118, 120, 290
fallacies, 69
Fallot's tetralogy, 38, 154, 160, 161, 162, 298
false, 24, 30, 44, 47, 48, 49, 57, 89, 177, 200, 290
false negative, 24
false positive, 24, 286
false-negative, 44
falsified, 17
Falstaff, 34
family history, 24, 25, 26, 89, 167, 168, 169, 173, 181, 193, 267, 291, 293, 294, 299, 302, 304, 326
   famous, 65
fat, 290
fatigue, 192, 213
fatty meat, 167
faux pas, 18, 20

fear of death, 15
febrile illness, 13, 33, 119, 121, 199
feedback, 2, 34, 38, 122, 130
feeling faint, 117
Feigenbaum, 84
felodipine, 262
femoral, 40, 91, 145, 168, 274
fenfluramine, 150, 222
fentanyl, 252
fertility, 208
FFR, 101, 102
fibrillation, 13, 14, 33, 36, 37, 68, 107, 108, 109, 115, 121, 123, 124, 125, 127, 128, 217, 225, 227, 239, 244, 251, 253, 273, 337, 338, 339, 340, 370, 371, 372, 373, 386, 391, 394, 395
fibrin, 202, 241
fibrinogen, 172, 218, 285, 290, 338
fibrinolysis, 235, 240, 241
fibroblasts, 145, 300, 302
fibroelastomas, 221
fibromas, 221
fibromuscular dysplasia, 211
fibrous tissue, 96
Fick Principle, 99
financial, 7, 52, 69, 240, 324
fingernail beds, 150
first heart sound, 39, 80, 137, 138
five pieces of fruit per day, 312
fixed splitting, 157
flavonoid, 175, 312, 317, 318, 371, 379
flecainide, 128, 186, 247, 248, 249
flow >, 159
flow ratio, 100, 101
flow-dependent, 148
flu-deoxyglucose (18F), 209
flummoxed, 29
fluoroscopy, 139, 272, 274
fly-on-the-wall, 18

foam cells, 301
focussed style, 16
foetal abnormalities, 217
foetor, 34
foetus, 156, 159, 210, 212, 217, 218
Fontan, 164
food, 28, 310
Food and Heart Disease, 310
foramen ovale, 115, 156, 163
foreign patients, 23
forensic, 30
forensic history taking, 30
forethought, 288, 289
form filling, 64
formulae, 82
Forssmann, 98
four-chamber view, 49
fourth heart sound, 145
Foxglove, 186, 226, 393
Fragile X syndrome, 298
frailty, 6, 34, 235
Framingham Heart Study, 88, 299, 384
Frances Crick, 296
Frank Blake, 69
Frank Partridge, 241
Frank Starling relationship, 140, 187, 189
Frank, O, 81, 378
fraudulent, 30
free oxygen radicals, 312
frightening, 119, 252
frontal plane, 104, 111
fully informed consent, 21
functional cardiac assessment, 87
functional flow reserve, 101
furosemide, 231, 233

G

Galvani, 103
Games People Play, 17
GARFIELD study, 339, 373
garlic, 28, 312, 316, 390
gas diffusion, 62
gender, 7
gene, 68, 197, 297, 295, 298, 302, 303, 304, 326
genetic, 24, 29, 69, 88, 113, 167, 168, 176, 191, 192, 195, 225, 283, 293, 295, 297, 298, 299, 300, 301, 302, 303, 304, 305, 310, 323, 375
   genetic genome research, 295
   genetic screening, 180
   genetic variants, 302
   genetically predisposed, 173
   genetics, 52, 300
   genotype, 25
Geordie, 20
Gerard Manley Hopkins, 18, 379
Gerhardt, 237
GFR, 192
GI bleeding, 125, 126
giant cell myocarditis, 200
ginseng, 28
glaucoma, 243, 246, 249
Global Burden of Disease Study, 165, 379
GLP-1 receptor, 174
GMC, 54, 55, 63
God, 65
gold standard, 59, 96, 196
golden needle, 69
Goswell, 103, 114
GP, 31, 174, 282
GPIIb/IIIa, 238
gradient, 147, 148, 154, 161, 195, 214
graft failure, 273, 336

grammar, 19
gram-negative bacteria, 202
granulomata, 209
grapes, 318
Greeks, 20, 237
growth factors, 300
growth hormone, 208
GTN, 264, 265
guesses, 49
guesswork, 49, 70
guideline, 22, 55, 267
gut biome, 52
GWAS (Genome-Wide Association Study), 301, 302, 367
gymnasium, 283

H

haemangiomas, 221
haematemesis, 205
haematology, 67
haematuria, 205
haemochromatosis, 193
haemoglobin, 67
haemorrhagic, 126, 238, 239, 304
halo bias, 2
hamartomas, 221
Harry Beck, 105
HbA1c, 183, 290
HCM, 37, 146, 193, 194, 195, 215, 216, 243, 265, 298, 367
HCN2, 113
HCN4, 113
HDL cholesterol, 169, 284, 302
headache, 57
healing, 311

health, 6, 27, 28, 166, 188, 212, 213, 235, 282, 307, 310, 311, 317, 319, 324
health, 51, 125, 206, 306, 311, 313, 315, 317, 318, 319, 368, 377, 384, 390, 391
heart attack, 294
heart attack prevention, 238
heart attack survivors, 330
heart attacks, 282
heart block, 21, 33, 79, 108, 118, 133, 196, 199, 200, 205, 209, 210, 227, 249, 250, 270
heart disease, 289, 291, 293, 294, 295
heart disease detection, 3
heart failure, 3, 11, 12, 15, 31, 32, 36, 37, 40, 41, 49, 61, 62, 79, 82, 83, 86, 108, 114, 123, 124, 128, 129, 137, 140, 141, 143, 150, 152, 153, 163, 179, 183, 185, 186, 188, 189, 191, 192, 193, 194, 195, 196, 197, 198, 199, 200, 201, 204, 207, 208, 213, 215, 222, 225, 228, 229, 230, 231, 232, 233, 234, 242, 243, 244, 246, 247, 260, 261, 262, 263, 287, 289, 291, 325, 330, 331, 341, 371, 373, 374, 375, 378, 385, 390, 391, 392
heart muscle, 74, 86, 87, 89, 94, 114, 140, 196, 291
heart shape, 111
heart sounds, 38, 39, 79, 85
heart transplantation, 101, 118, 231, 269
heart valve, 21, 43, 55, 85, 125, 289, 340
Heberden, 8, 9, 33, 60
heparin, 218, 239
hepatic failure, 34, 53
hepatitis, 52
herpes 6, 198
Herxheimer reaction, 76, 206
heuristic, 69
HFpEF, 188, 196, 234
hibernating, 86, 95, 96
hibernating myocardium, 86
hidden in plain sight, 66
Hierarchical Phase Contrast Tomography, 72, 380
high blood pressure, 293
Hispanic, 166

histological, 3, 45, 194
historic data, 34
history, 6, 8, 18, 19, 20, 23, 24, 26, 27, 28, 29, 30, 32, 34, 44, 53, 65, 67, 104, 276, 237, 269, 284, 286, 293, 371, 378
history taking, 8, 18, 32
HIV and AIDS, 52
HMG-CoA, 174, 266, 320
HMG-CoA Reductase Inhibitors, 266
hoarseness, 179
hobby horse, 52
Hodgkin, 106, 381
Holmes and Rahe, 26
Holt-Oram syndrome, 298
Homan's sign, 42
homeostasis, 96
homocysteine, 172, 290, 314
HOPE trial, 332
horizontal plane, 104, 111
horses for courses, 29
hospital ward, 64
hsCRP, 172
hubris, 67
human error, 48
human wisdom, 69
hydralazine, 210, 215
hydrochlorothiazide, 233
hydrodynamics, 74
Hygieia, 311
hyoscine, 120
hyperactivity, 29
hypercholesterolaemia, 267, 326
hypercoagulable state, 217, 218, 338, 340
hyperinflation, 40
hyperkalaemia, 48, 57, 113, 234, 260, 261
hyperlipidaemia, 167, 172, 173, 174, 268, 284, 329, 330
hypertension, 3, 24, 25, 37, 39, 50, 67, 87, 88, 89, 92, 101, 112, 125, 137, 139, 143, 150, 152, 153, 157, 159, 161, 167, 168, 174, 175, 176, 177, 181,

187, 193, 194, 207, 208, 211, 212, 215, 228, 229, 232, 233, 234, 242, 243, 244, 245, 246, 258, 260, 261, 262, 263, 268, 286, 291, 294, 299, 304, 315, 317, 321, 323, 326, 328, 331, 341, 374, 375, 380, 384, 386
   hypertensive genes, 89
   hyperthyroidism, 207, 243
   hypertriglyceridaemia, 175
   hypertrophic, 193, 253, 367
   hypertrophied, 89, 110, 143, 146
   hypertrophy, 40, 78, 79, 87, 88, 89, 107, 145, 147, 149, 150, 185, 191, 194, 195, 242, 246, 249, 258, 291, 317
   hyperventilation, 11, 35, 61, 62
   hypokalaemia, 113, 208, 248
   hypomagnesemia, 248
   hyponatraemia, 31, 192
   hypoperfusion, 192
   hypotension, 14, 178, 188, 208, 228, 245, 246, 250, 260, 265
   hypothyroidism, 34, 207

I

   IBS, 66
   ICD, 162, 257
   idarucizumab, 239
   idiopathic, 79, 107, 108, 236, 257
   ignorance, 326
   IHD, 61, 87, 91, 159, 170, 171, 284, 291, 367
   IL-6, 168, 170, 191, 300
   iliac endofibrosis, 183
   iliac stenoses, 183
   ill-educated, 166
   illiterate, 19
   Iloprost, 265
   IMAGINE trial, 332
   imaging, 44, 58, 84, 85, 91, 94, 96, 151, 209, 383
   immune system, 300

immunosuppression, 200, 271
impedance, 229
implantable cardioverter-defibrillator, 13, 134, 257
in utero, 156, 159, 210
incompatibility, 29
indapamide, 232
index of suspicion, 283
indigestion, 9, 10, 60, 228
individual, 17, 18, 29, 45, 47, 69, 77, 91, 92, 110, 167, 172, 173, 194, 203, 225, 235, 243, 248, 295, 299, 306, 307, 323
inductive, 69, 312
ineptitude, 54
inexperience, 59
infarct size, 250
infection, 136, 192, 198, 201, 202, 206, 207, 271, 273
infective endocarditis, 141
inferior infarction, 76, 171
inflammation, 57, 58
inflammatory cells, 168
inflammatory heart disease, 33
inflammatory markers, 285
infliximab, 284
infusion, 64
ingested fat, 167
inheritance, 3, 25, 154, 294, 298, 304
inherited haemochromatosis, 297
inlet valve, 73
INR target, 341
inscape, 18
insight, 56, 326
inspiration, 10, 39, 62, 81, 98
insulin, 169, 174, 175, 213, 283, 290
integrated circuit, 50
intelligence, 19, 203
interatrial septum, 98
intercellular adhesion molecule 1 (ICAM-1), 300
intercostal muscles, 35, 40

interleukin 4, 300
interlobular septae, 92
internal jugular, 36
internal mammary artery bypass, 270
inter-personal game, 17
interplay style, 17
interpretation, 3, 48, 50, 57, 59, 92, 134, 236
interpretation, 45, 50
interstitial fluid, 229, 230
interstitial lung fluid, 230
intervention, 1, 9, 33, 64, 79, 85, 96, 139, 144, 149, 150, 153, 154, 157, 158, 160, 161, 162, 163, 164, 166, 176, 180, 181, 183, 224, 240, 256, 269, 271, 273, 276, 282, 283, 285, 286, 288, 323, 331
interventricular septum, 76, 86
interview, 13, 17, 66, 241, 271
intima, 56, 91, 98, 169, 177, 266, 301, 387
intimal metabolism, 167
intimal tearing, 93
intracardiac pacing, 134
intracellular action potential, 112
intracranial bleeding, 126
intramural haematoma, 176
intransigence, 64
intrauterine steroids, 210
intravascular ultrasound images, 97
intraventricular septum, 115, 170
intrinsic rate, 77, 78, 108, 121
invasive, 1, 44, 55, 56, 84, 85, 93, 254, 272, 276, 277
investigation, 23, 41, 43, 44, 45, 46, 47, 50, 51, 52, 55, 56, 57, 65, 91, 133, 161, 162, 194, 196, 199, 204, 214, 283, 284, 286, 290
ionotropic, 128, 228, 262
ionotropy, 187
irbesartan, 261, 262, 374, 386
iron storage, 194, 196
irregular, 16, 33, 107, 108, 109, 117, 127, 135
irregularly irregular, 33

ischaemia, 9, 12, 33, 82, 87, 91, 108, 122, 123, 125, 130, 140, 145, 146, 172, 182, 184, 185, 188, 199, 231, 246, 250, 275, 276, 282

ischaemic, 41, 55, 87, 91, 96, 123, 142, 182, 192, 204, 205, 225, 228, 239, 241, 253, 254, 273, 285, 287, 290, 313, 328, 333, 334, 335, 337, 338, 339, 371, 373, 387, 391

ischaemic heart disease, 91, 123, 225, 285, 287, 313, 334, 371, 391

ISIS-2 trial, 327

isoprenaline, 53, 119

isosorbide mononitrate, 264

IT, 23

ivabradine, 118, 189, 227, 228, 378, 390

IVUS, 97, 99, 173, 183, 333, 367

J

James Black, 242

James Watson, 296

Janeway lesions, 204

jaundice, 32

jet, 141, 158

John Duns Scotus, 18, 379

Judge Judy, 17

judgement, 1, 22, 30, 42, 50, 54, 55, 56, 65

jugular venous pressure, 31

junior doctor, 1, 51, 52, 54, 58, 63, 183, 241, 334

justification, 46, 57

JVP, 31, 35, 36, 37, 61, 62, 86, 97, 98, 127, 153, 154, 191, 229, 230, 325

K

karyotyping, 68

Kawasaki disease, 303, 374

KCNH2, 122, 298

Keith Jefferson, 73, 92
Kerley B lines, 92, 93, 138
ketosis, 34
koróni, 74

L

labetalol, 215
labile, 25, 37, 87, 88
labile hypertension, 25, 37, 88
laboratory, 48, 49
laboratory results, 63
LAMB study, 220
language, 20
Laplace, 82, 145, 150, 188, 190, 383
late diastolic murmur, 40
late systolic murmur, 141
LATS, 207, 372
lawnmower engine, 184
lawyers, 17, 23
LBBB, 39, 80, 81, 111, 112, 132
LDL, 91, 168, 169, 173, 192, 266, 290, 301, 302, 312, 313, 315, 316, 317, 320, 367
leadless pacemakers, 278
left atrial ablation, 256
left atrial appendage, 139, 143, 338
left atrial hypertrophy, 107
left atrium, 73, 76, 93, 97, 107, 140, 142, 143, 157, 220
left main stem, 74
left ventricular hypertrophy (LVH), 3, 37, 50, 73, 87, 88, 110, 111, 158, 383
Leiden factor, 290
leiomyosarcomas, 221
Leonardo da Vinci, 74
leukocyte chemotaxis, 169

Levine-Lown-Ganong, 109
lidocaine, 134, 249
life support, 255
LIFE trial, 261
lifestyle, 26, 27, 28, 61, 70, 167, 256, 288, 294, 295, 305, 307, 310, 323
life-threatening, 43, 62
lignocaine, 249
lily of the valley, 26
LIMA graft, 272
limb ischaemia, 182
limus, 275
linguistic ineptitude, 20
Linus Pauling, 311
lipomas, 221
liraglutide, 174
lisinopril, 260
livedo reticularis, 181
liver function, 53
liver pulsation, 153
Liverpudlian, 20
LLG, 109, 115
Loeys–Dietz syndrome, 303
logical, 282
long QT syndrome (LQTS), 115, 134, 217, 244, 250, 254, 298
longer QRS, 110
longevity, 26
losartan, 261, 262, 374, 386
loss, 6, 34, 53, 64
loss of face, 64
Louis Washkansky, 271
low BP, 9, 13, 118, 119
low frequency, 85, 146
low-dose aspirin, 334, 341
Lp-PLA$_2$, 172
luck, 54
lung auscultation, 40
lung disease, 62, 182, 208

lung function, 62, 129, 228
lupus erythematosis, 74
lusitropy, 187, 188
LV dysfunction, 49, 86, 87, 124, 144, 151, 152, 253, 340
L-version), 163
LVH, 35, 37, 38, 50, 73, 79, 87, 88, 89, 90, 144, 145, 146, 147, 151, 161, 194, 261, 297, 304, 367, 395
lysis, 48, 169
lysosomal storage disorder, 297

M

M_1T_1, 80
machine output, 50
machine-generated, 47
machinery murmur, 159
macitentan, 265
macrophages, 145, 168, 300, 302, 302, 303
Magdi Yacoub, 270
magnesium, 134, 175, 244, 315, 382
magnetic field sensor, 257
main stem, 170
major bleeding, 126, 336, 338, 339
management, 3, 6, 7, 8, 16, 22, 27, 43, 45, 46, 48, 49, 51, 54, 62, 63, 65, 82, 89, 120, 123, 129, 140, 148, 152, 161, 164, 176, 178, 179, 181, 194, 203, 215, 224, 225, 235, 249, 250, 256, 260, 267, 291, 306, 324, 325, 326, 3 41, 372, 373
manganese, 175
mannitol, 231
mannitol, 233
manslaughter, 63
mapping, 102, 195, 256, 257, 301
Marfan Syndrome, 141, 143, 150, 177, 180, 215, 298, 303
Mason Sones, 98, 270
master, 20

MAT, 130
mathematics, 70
matriarch, 66
matrix, 25, 302
mean, 19, 284
measurement, 48
mechanical assist, 200
mechanical functioning, 8, 99
mechanical hearts, 269
mechanical valves, 140, 216, 218, 340
medial hypertrophy, 89
mediastinal obstruction, 36, 37
mediastinal space, 73
medical experience, 54
medical expertise, 54, 64
medical history, 8, 20
medical insurance, 52
medical knowledge, 24, 325
medical school, 67
medical screening, 58
medical student, 106, 175, 206
medical style, 19
medication, 10, 29, 33, 67, 231, 248, 266, 325, 329
medico-legal, 29, 51, 58
Mediterranean, 294, 307, 313
MEDLINE, 331
melaena, 205, 236
membranous septum, 156
men, 54, 55, 66
Mendelian, 25
meningitis, 42, 205
menopause, 242, 326
mental breakdown, 83
MERIT-HF study, 189, 385
mesothelioma, 221
meta, 7, 18

meta-analysis, 239, 240, 243, 256, 260, 273, 301, 302, 307, 318, 320, 327, 328, 329, 330, 332, 369, 370, 371, 385, 386, 388, 389, 394
metabolic, 15, 169, 175, 213, 278
metabolic syndrome, 169, 175, 290
meta-data, 18
meta-information, 7, 8, 55
metalloproteinases, 300, 301
metaphor, 30
metastases, 41
methoxamine, 246
methylated spirit drinkers, 175
methyldopa, 215
metoprolol, 243, 244, 385
mexiletine, 249
Michael J. Davies, 93
micro re-entry, 130
micro-behaviour, 7
microbiome, 225
microcalcification, 94
microelectrode, 106
midazolam, 252
mid-diastolic 'plop', 220
mid-diastolic murmur, 36, 68, 137, 138, 220
middle-aged deaths, 284
middle-class, 306
migraine, 24, 57, 243, 284
milrinone, 228
mind-set, 28
minerals, 28, 295, 305, 312, 315
minerals, 315
misinterpreted, 58
misleading, 30, 282
missed beats, 14
mitochondria, 192
mitochondrial DNA, 296
mitochondrial function, 296
mitogen-activated protein (MAP), 258

MitraClip, 144
mitral area, 40, 68, 149
mitral incompetence, 39, 43, 142
mitral murmurs, 35, 142
mitral stenosis, 16, 37, 39, 41, 68
mitral valve, 28, 40, 73, 80, 98, 130, 137, 138, 139, 140, 141, 149, 153, 156, 160, 195, 202, 214, 220, 269, 298
mitral valve prolapse, 141
mitral valve prolapse syndrome, 141
MLC, 242
modulation, 191
molecular, 3, 69, 218
molecular guided therapy, 69
money, 51, 288, 324
monoclonal antibodies, 201
mono-morphic, 135
monophasic devices, 252
monosaturated fatty acids, 318
Montgomery, 310
Monty, 63, 310
morbidity, 3, 26, 27, 76, 77, 87, 92, 126, 166, 171, 176, 237, 261, 267, 281, 282, 285, 287, 288, 289, 312, 316, 323, 325, 375
mortality, 26, 27, 77, 87, 88, 92, 99, 123, 124, 126, 152, 153, 165, 166, 171, 176, 178, 199, 201, 216, 228, 229, 237, 239, 243, 260, 261, 273, 275, 276, 281, 282, 285, 287, 288, 289, 307, 312, 313, 316, 318, 320, 321, 323, 325, 328, 329, 330, 331, 332, 335, 371, 375, 377, 378, 388
MPTS, 54, 55, 63
MRI, 2, 49, 58, 84, 87, 96, 142, 161, 162, 178, 181, 182, 195, 196, 200, 205, 209, 367
MRSA, 202
multi-morphic, 135
multiple diagnoses, 43
multiple wavelet hypothesis, 127, 370
multislice CT, 172
multivitamins, 311
Munger, Charlie, 52
murmurs, 38, 39, 291

muscle hypertrophy, 145
mutation, 154
mutations, 293
MVP, 140, 141, 368
mycotic, 204, 205
myeloma, 197
myeloperoxidase (MPO), 192
myocardial, 49, 57
myocardial biopsy, 200
myocardial excitability, 227
myocardial fibrosis, 137, 234
myocardial function, 128, 234, 290
myocardial oxygen, 9, 12, 86, 188
myocardial viability, 95, 96
myocarditis, 3, 42, 122, 192, 193, 194, 198, 199, 200, 201, 209, 210, 254, 296
myocardium, 79, 81, 86, 94, 95, 96, 109, 127, 193, 194, 228, 245, 246, 290
myocyte, 114, 134, 189, 190, 226, 248
myofibril, 82
myosin, 195, 200, 242, 264, 298
myosin inhibitors, 195
myxoedema, 32, 119, 207

N

NAME study, 220
nausea, 121
NE, 119, 241, 242, 245
nebula thinking, 65
neck, 9, 33, 34, 35, 36, 40, 98, 144, 154, 158, 177, 185
necrosis, 189, 200
necrotising arteritis, 209
negative family history, 25, 293
negligence, 63

neoplasm, 220
neoplastic disease, 123, 198
neprilysin inhibitors, 190
nerve root compression, 42
neural network, 69
neuropathy, 25, 42
neuroregulin (NRG, 191
neurotic, 46, 70
neuroticism, 70
New York Heart Association, 11, 12, 253, 375, 377
news media, 68
NHS, 22, 23, 51, 52, 53, 69, 90, 172, 173, 240, 249, 268, 282, 285, 324
nicardipine, 263
NICE, 89, 239, 247, 268, 328, 341, 368
nicotine, 40, 241
nifedipine, 262
nitrates, 215, 262, 263, 264, 265, 385
nitric oxide (NO), 169, 229, 242, 259
nitrites, 263, 265
nitroglycerin, 264
NO synthetase (eNOS), 169
NOAC, 239, 240, 335, 338, 339, 368
Nobel Prize, 103, 269
nocturnal dyspnoea, 37, 185, 214
non-compliance, 18
non-fatal MI, 328
non-ischaemic, 192
non-valvular atrial fibrillation, 338
Noonan syndrome, 154, 160, 298
normal pulses, 182
nostalgic, 41
NSAID, 21
NSTEMI, 112, 395
NT-proBNP, 191
numerical analysis, 46
nurses, 1, 18, 22, 27, 63, 64, 66, 118, 292
NYHA, 11, 253, 254

O

O_2 saturation, 99, 100
obese, 10, 35, 36, 106, 111, 142, 174, 184, 308
obesity, 11, 61, 86, 147, 167, 175, 184, 213, 258, 273, 298, 308
obligatory, 59
obsequious, 20
obsessive, 64, 70, 89
occlusion, 76, 82, 171, 206, 303
ODI, 239
oedema, 15, 31, 37, 40, 41, 58, 61, 68, 92, 137, 142, 150, 152, 185, 200, 207, 213, 214, 215, 216, 229, 230, 231, 232, 233, 263, 325
oedematous feet, 185
oesophageal probe, 85
oesophagus, 73
offspring, 25, 213
older adults, 13, 15, 91, 99, 129, 137, 194, 201, 225, 231, 236, 237, 314, 329, 337, 338, 387
olive oil, 318
olmesartan, 261, 262, 374, 386
omega, 175, 312
omega 3, 6 and 9, 295
omega-3, 6 and 9, 175
omission, 18, 24, 48
Onyx, 275
opening snap, 40, 68, 137, 138
operations, 68, 271
ORBIT score, 126
organ failure, 188
organic heart disease, 112
orthodromic, 132
orthogonal axes, 257
orthopnoea, 11, 37, 137, 185, 192
orthopnoea, 214

oscilloscope, 85
Osler's nodes, 201, 204
osteopathy, 27
outlet, 73, 74, 395
outsiders, 26, 27
overheating, 119
overriding aorta, 298
overweight, 289
oxygen consumption, 99, 188, 192, 228, 264
oxygen delivery, 74, 192
oxymetazoline, 246

P

P_2, 39, 81, 137, 153, 157, 160
$P2Y_{12}$ inhibitor, 335
$P2Y_{12}$ receptors, 238
pacemaker, 13, 28, 52, 64, 107, 113, 119, 120, 130, 153, 200, 227, 241, 251, 256, 270, 272, 277, 279, 395
pacemaker implantation, 64, 277
pain, 7, 9, 10, 13, 25, 41, 42, 43, 55, 57, 60, 66, 119, 120, 141, 169, 177, 179, 181, 182, 194, 196, 199, 204, 209, 249, 263, 281, 395
pallor, 9, 12, 13, 33, 35, 40, 118, 119, 240
Palmaz-Schatz®, 274
palpitation, 7, 9, 12, 14, 15, 29, 40, 68, 117, 122, 135, 194, 196, 207, 228, 247, 265, 291
pan-systolic murmur, 39, 142, 153
paper speed, 108
papillary muscle, 142, 184, 216
PARADIGM-HF trial), 190, 371
paradox, 91, 333
paradoxical embolisation, 163
parasympathetic, 78, 79, 118, 119, 131, 241, 242, 248, 303
parent, 25
Parkinsonism, 32

paroxysmal, 117, 214
paroxysmal AF, 256
PARTNERS STUDY, 182, 380
parvovirus 19 (PVB19), 198
passive atrial filling, 78
past history, 21
Pasteur, 32
patent foramen, 97, 163
pathognomonic, 21
pathological anatomy, 46, 72
pathologist, 175
pathology, 47, 58, 286
pathophysiological, 16, 184, 284
patho-physiologically, 45
pathophysiology, 16
patience, 7
patient confidence, 29
patient involvement, 324
patient management, 8, 45
patient satisfaction, 292
patients, 1, 2, 6, 7, 8, 9, 11, 12, 13, 14, 15, 16, 17, 18, 19, 20, 21, 22, 23, 24, 25, 26, 27, 28, 30, 31, 33, 35, 39, 42, 43, 44, 46, 49, 50, 51, 52, 53, 55, 57, 58, 63, 64, 69, 76, 80, 86, 87, 89, 96, 101, 107, 108, 109, 110, 118, 119, 120, 121, 122, 123, 124, 125, 128, 129, 131, 135, 136, 139, 140, 141, 142, 145, 146, 150, 151, 152, 153, 159, 160, 163, 168, 169, 172, 173, 174, 175, 176, 177, 178, 179, 181, 182, 183, 185, 193, 194, 196, 197, 199, 200, 201, 203, 204, 207, 208, 210, 212, 215, 217, 218, 224, 225, 226, 230, 231, 232, 234, 235, 236, 237, 239, 240, 241, 242, 243, 244, 247, 248, 250, 251, 253, 254, 256, 260, 261, 262, 263, 264, 265, 266, 267, 268, 269, 270, 271, 272, 273, 274, 275, 276, 277, 278, 281, 282, 283, 284, 285, 286, 287, 288, 289, 290, 291, 292, 293, 297, 302, 311, 313, 315, 316, 319, 320, 321, 323, 324, 325, 326, 327, 329, 330, 332, 333, 335, 336, 337, 338, 339, 340, 341, 369, 370, 373, 376, 378, 381, 385, 387, 388, 389, 391, 394
patronising, 20
Paul Kligfield, 32
Paul Wood, 2
P-cell, 107

PCI, 60, 96, 101, 174, 271, 274, 275, 276, 277, 368, 372, 375, 391
PCR, 49
PCSK9, 302
PCWP, 97
PDE inhibitors, 265
PDE) Inhibitors, 228
PDE-3 inhibitors, 228
PDE-4 inhibitor, 229
PEAR1 and SVEP1, 300
penetrance, 25
penicillin, 26, 49, 76, 141, 202, 203, 206
percuss, 41
percussion, 61
percutaneous, 53
Percutaneous Coronary Intervention (PCI), 274, 368, 389, 391
perfusion scan, 96
pericardial effusion, 193, 209, 221
pericardial space, 178
pericarditis, 10, 62, 122, 198, 208, 209, 210
pericardium, 73, 221, 222
peripheral blood flow, 34, 183
peripheral perfusion, 34, 118, 230, 240
peripheral tissue perfusion, 230
peripheral vascular disease, 25, 165, 181, 368
personal circumstances, 291
personal data, 70
personal issues, 7, 15
personalisation, 69
personalised medicine, 69
perspective, 18, 42, 43, 48, 49
PET, 2, 49, 94, 95, 96, 209
petit mal, 13
phaeochromocytoma, 245
pharmaceutical industry, 28
pharmacological, 27, 87, 242, 276, 285
phenotype, 25
phenotypic expression, 68

phenoxybenzamine, 245
phentermine, 174
phentolamine, 245
phenylephrine, 246
phonocardiography, 39, 85
phosphodiesterase III inhibitors, 229
phosphodiesterase inhibitors, 227, 228, 229
physical activity, 11, 12, 309, 385
physical processes, 8
physical signs, 32, 33, 38, 42, 47, 61, 66, 145, 286
physician, 3, 10, 47, 186, 226, 283, 324
physics, 48, 106
physiological murmurs, 213
physiology, 3, 46, 72, 77, 213
physiotherapy, 27
phytosterols, 316
Pim de Feyter, 171, 240
placebo, 28
placenta, 156, 218, 258
plaque morphology, 168
plasma volume, 212
plasmin, 241
platelet adhesion, 169, 264
platelet adhesiveness, 313
platelet aggregation inhibitor, 183
pleura, 41
pleural effusion, 41, 62
pleuritic, 10
pliability, 85, 137
pneumonia, 26, 41, 58, 61, 63, 204, 283
pneumothorax, 41, 61, 62, 63
Poiseuille, 82, 285
Poles, 20
political, 22, 27, 171, 284, 307
political correctness, 27
polyangiitis, 209
polyarteritis, 170

polyarteritis nodosa, 211
polycythaemia rubra vera (PRV), 67
polymerase chain reaction, 49
polymorphic VT, 254
polymorphism, 168, 301, 302
polymyositis, 210
polypeptides, 191
polyphenols, 312, 314, 317, 318
Polypill, 282, 319, 320, 327, 328, 373, 382
polyunsaturated fat, 312
polyunsaturated Fats, 314
poor, 17, 26, 34, 50, 67, 75, 89, 91, 118, 140, 145, 150, 158, 166, 199, 200, 230, 306, 307, 316
popliteal, 168
population-based, 173, 314, 381
posterior descending artery, 75, 86
posterior infarction, 76, 171
posterior thoracic pain, 179
posterior tibial, 182
post-infarction, 229, 237, 282
post-operative, 273
potassium, 48, 57, 106, 113, 128, 134, 226, 230, 232, 233, 234, 244, 247, 248, 258, 341
potassium channel current, 128
potassium depletion, 230, 232, 233
potassium efflux, 106, 247, 248
potassium loss, 226
power, 19, 28
PR interval, 79, 108, 109, 120, 121, 132, 163, 164, 248
prasugrel, 335
pravastatin, 320, 328, 329
prazosin, 246
pre-determined, 25, 310
prediction, 91, 299, 301, 387
predictive, 44
predisposition, 293
pre-eclampsia, 215

pre-excitation, 109, 395
pregnancy, 3, 13, 46, 119, 212, 213, 214, 215, 216, 217, 218, 225, 248, 263, 373
pregnant women, 212
pre-load, 187
premonition, 13
pressure gauge, 131
pressure gradient, 38, 138
presumptions, 27, 49, 278
presumptive diagnosis, 32
pre-symptomatic, 172
presystolic accentuation, 137, 138
pre-systolic sound, 146
Prevention of Heart Disease, 305
pride, 30
primary hypertension, 88, 89, 194
primary outcome, 336
PRIMARY PREVENTION, 305
primitive receptor (P)RR, 259
primum septum, 156
principle, 29, 43, 105, 171
private hospital, 21, 53
pro-arrhythmic, 114, 128, 243, 244, 247
probability, 24, 283
procainamide, 210, 247, 249
procainemide, 227
pro-drug, 238
professionalism, 23
professionals, 17, 23, 63, 326
profibrotic, 258
prognosis, 6, 67, 75, 76, 77, 88, 89, 96, 140, 145, 150, 158, 192, 197, 199, 201
prognostic significance, 90
programming, 69
progression, 12, 45, 46, 78, 89, 148, 152, 170, 179, 180, 183, 266, 267, 314, 315, 333, 389
pro-inflammatory cytokines, 300

proof, 283, 288, 311, 317
prophylactic, 92, 126, 141, 202, 203, 238, 268, 281, 285, 311, 318, 319, 323, 325, 375, 391
prophylaxis, 141, 148, 216, 244, 327
propofol, 252
propranolol, 53, 129, 242, 243, 244
prosthetic heart valve risk, 340
prosthetic valves, 141, 201, 202, 203, 236, 340
protein C, 218
protein S, 218
prothrombin time, 235, 239
protocol, 53, 54, 55, 200
pseudoephedrine, 246
pseudo-infarct, 199
psoriasis, 24, 170
psychiatry, 27
psychosomatic research, 26, 381
psychotherapy, 27
PT ratio >, 236
PTSD, 11
public interest, 288
public opinion, 68
pulmonary arteries, 39, 74, 92, 97, 98, 100, 138, 157, 159, 162, 164, 265
pulmonary capillary wedge pressure, 97, 101
pulmonary disease, 40, 101, 130, 152, 273, 330
pulmonary emboli, 37, 61, 62, 68, 122, 205
pulmonary hypertension, 37, 38, 39, 265
pulmonary regurgitation (PI), 154
pulmonary valve, 39, 81, 115, 160
pulmonary valve closure (P_2), 39
pulmonary vein ablation, 256
pulsatile mass, 181, 205
pulse, 8, 12, 33, 34, 35, 36, 37, 40, 61, 70, 78, 88, 107, 108, 109, 117, 118, 119, 120, 121, 127, 140, 145, 146, 150, 151, 213, 326
pulse waveform, 33, 150
pulsus parvus, 145
Purkinje, 77, 78, 106, 109, 186

PVD, 181, 182, 183, 368

Q

QRisk3, 285
QRS complex, 37, 50, 78, 109, 112, 251, 252
Q-T interval, 227
QTc, 134, 248, 249, 250
quercetin, 317
questioning, 18, 29, 30, 287
questions, 7, 15, 16, 19, 23, 24, 33, 51, 61, 80, 94, 112, 118, 286
queue, 51
QUIET Trial, 333
quinidine, 186, 247, 248, 249
quinine, 247

R

RAAS, 189, 190, 257, 259, 315, 368
race, 7, 26
RACE II Trial, 126, 387, 392
races, 25
radiation, 44, 45, 51, 84, 95, 135, 145, 217, 257
radioactive, 67, 95, 208
radiofrequency, 255
raised lipids, 91
raison d'être, 311
RALES trial, 190, 234, 387
ramipril, 260, 328, 331
random selection, 67
rare, 14, 52, 67, 74, 134, 140, 141, 150, 151, 169, 194, 208, 209, 210, 211, 217, 220, 322
RBBB, 81, 110, 111, 157, 162, 163, 395

re-assurance, 286
recidivism, 64
reciprocating, 122, 132, 249
recognition, 6, 26, 50
recovery, 120, 121, 133, 139, 201, 252, 375
red cell mass, 212
red tape, 64
red wine, 175, 313, 314, 375
red yeast rice, 266, 320
re-entry, 14, 102, 108, 109, 121, 122, 127, 130, 131, 134, 248, 251, 255, 256
reference, 15
referral, 7
refractoriness, 227
refractory period, 122, 128, 130, 134, 186, 191, 227, 248
regulators, 22, 53, 54, 55, 179, 235, 237
regulatory action, 22
regulatory authorities, 178
regurgitant flow, 85, 151
regurgitation, 38, 142, 149, 153, 206, 386
reinfarction, 275, 335
relationships, 26, 292
relative, 30, 59, 224, 260, 287, 288, 321, 328
relaxation, 34, 80, 187, 242, 264
RE-LY study, 240, 388
remodelling, 128, 190, 191, 303
renal dysfunction, 15, 244, 273
renal failure, 48, 61, 178, 211, 231, 233, 234, 260, 273
renal impairment, 129, 322
René Favaloro, 271
René Thom, 82
renin, 189, 233, 243, 257, 258, 259, 368, 383
renin-angiotensin-aldosterone (RAAS), 189
re-perfusion, 96
repolarisation, 113, 134
reports, 47, 54
research, 45, 47, 52

resistance, 21, 79, 81, 89, 101, 157, 159, 175, 185, 187, 192, 212, 213, 218, 242, 257, 290

restenosis, 274

resveratrol, 314, 317

reteplase, 241

retirement, 288

retrograde, 98, 102, 131, 132

retrospective analysis, 286

revascularisation, 86, 182, 244, 262, 275, 328

Reye's Syndrome, 21, 22

rhabdomyoma, 221

rhabdomyosarcomas, 221

rhetorical question, 52

rheumatic fever, 16, 21, 68, 93, 136, 142, 148, 193, 199

rheumatic heart disease, 39, 68, 123, 124, 136, 143, 145, 212, 340

rheumatoid arthritis, 136, 170

rhonchi, 40

rhythmicity, 250

ribosomal RNA (rRNA), 296

rich (wealthy), 27, 91, 93, 129, 168, 169, 172, 266, 282, 306, 307, 318, 371

richest subgroups, 166

right atrium, 35, 36, 73, 74, 75, 77, 78, 97, 98, 132, 156, 157, 170

right coronary, 12, 75, 76, 77, 86, 170, 171, 270

right heart failure, 31, 37, 186, 229

right ventricular hypertrophy, 38, 73, 112

risk, 2, 7, 22, 24, 25, 26, 27, 45, 52, 53, 54, 55, 56, 63, 83, 87, 88, 89, 90, 91, 92, 96, 98, 99, 117, 123, 124, 125, 126, 127, 128, 132, 135, 141, 146, 148, 161, 162, 167, 172, 173, 174, 178, 179, 180, 181, 182, 195, 199, 202, 203, 206, 210, 214, 215, 216, 217, 241, 243, 245, 246, 253, 254, 255, 258, 260, 266, 267, 268, 271, 273, 275, 282, 284, 285, 287, 290, 299, 300, 301, 302, 306, 307, 308, 309, 311, 312, 313, 314, 318, 319, 320, 321, 324, 327, 328, 330, 334, 335, 337, 338, 339, 340, 341, 370, 373, 378, 380, 381, 385, 386, 387, 391, 392

rivaroxaban, 183, 218, 239, 335, 336, 337, 339, 388, 389

RNA, 296

ROCKET trial, 240, 388

roflumilast, 229
Romans, 311
Romhilt-Estes LVH Point Score System, 146
ROSC, 251
rosuvastatin, 266, 320, 322, 329
Roth spots, 201
Royal College of Physicians, 8, 32, 282
Royal Family., 27
Royal Navy, 76
R-Rate Response Pacing, 278, 279
rs1800470 genotype, 302
rubidium PET scan, 49
rules, 8, 21, 22, 55, 56, 67, 235
rupture, 93, 94, 180, 181, 184, 205, 206, 216, 266, 301
Russian, 20
RVH (Right Ventricular Hypertrophy), 36, 38, 62, 73, 153, 157, 160, 162, 298

S

safety, 28, 283, 285
sagittal plane, 104, 111
salt and water retention, 192
sampling error, 48
saphenous vein graft, 273
sarcoidosis, 193, 208
sarcolemma, 114, 189
sarcomere disruption, 194
SARS-CoV-2, 198
saturated fat, 167, 175, 295
scale, 282
scanning, 2, 51, 58, 72, 92, 93, 94, 95, 96, 99, 137, 161, 172, 180, 181, 182, 196, 209, 330, 380
science, 28, 57, 68

screening, 3, 52, 58, 75, 276, 283, 284, 285, 286, 287, 288, 289, 291, 323, 324, 392
screening procedures, 291
second heart sound (S2), 214
secondary amyloidosis, 197
secondary heart disease prevention, 3
secondary prevention, 3, 176, 238, 243, 305, 309, 319, 320, 323, 325, 327, 328, 329, 332, 334, 335, 337, 341, 369, 371, 373, 377, 382, 389
secondary tumours, 221
secundum septum, 156
security, 23
sedation, 120, 252
selenium, 175, 315
semaglutide, 174
semi-conscious, 13
sensitive, 16, 17, 46, 47, 58
sensitivity, 13, 27, 44, 47, 49, 50, 56, 119, 225, 268
sepsis, 62, 122, 134, 201
septal defect, 34, 155, 270
septal muscle, 156
septicaemia, 42, 184
serine endopeptidase, 125
serum TGFβ1 protein, 302
Shakespeare, 34, 390
Sherlock Holmes, 66
SHIFT study, 189, 228, 390, 391
shock, 101, 134, 188, 189, 246, 251, 252, 255
shortness of breath, 3, 9, 11, 23, 60, 68, 93, 137, 139, 184, 185, 186, 213, 284, 287, 290, 291
shunt, 62, 100, 156, 157, 158, 162, 163
sickness, 311
sildenafil, 228, 265, 321
simvastatin, 225, 266, 320, 328, 329
single ventricle, 164
single-nucleotide polymorphism (SNP), 301
sino-atrial block, 21, 107, 120, 278
sinoatrial block, 118

sinus, 11, 13, 14, 74, 76, 77, 78, 79, 102, 106, 107, 108, 115, 118, 119, 121, 122, 123, 124, 127, 128, 130, 131, 132, 137, 138, 153, 157, 170, 186, 199, 208, 227, 247, 251, 256, 277, 278, 340
  Sinus arrest, 13
  sinus node, 77, 78, 79, 102, 106, 107, 121, 132, 133, 227, 278
  sinus tachycardia, 11, 122, 123, 199
  sinus venosus defects, 157
  skeletal atrophy, 192
  SLCO1B1, 225
  SLE, 33, 136, 209, 210
  sleep, 14, 35, 36, 134, 185, 233, 244, 311
  sleep apnoea, 35, 36, 134, 192
  slow heart rates, 117
  slurred R-wave, 109
  small vessel, 193
  smokers, 276, 282
  smoking, 26, 61, 79, 167, 173, 182, 183, 302, 306, 307, 308, 309, 310, 321, 378
  smoking cessation, 173, 183, 283, 294, 307, 309, 378, 393
  social, 7, 8, 15, 18, 23, 26, 27, 30, 311, 381
  social status, 26, 27
  society, 26, 27
  socio-economic class, 282
  socio-economic group, 268
  SODA, 239
  sodium influx, 106, 107
  sodium reabsorption, 232, 234
  Sokolow-Lyon Criterion, 146
  soldiers' heart, 10
  sotalol, 128, 129, 135, 217, 244, 247
  soya, 28
  spanophiliac, 52
  sparks (myofibrillar), 103, 134
  specialist centre, 179
  specific, 15, 23, 46, 47, 49, 54
  specificity, 44, 47, 50, 56, 58, 69, 146, 200, 243, 268, 326
  sphygmomanometer, 131

spider naevi, 34, 53
spironolactone, 190, 233, 234, 387
splenic infarcts, 204
splinter haemorrhages, 201, 204
splitability index, 137
sputum, 40
squeezing force (contractility), 188
ST segment, 112
St. George's Hospital, 32, 79, 92, 93, 220, 278
stable plaque, 169
staff, 18, 64
stainless-steel, tubular stent, 274
standardisation, 22
Stanford, 177, 180, 210, 271
staphylococcus aureus, 201
Starling, EH, 81, 390
Starr-Edwards valves, 270
statin, 30, 170, 173, 174, 183, 210, 217, 266, 267, 268, 282, 285, 300, 319, 320, 321, 322, 328, 329, 330, 369, 388
statistical analysis, 92, 168
statistical appraisal, 281
statistical paradox, 284
statistically unreliable, 89
STEMI, 112, 395
stenoses, 12, 82, 101, 102, 183, 185, 272, 274, 285
stenosis, 11, 33, 37, 38, 40, 62, 68, 78, 86, 93, 101, 107, 136, 137, 138, 139, 140, 144, 145, 147, 148, 149, 152, 160, 195, 214, 216, 229, 256, 270, 274, 276, 302, 375, 395
stent-grafting, 181
stenting, 26, 60, 94, 161, 183, 217, 231, 269, 272, 275, 276, 281, 284, 291
Stephen Edmondson, 281
stethoscope, 36, 39, 41
Steve Harrison, 288
stimulus inhibition, 79
strategic judgements, 288
sstrategy, 29, 52, 283, 284, 285

stratification, 68, 125, 302, 391
streptococcus, 49, 201
streptokinase, 241
stress, 7, 10, 13, 49, 60, 83, 86, 87, 96, 122, 142, 167, 169, 190, 191, 192, 242, 245, 291, 303, 306
stressed, 9, 94
stressful, 119, 303
stridor, 179
stroke, 14, 42, 81, 88, 89, 90, 123, 124, 125, 126, 149, 161, 174, 188, 189, 194, 204, 213, 237, 238, 239, 241, 251, 256, 258, 260, 261, 268, 273, 275, 304, 309, 312, 313, 318, 319, 321, 325, 327, 328, 331, 332, 334, 335, 336, 337, 338, 340, 370, 373, 379, 391, 392
stroke rehabilitation, 14
stroke statistics, 166
structural heart disease, 14, 133, 254, 257
ST-segment elevation, 134
style, 17, 18, 20
sub-aortic, 195
subcutaneous nodules, 199
subendocardial, 33
subendothelial accumulation, 300
subtraction angiography, 182
sudden death, 76, 79, 132, 135, 152, 162, 193, 194, 206, 254, 298, 314, 324
suffering, 64
suggestion, 42, 51, 53, 61, 333
summary, 23, 336
sunlight and vitamin D status, 314
superior vena cava, 100, 179
supplement, 312
supplements, 28, 311, 312, 384
supra-cristal, 158
supraventricular arrhythmias, 250
surgeon, 281
surgery, 1, 2, 46, 55, 85, 136, 137, 139, 141, 143, 144, 147, 151, 152, 158, 159, 172, 178, 206, 269, 270, 273, 291, 336
surgical developments, 269, 270

survival advantage, 329
sustained VT, 253, 254
SVC, 100
SVT, 132, 135, 256
sweating, 9, 243
sympathetic, 20
sympathetic fibres, 119, 241
sympathetic stimulation, 131
symptom, 3, 9, 10, 45, 47, 60, 61, 68, 185, 186, 241, 263, 286, 289
sync, 33
synchronization, 252
synchrotron, 72
syncope, 7, 9, 12, 13, 16, 78, 79, 118, 119, 120, 121, 123, 141, 145, 147, 194, 196, 197, 199, 200, 220, 253, 254, 265, 277, 298, 341
SYNTAX trial, 271, 391
syphilis, 24, 52, 74, 75, 76, 206
systemic embolisation, 205
systemic lupus (SLE), 170
systolic pressures, 90

T

T waves, 57
T3, 34, 119, 241
T4, 34, 119, 241
tachycardia, 13, 61, 62, 86, 102, 109, 119, 121, 122, 123, 130, 131, 132, 186, 199, 208, 241, 243, 251, 253, 255, 256, 257, 393
Takayasu, 150, 177, 209
talent, 1, 54, 65
talking therapies, 27
tamponade, 178
tamsulosin, 246
Tarot, 50
TAVI, 149, 152, 269, 272, 341, 368
T-cells, 145, 275

teacher, 27, 32
teachers, 2, 27, 32, 39, 64
teaching exercise, 53
technical excellence, 16
technique, 17, 18, 19, 37, 48, 54
technology, 69, 257
telephone, 26
temperature, 33, 100, 123, 207, 314
tempting providence, 64
tenderness, 41, 60, 209, 236
tendon reflexes, 34
tenecteplase, 241
terazosin, 246
tertiary medical centres, 16
TEVAR, 178
TGFβ1 signalling, 302
TGFβR2 receptor, 303
theoretical, 66
therapeutic, 1, 8, 22, 23, 27, 28, 29, 129, 191, 226, 230, 233, 240, 266, 276, 303
thermionic valves, 104
thermodilution method, 100
thiazides, 231, 232
third party 'referees', 54
Thomas Lewis, 114
thoracotomy, 252
three wise men, 54
thrill, 158
thromboembolism, 129, 217, 339, 340, 341, 372
thrombophlebitis, 15
thrombus, 139, 266
thyroid hormones, 34, 119, 241
thyrotoxicosis, 33, 34, 62, 121, 122, 123, 125, 184, 217
TIA, 125
ticagrelor, 335
Tietze, 10, 60
tightness, 9, 181

time dimension, 42
TIMIC study, 200, 378
timing, 46, 138
tiredness, 213
tissue necrosis, 167
TNF-α, 170, 191, 300
tobacco, 173
tonsillar abscess, 21
toothbrushing (and SBE), 202
torsade de pointes, 244, 249
total cholesterol, 91, 266
toxic, 28, 53, 63, 119, 122, 175
toxic effect, 119
toxic state, 53, 122
toxicity, 57, 130, 190, 224, 227
TR, 152, 153
tRAAS, 258
training, 1, 54, 118, 255, 309
trandolapril, 260
transcatheter aortic valve implantation, 149
transcatheter repair, 144
transcatheter valve replacement, 149
transfer RNA (tRNA), 296
transforming growth factor β (TGF-β), 301
transfusion, 127
transoesophageal, 85, 139, 205
transplanted heart, 197
transposition, 158, 161
transverse axial tubular system, 114
trastuzamab, 191
travel, 21
travel sickness, 13
treadmill, 86
treatment, 8, 14, 21, 23, 26, 28, 31, 33, 34, 43, 66, 87, 88, 89, 92, 134, 135, 144, 155, 173, 174, 181, 195, 197, 200, 203, 205, 206, 208, 215, 222, 225, 229, 235, 241, 243, 244, 256, 263, 264, 265, 268, 270, 273, 275, 276, 281, 283, 286, 315, 319, 323, 329, 333, 334, 382, 384

tremor, 13, 32, 243
tribunal, 63
TriClip, 153
tricuspid valve incompetence, 37
triggering, 114
triglyceride, 91, 290
TRILUMINATE single-arm study, 153, 386
Trisomy 21 (Down's syndrome), 297, 298
troponin, 9, 82, 199, 290, 298, 322
trust, 17, 23, 24, 29
Trypanosoma cruzi, 199
TSOAC, 239
T-tubule, 114
turbulent flow, 168
Turkish, 20
Turner syndrome, 160, 298
T-wave inversion, 38, 146
T-waves, 31, 57, 103, 113, 129, 194, 208, 226
Type A personality, 306

U

Uist, 66
UK, 19, 21, 27, 51, 55, 63, 67, 84, 129, 133, 174, 175, 177, 193, 220, 226, 232, 237, 247, 249, 267, 268, 283, 291, 313, 323
Ukrainians, 20
ultrasound, 326
ultrasound transducer, 285
umirolimus, 275
uncommunicative, 19
unconscious, 29, 117
uncooperative, 18
underactive thyroid, 118
Underground Map (London), 105
understanding, 325

unfitness, 11, 61, 86, 182
unipolar, 104
unpleasant, 44
unstable angina, 169
unstented, 274, 275
unworthy test, 51
urgency, 11, 45, 62, 236, 306
urgency, 45
URTI, 199, 368
US, cardiovascular deaths, 308
USA, 127, 139, 175, 274, 370, 375

V

vagal nerve stimulation, 14
vagal reflexes, 13
vagal stimulation, 128
valsalva manoeuvre, 52, 131, 146
valsartan, 190, 261, 262, 333, 374, 386, 392
VALUE trial, 333
values, 27
valve annulus, 141, 147, 205
valve defects, 40, 122, 154
valve implantation, 341
valve leaflet, 149, 153
valve obstruction, 147
valve problems, 289
valve repair, 139, 152, 222, 270
valvotomy, 136, 139
valvular heart disease, 193, 340
Vane, 237, 392
variations, 3, 50, 56, 74, 77, 225, 297
vasa vasorum, 177
vascular cell adhesion molecule 1 (VCAM-1), 300
vascular resistance, 101

vasoconstriction, 101, 192, 229, 242, 244, 245, 246, 258, 265
vasomotor syncope, 13, 121, 341
vasovagal reflex, 9
vector, 50, 79, 105, 107, 109, 111, 395
vegetables, 28, 307, 312, 313, 318, 379, 392
vegetarian diet, 313
vegetations, 85, 151, 202, 204, 205, 206, 210
veins, 15, 37, 42, 61, 74, 92, 97, 107, 127, 157, 183, 272, 273
vena cava, 53, 100, 157, 179
vena caval pressure, 185
venesection, 64
venous blood, 74, 163
venous cannulation, 64
venous compression, 218
ventricular, 3, 11, 12, 13, 14, 36, 37, 38, 40, 49, 50, 73, 74, 76, 77, 78, 79, 80, 81, 85, 86, 87, 88, 97, 105, 108, 109, 110, 111, 112, 121, 122, 123, 126, 127, 128, 129, 130, 132, 133, 137, 139, 140, 142, 143, 145, 150, 153, 156, 158, 160, 162, 164, 178, 185, 187,188, 191, 192, 194, 196, 200, 206, 209, 216, 217, 221, 226, 227, 229, 230, 234, 243, 244, 249, 250, 251, 253, 254, 256, 257, 258, 272, 277, 332, 333, 378, 383, 387, 389
  ventricular contractility, 86, 128, 226, 243
  ventricular tachycardia, 135
verapamil, 118, 178, 227, 247, 262, 321
verification, 24, 32, 47, 141, 287
VF, 79, 117, 133, 134, 135, 241, 243, 250, 251, 252, 253, 255, 298
vicious spiral, 188
village priest, 27
viral myocarditis, 198
viruses, 198, 296
Visscher, 81, 378, 390
vitamin(s), 28, 235, 239, 282, 283, 295, 305, 311, 312, 314, 315, 341, 367, 368, 371, 379, 384, 391
  vitamin D blood levels, 315
  vitamin D supplementation, 315
  vitamin K, 125, 235, 236, 239, 282, 312, 339, 340
  vitamin K antagonists, 125, 282, 339
  vitamin K epoxide reductase, 235

VKORC1, 225
Vladimir Demikhov, 272
vocabulary, 19
vogue, 69
voltage wave, 79
voltages, 37, 50
volume overload, 149
VOO, 278
VOYAGER PAD trial, 183
VSD, 39, 100, 154, 156, 157, 158, 159, 160, 162, 215, 216, 298, 368
vulnerable period, 134, 251
vulnerable plaque, 93, 94, 173, 266
VVI, 277, 278
VVIR, 278

W

walking distance, 183
Waller, 103, 104, 114, 372
Walsh, 103
wandering, 107
warfarin, 125, 129, 218, 225, 235, 236, 237, 239, 240, 312, 334, 335, 337, 338, 340, 368, 370, 371, 377, 381, 388, 393, 394
wartime, 10, 64
wave front, 78, 79, 80, 102, 105, 106, 107, 108, 109, 110, 111, 112, 114, 122, 132, 257
wave function, 50
waveform, 33, 35, 36, 40, 104, 106, 107, 154, 251
wealth, 27
weight gain, 185
weight loss, 150, 175, 204, 209, 220, 231
weight of clinical significance, 65
weighting factor, 61
Wenckebach, 109, 121, 247, 395
western world, 91, 193, 284

wheezing, 40
Whipple disease, 136
white coat, 25, 88, 384
WHO, 166, 194, 215, 313
Wilkin score, 137
William Harvey, 8, 114
William Withering, 129, 186, 226
wisdom, 24, 247
witnesses, 13
Wolff L, Parkinson J, and White P.D (see WPW), 109, 131, 393
women, 41, 55, 126, 166, 174, 181, 208, 210, 212, 213, 216, 218, 232, 233, 285, 326, 337, 338
World Heart Federation, 305
worthy test, 51, 52
WOSCOPS (West of Scotland Coronary Prevention Study, 321
WPW, 14, 109, 115, 131, 132, 163, 249, 250

X

xiphisternum, 36, 73, 153, 160
X-linked abnormalities, 297
X-ray, 291

Z

zatarolimus, 275
zinc, 175
Zutphen Elderly Study, 318, 380

www.ingramcontent.com/pod-product-compliance
Lightning Source LLC
Chambersburg PA
CBHW070520010526
44118CB00012B/1038